T0277487

CELEBRATING

50 YEARS

Texas A&M University Press
publishing since 1974

SEA
CHANGE

Larry McKinney, General Editor
John W. Tunnell Jr., Founding Editor

Sponsored by the Harte Research Institute for Gulf of Mexico Studies,
Texas A&M University–Corpus Christi

SEA CHANGE
A Message of the Oceans

SYLVIA ALICE EARLE

Texas A&M University Press
College Station

Credits for poetry quoted: Ogden Nash. "A Beginner's Guide to the Oceans" © 1937, in *Good Intentions*. Boston: Little, Brown and Company, p. 53. Dr. Seuss. 1971 *The Lorax*. © 1971 by Theodor S. Geisel and Audrey S. Geisel. New York: Random House. Scott McVay, in Osaka, August 18, 1970, "The Alternative?"

Jellyfish by parkjisun from the Noun Project.

♾ This paper meets the requirements of ANSI/NISO Z39.48-1992 (Permanence of Paper). Binding materials have been chosen for durability. Manufactured in the United States of America

Library of Congress Control Number: 2020949624

ISBN-13: 978-1-62349-904-4 (cloth: alk. paper)
ISBN-13: 978-1-62349-905-1 (ebook)
978-1-64843-272-9 (paper)

A list of titles in this series is available at the end of this book.

I dedicate this to my best friends, Alice Richie Earle and Lewis Reade Earle, whose caring ethic is reflected on every page; to my irreverent, wonderfully tolerant and loving offspring, Elizabeth, Richie, and Gale; to my lifelong hero, Harold J. Humm; and to all who seek to understand and protect the wild ocean.

CONTENTS

Part III SEA CHANGE

A gallery of images follows page 108.

PREFACE TO THE NEW EDITION

So much has changed in the twenty-five years since this book was published that it is hard to know where to begin to highlight new discoveries, insights, and shifting attitudes about the importance of the ocean to everyone, everywhere, all of the time. A good place to start might be where the book ended, where I articulated the urgent concerns that motivated its writing:

> This is the time as never before and perhaps never again to establish policies—on a small personal scale as well as on a broad public scale—to protect and maintain planetary health. To be effective, actions must be taken before ecosystems are further traumatized or destroyed, before vested interests become too firmly established, before the arrival of that worrisome "point of no return."

Five years later, in 2000, I attended an Explorers Club dinner in Houston celebrating the beginning of the new millennium. My host, Brian Hanson, seated me next to a tall, lean Texan, William "Will" Harte, who introduced himself as a "rancher" passionate about wildlife and wild places. Recognizing a kindred spirit, I gave him a copy of *Sea Change*, a small act that set in motion events that have had large, enduring consequences. Not only did Will read the book; he passed it along to his father, Edward H. Harte, a journalist, newspaper publisher, philanthropist, and pioneer in environmental conservation and also a beloved and respected resident of Corpus Christi, Texas. I was unaware that family discussions were underway at the time about Ed Harte's desire to "give back" to his community, state, and country with a major gift. Already, Ed and his late wife, Janet, had supported major cultural institutions in Corpus Christi, and together with his brother, Houston Harte, they had contributed their 66,000-acre ranch to the Nature Conservancy, leading to a significant expansion of Big Bend National Park. He also was instrumental in protection of Padre Island and the designation of Mustang Island as a state park. Owing to his love of wild birds and wild places, he served on the board of the Audubon Society for many years and was chair from 1974–79.

The then-president of Texas A&M University–Corpus Christi, Robert Furgason, was encouraging Ed to consider supporting a new research facility to be based at the university. As a catalyst, *Sea Change* apparently helped bring into focus the potential for creating something that would bring the United States, Mexico, and Cuba together to have a tri-national impact on exploring and caring for a part of the ocean important to the Harte family, to me, and to the world: the Gulf of Mexico, Earth's ninth-largest body of water.

As the year 2000 progressed, I was asked to join discussions about how the $46 million gift Ed Harte would make could be most effectively used, where it would be based, and who might be involved. Few things are as exhilarating as being in at the beginning of something that promises to deliver on the one condition Ed Harte mandated: make a difference. To help guide the process, a brain trust was assembled of nineteen scientists, business leaders, explorers, conservationists, writers, and educators who deliberated with members of the Harte family and university officials at meetings in various Gulf states, in Mexico, and in Havana, Cuba.

Several years later, at a meeting in Monterrey, Mexico, someone asked how the Harte Research Institute for Gulf of Mexico Studies came to be. The answer is that many people put their heads together over many months and worked very hard to turn a vision into reality under the leadership of President Furgason and John Wesley "Wes" Tunnell, a marine scientist with decades of experience exploring the Gulf of Mexico and mentoring hundreds of students. But in a moment of generous overstatement, Ed Harte nodded to me and said, "Well, she wrote the book, and I wrote the check."

In the twenty-year history of the Harte Research Institute (HRI), many books have been written as a consequence of the energy, resources, and intellect HRI has attracted.

In 2001, Wes Tunnell and I were reflecting on how much has been learned about the Gulf of Mexico and how much has changed since the publication in 1954 of US Fish and Wildlife Service Bulletin 89, *The Gulf of Mexico, Origin, Waters, and Biota*, edited by Paul Galtsoff. Both of us had relied on this inch-thick volume as the baseline for what was known about the Gulf during our early years of research—for Wes on micromollusks, for me on marine algae. We agreed that a fifty-year update on the state of the Gulf was overdue. In his typical "let's get it done!" fashion, Wes began implementing an assessment of the biology, geology, physical oceanography, and social and economic status of the Gulf, a monumental vision that he shared with Darryl Felder, a University of Louisiana biologist with a passion for decapods, especially the enormously diverse ten-legged creatures most people know as crabs, shrimp, and lobsters. Owing to the commitment of Tunnell, Felder, and hundreds of contributors, as of this writing, seven updated volumes of *The Gulf of Mexico, Origin, Waters, and Biota* have been published, with *Volume 1, Biodiversity* alone comprising 1,400 pages of text and photographs and weighing in at a hefty 7.5 pounds.

In 2005, Robert Furgason retired as president of the university and became the first executive director of HRI, with Wes Tunnell serving as associate director. They spearheaded what has become known as the "Harte Model," an interdisciplinary center organized into areas headed by endowed chairs supporting leading experts in coastal and marine geospatial sciences, ecosystem studies and modeling, biodiversity and conservation science, fisheries and ocean health, marine policy and law, and socioeconomics. Two years later, Furgason retired again and Larry McKinney, an energetic and widely esteemed conservation biologist, was lured out of retirement from his former role as senior director of aquatic resources for the Texas Parks and Wildlife Department to lead HRI.

Early on, HRI initiated the GulfBase website, a searchable and sortable database that includes data on bays and lagoons, reefs and islands, and events to help researchers, institutions, policy-makers, and the general public work together in ways that are beneficial to the Gulf. Later, the Gulf of Mexico Research Initiative Information and Data Cooperative (GRIIDC) began implementing a research database to receive and process data collected throughout the Gulf. And, in January 2015, under Larry McKinney's leadership, Texas OneGulf, a consortium of nine institutions led by HRI, was designated by the Texas Commission on Environmental Quality (TCEQ) as a Center of Excellence.

When Hurricane Katrina struck the Gulf coast in 2005, a then-proposed "State of the Gulf" summit hosted by HRI had to be postponed for a year, but the delay provided an opportunity to bring together an unprecedented cross section of government, industry, science, conservation, and public participants strongly motivated to work together to deal with the impact of recurrent megastorms and of the destructive consequences of human activities that are causing relentless degradation of the ocean in the Gulf and globally. Interest was especially high since ocean health has been declining globally, and solutions that could be applied in the Gulf could inspire positive actions elsewhere.

The importance of the second HRI-hosted State of the Gulf summit, in 2011, was magnified by the aftermath of the catastrophic Deepwater Horizon oil spill that occurred in April 2010, a disaster costly in terms of human lives lost, families and communities disrupted, and immense damage to wildlife and ecosystems from the edge of the Gulf to the great depths of the ocean as well as far-reaching economic losses. HRI launched the comprehensive "Gulf of Mexico Ecosystem Report Card Prototype" to inspire actions to move from sweeping losses to recovery and long-term stability.

At the time of the Explorers Club dinner in Houston in 2000, I was two years into my appointment as a National Geographic Society Explorer in Residence, where I was charged with creating a project that they would help support. Coincidentally, I had the opportunity to apply for and receive a $5 million grant from the Richard and Rhoda Goldman foundation to use small submersibles to explore and document under the banner "The Sustainable Seas Expeditions," the nature of the thirteen United States National Marine Sanctuaries, managed by the NOAA. Details are given in my 2009 book, *The World Is Blue*, about how the five-year project attracted more than a hundred institutions, agencies, and industry partners with numerous volunteer scientists, teachers, students, and others.

In collaboration with HRI, one segment proved to be especially memorable: Wes Tunnell, Robert Furgason, and other HRI scientists trained to be pilots of the one-person subs DeepWorker and Deep Rover using an indoor test facility at Texas A&M University. With institute support, an expedition to Mexico's Gulf Coast was organized, and a Mexican Navy ship, Mexican Navy Admiral (and HRI advisor) Alberto Vasques, and forty cadets were enlisted to dive, document, and communicate results of explorations around protected reefs near Vera Cruz. Admiral Vasques and I share the same birthday, August 30, which we celebrated aboard ship and underwater in the reefs offshore from Vera Cruz in 2003, when coral spawning was documented for the first time in the Western Gulf.

The 1998 to 2003 Sustainable Seas Expeditions came at a time when realization was growing that actions were needed to scale up protections for places in the ocean equivalent to those for parks and other areas on the land. Populations of many large fish, including sharks, had declined by as much as 90 percent. Coral reefs, sea grass meadows, coastal marshes, and mangroves were disappearing through deliberate displacement by human activities or as a consequence of climate change, loss of diversity, ocean acidification, and other pressures. At the time, a fraction of a percent of the ocean was being deliberately safeguarded for fish and other wildlife. Gradually, momentum has grown nationally and globally for protection of places in the sea that are helping depleted species and ecosystems recover from a long history of exploitation.

In 2006, President George W. Bush designated the world's largest fully protected park, 139,818 square miles of the ocean surrounding the northwestern Hawaiian Islands, as the Papahānaumokuākea Marine National Monument. This action, coupled with growing evidence that protected areas "work," offering economic benefits as well as serving as effective sources of renewal for degraded areas, helped inspire other leaders to approve enhanced protection. Australia, New Zealand, the UK, Chile, Mexico, Costa Rica, and many other nations have made unprecedented commitments for marine protected areas. Tommy Remengesau, president of the Micronesian island nation of Palau, established 80 percent of the country's entire Exclusive Economic Zone as a fully protected region, maintaining the inshore 20 percent for managed use. In 2016, President Barack Obama expanded the Papahānaumokuākea Marine National Monument to the edge of the two-hundred-mile Exclusive Economic Zone, encompassing 583,000 square miles of ocean where even fish are safe.

As of 2020, about 3 percent of the ocean within the jurisdiction of individual countries had full or highly protected status, a significant increase since the first marine parks were created—Australia's Great Barrier Reef Marine Park Authority and, in the United States, a shipwreck site off the coast of North Carolina and a small area near Key Largo, Florida. An additional 4 or 5 percent of ocean is being managed with the intent of sustaining healthy ecosystems while continuing to extract wildlife and otherwise subject the areas to various human uses. But the combined 7 to 8 percent is still well below the United Nations Sustainable Development Goal of 10 percent by 2020. A new target, 30 percent by 2030, is gaining traction as evidence grows that protection, especially full protection, of critical areas in the sea is one of the most effective, powerful, and realistic ways for people to do something to restore and protect ocean health.

In April 2010, a few days before the Deepwater Horizon oil spill, I was at sea in the Galapagos Islands in the company of more than a hundred guests of Technology Entertainment Design, TED, the organization that famously gives individuals opportunities at an annual conference to present a three-to-eighteen-minute "talk of their lives" that will then be communicated to a global audience. From time to time, they also bestow a prize of money and fulfillment of a wish "big enough to change the world" to individuals who are then helped by the TED community to make the wish come true. In 2009, I was one of three chosen to receive the prize, and the Galapagos expedition aboard the National Geographic–Lindblad ship, *Endeavor*, was made possible by the contribution of an individual who wanted to help implement this wish:

> *I wish you would use all means at your disposal—Films! Expeditions! The Web! New submarines!—to create a campaign to ignite public support for a global network of Marine Protected Areas, "hope spots" large enough to save and restore the ocean, the blue heart of the planet.*
>
> *How much? Some say ten percent, some say thirty percent. You decide: How much of your heart do you want to protect? Whatever it is, a fraction of one percent is not enough.*

The outcomes from the expedition included $16 million in contributions from participants made to various organizations that were working to establish ocean education, research, and protection.

One inspired individual backed the production of a film that was acquired by Netflix and aired globally in 2014 as *Mission Blue*. After the expedition, a nonprofit organization that I had recently started in order to develop and produce manned submarines for research and exploration morphed into the Mission Blue Foundation, dedicated to developing the Hope Spot concept.

Now, ten years later, Mission Blue has more than 200 partners, and working with the International Union for Conservation of Nature (IUCN) and local champions and communities around the world, more than 120 Hope Spots have been designated with the goal of winning enhanced and ultimately full protection for a network of protected areas. An early partnership with Google to highlight Hope Spots on their Google Earth system continues, and a relationship with Environmental Systems Research Institute (ESRI) is underway to incorporate data, images, and stories into the ESRI Story Map program that can be widely shared.

Some Hope Spots already have significant protection, such as the waters of Palau, the Galapagos Islands, and Mexico's Cabo Pulmo in the Gulf of California. Others such as San Francisco Bay and Florida's Gulf Coast have major opportunities for significant improvement.

In every case, partners are involved from other conservation organizations, universities, and local schools and businesses. In keeping with "the wish," a vigorous communication program reaches millions of people with stories and information about ocean issues.

The vision of developing and using submersibles for exploration and ocean care has a new home, an organization called Deep Hope. My daughter, Elizabeth Taylor, and her husband, Ian Griffith, have owned since 1997 a venture that I started in 1992, Deep Ocean Exploration and Research (DOER). With a team of engineers, they have created and operated dozens of innovative systems, including the manipulator arm and other equipment for the *Deepsea Challenger* sub piloted by James Cameron to the deepest part of the ocean, a 6,500-meter remotely operated system used by the University of Hawaii, and camera systems that can document the presence of oil in deep-sea sediments. Their design for a pair of 1,000-meter subs that can accommodate a seasoned pilot and two passengers, any of whom can drive the sub and operate the arms, cameras, and other equipment, is at the core of Deep Hope's plans for engaging young explorers as well as scientists, artists, and many others who are making decisions about the ocean but who have never experienced what it is like to be under the sea.

With a growing sense of urgency, and with new technologies to help, exploration of the ocean is rapidly advancing, just in time to inform critical decisions about the way forward.

Consider: The first circumnavigation of the world occurred 500 years ago, at a time when it was widely accepted that Earth was at the center of the universe. The discovery of oxygen as a basic element did not occur until 1772, and soon thereafter, the composition of water—H_2O—was verified. In 1872, the first global expedition dedicated to ocean exploration took place, the voyage of the HMS *Challenger*. When my father was born in 1900, rapid transit meant riding a fast horse, but within twenty years, cars, especially Henry Ford's Model T, had become common. Air travel and access to space followed, with the first circumnavigation of the moon occurring in 1968. As the *Apollo 8* spacecraft emerged from the far side of the moon, astronaut William Anders recorded the iconic photograph *Earthrise*, an image of a small blue sphere suspended within an infinity of sparkling planets, moons, and stars devoid of trees, birds, or any other life as we know it.

In 2000, thousands of scientists set out to inventory life in the ocean, top to bottom, in an ambitious ten-year "Census of Marine Life." Museum specimens were examined; publications, logbooks, and expedition journals searched; and new expeditions launched. They could account for about 260,000 species but estimated that 1 to 10 million remain to be discovered, not including many millions of microbes. A new International "Decade of Ocean Research" was launched in 2020, and an international consortium of researchers plans to produce an accurate map of the seafloor within the next ten years.

There are many reasons to feel a sense of despair about the state of the world since the publication of *Sea Change* twenty-five years ago. After all, it took 4.5 billion years for Earth's living systems to transform barren rocks and water into the vibrant planet we know as home, and it has taken humans about 4.5 decades to unravel those systems, putting our existence in jeopardy. Ten years ago, in *The World Is Blue*, I wrote:

> *The next ten years may be the most important in the next 10,000 years—the best chance our species will have to protect what remains of the natural systems that give us life. To cope with climate change we need new ways to generate power. We need better ways to cope with poverty, wars, and disease. We need many things to keep and maintain the world as a better place. But nothing else will matter if we fail to protect the ocean. Our fate, and the ocean's, are one.*

Now, ten years later, a January 2020 report by NOAA and NASA confirmed that 2010 to 2019 was the hottest decade since record keeping began 140 years ago, that 2019 was the hottest year ever recorded, and that ocean temperatures are the highest they have ever been. The UN Intergovernmental Panel on Climate Change (IPCC) reported in 2018 that the world has just over a decade to get climate change under control before irreversible "tipping points" are reached with catastrophic consequences for life on Earth.

But as grim as these facts are, there are also reasons for optimism, and knowledge is the key. In ignorance, humans have unwittingly consumed much of the natural world that not only has fostered our prosperity but also maintains Earth as a safe haven in a formidably inhospitable universe. Now we know what could not be known during most of the time humans have existed, and the pace of discoveries is accelerating. The ocean is the cornerstone of Earth's life-support system, and we must take care of it as if our lives depend on it, because they do.

ACKNOWLEDGMENTS

Several special people made this book possible, especially Dan Strone, whose persistent calm encouragement provoked me to write it; Phyllis Grann, whose endorsement inspired me to believe that I could—and should—get on with it; Laura Yorke, who championed the project, remained patient through its long incubation, and, as my editor, skillfully nudged the rough-edged manuscript into smoother prose; and to Eileen Cope for her steady, good-natured help from start to finish. Special thanks are due Francesca Cava, Linda Glover, Elaine Harrison, Joan Membery, Marian Rivman, and John Robinson for readily sharing their time, friendship, and valuable insights. My daughter Elizabeth has been a cherished partner in the endeavor, bringing a never-run-out supply of ingenuity, common sense, humor, and help. I am grateful to the rest of my family and friends, and especially to my mother and to Richie and Gale, who live many miles away and have been cheerfully understanding during months of serious neglect while I took the necessary long, deep dive into piles of notebooks, papers, libraries, and records of personal experiences.

The book has been strengthened by advice provided by those who willingly waded through all or part of various versions: Harold Humm and Suzie Humm, David Attenborough, Lynne Carter, Arthur C. Clarke, John Craven, Charlotte Delahay, Hugh Downs, William Eichbaum, Kathryn Fuller, Gale Hey, Graeme Kelleher, Reeve Lindbergh, John Loret, Roger McManus and the staff of the Center for Marine Conservation, Scott McVay, Sir John Rawlins, Bruce Robison, Gregory Stone, Kathryn Sullivan, Elizabeth Taylor, Leighton Taylor, Don Walsh, Geraldine Wenz, Robert Wicklund, and Edward O. Wilson.

I especially thank Al Giddings for the many one-of-a-kind adventures we have shared, and for use of several of his memorable photographs. Others who have contributed photographs, and to whom I am very grateful, are Ibraham Alam, Dale Andersen, John J. Domont, Charles Fowler, William Harrigan, Peter Hill, Tom Iliffe, Bates Littlehales, Amos Nachoum, Phil Nuytten, John Robinson, Dirk Rosen, Ken Sakamoto, Flip Schulke, Harold Vokes, and Robert Wicklund.

I want to acknowledge the wisdom imparted by Sir Peter Scott, Maurice Strong, Russell Train, Paul MacCready, and my friends at Deep Ocean Engineering, Inc., the California Academy of Sciences, Duke University, the Radcliffe Institute, the Farlow Herbarium at Harvard University, the World Wildlife Fund-U.S., World Wide Fund for Nature-International, the Charles A. and Anne Morrow Lindbergh Foundation, the National Geographic Society, and the Explorers Club for many inspirational years of association during development of my personal spin on the concept of finding a balance between technology and nature, and the need to harmonize policies for both a sound environment and a sound economy.

I am grateful to the Center for Marine Conservation for contributing many vital statistics, and, together with the Caribbean Marine Research Center and their Sea Change Trust, for providing a "spiritual home" for the *Sea Change* ethic. And I would be remiss not to acknowledge the vital part played by my impatience with certain aspects of the U.S. government bureaucracy. I treasure the experiences, insights, and friendships gained while serving as Chief Scientist of the National Oceanic and Atmospheric Administration (NOAA) but this book could be written only with the priceless independence and freedom that comes with being a private U.S. citizen.

Finally, I want to thank the fish and countless other sea creatures who first lured me into the sea, and who have inspired a fierce sense of urgency about understanding their realm—and caring for it.

INTRODUCTION

"Suppose the oceans dried up tomorrow. Why should I care? I don't swim. I hate boats. I get *seasick!* I don't even like to eat fish. Why should I object if some of them—or all of them—go extinct? Who needs the ocean?"

The questions were fired at me without pause by an impish young Australian standing along the shore near Melbourne. She thrust a microphone under my chin, and smiled expectantly. I was a featured speaker at a 1976 conference about the oceans, and anticipated answering questions about what it was like to live underwater, meet sharks, dive on shipwrecks, or perhaps to tell about exploring unknown depths, about going where no woman—or man—had gone before.

Groaning silently, I thought, *Good grief! Can she be serious? Suppose the oceans dried up tomorrow! What a concept! Who needs the ocean? Who doesn't need the ocean!* I glanced at the rippled edge of the vast, sparkling blueness that dominates the planet, embraces islands and continents, shapes the character of climate and weather, and from the sunlit surface to the greatest, darkest depths seven miles down is home for most of life on earth. Then I said, with a sweep of my arms:

"Right, dry up the oceans. Think of all the good stuff lost at sea that you could just scoop up. The trouble is, there wouldn't be anybody around to do that. Without an ocean, there would be no life—no people, anyway."

"Well, how so?" she prodded. "People don't *drink* saltwater."

"Okay," I began. "Get rid of the ocean, and Earth would be a lot like Mars. Cold, barren, inhospitable. Ask those who are trying to figure out how astronauts can live there. Or, how about the moon. There's a place with no bothersome ocean. And no life. Or Venus. Yes, the beautiful—and lifeless—hot planet with no ocean. It doesn't matter where on Earth you live, everyone is utterly dependent on the existence of that lovely, living saltwater soup. There's plenty of water in the universe without life, but nowhere is there life without water."

I paused, looked back at the incoming waves, then added—hoping I was in tune with the reporter and her audience—"The *living ocean* drives planetary chemistry, governs climate and weather, and otherwise provides the cornerstone of the life-support system for all creatures on our planet, from deep-sea starfish to desert sagebrush. *That's* why the ocean matters. If the sea is sick, we'll feel it. If it dies, we die. Our future and the state of the oceans are one."

The importance of the ocean to life on Earth seemed so obvious to me, a thoroughly academic but perpetually waterlogged marine scientist who had spent years at universities and research laboratories as well as at sea, and thousands of hours diving in the Atlantic, Pacific, and Indian oceans. That and weeks of living underwater in submerged quarters and many hours cruising the depths in small submersibles were practical "fins-on" preparation for a career that would include starting and running a company to design and operate underwater equipment and traveling to more than sixty countries for lectures, research, and meetings on ocean policy.

In retrospect, the questions posed by the tongue-in-cheeky Australian were mild compared to later grilling by many who really do *not* see the relevance of the ocean to their everyday lives. Years later, as Chief Scientist of the National

Oceanic and Atmospheric Administration (NOAA)* from 1990 to 1992—where I was known to my friends as the U.S. "sturgeon general," charged with looking after planetary health—I came into contact with a startling number of bright and influential people who seemed not to know or care that the sea is changing, and that the fundamental underpinnings of all that we hold dear as humans is jeopardized by such ignorance and indifference. I wondered what could jolt people from their complacency, could make them notice the dangers of overfishing, of poisoning the sea with toxic chemicals that return to us in intricate but inevitable pathways, and of blindly tinkering with Earth's life-support mechanisms.

I have tried to think of other responses to those simple-sounding questions about why the oceans matter. Sometimes, I try to imagine what intelligent aliens, viewing Earth from afar, might think about the sea. From their perch in the sky, they could immediately see what many earthlings never seem to grasp: that this is a planet dominated by saltwater! In fact, the ocean is the cornerstone of the systems that sustain us: every breath we take is linked to the sea. Clouds of freshwater are lofted from the sea surface to the atmosphere as vapor and return there, via the land, as fog, rain, sleet, and snow. Swirling hurricanes, billowing thunderheads, wispy cirrus clouds, are born in the ocean. This vast, three-dimensional realm, accounting for 97 percent of Earth's water, also makes up more than 95 percent of the biosphere, the planet's "living space." NOAA biologist Nancy Foster says it succinctly: "*Earth* is a *marine* habitat."

Sometimes I try a poetic approach and describe how luminous, rainbow-colored jellies, starlike planktonic creatures, giant squid, translucent pink prawns, gray dolphins, brown lizards, spotted giraffes, emerald mosses, rustling grasses, every leaf on every tree and all people everywhere, even residents of inland cities and deserts who may never see the sea, are nonetheless dependent upon it.

---

* NOAA accounts for about two thirds of the budget of the Department of Commerce and has five divisions: the National Weather Service, the National Ocean Service (incorporating the U.S. Coast and Geodetic Survey), the National Environmental Satellite and Data Information System, Oceanic and Atmospheric Research (including a network of national laboratories), and the National Marine Fisheries Service. It is managed by a team of four, including the Administrator (the Undersecretary for Oceans and Atmosphere) and the Chief Scientist. No preparation is adequate for such a job, but years of direct contact with the subject matter helped.

Often I quote the eloquent bard of science Loren Eiseley in *The Immense Journey*:

> *If there is magic on this planet, it is contained in water. . . . Its substance reaches everywhere; it touches the past and prepares the future; it moves under the poles and wanders thinly in the heights of air. It can assume forms of exquisite perfection in a snowflake, or strip the living to a single shining bone cast upon the sea.*

Occasionally I draw upon Ogden Nash's "A Beginner's Guide to the Ocean":

> *It is generally understood to be the source of much of our rain,*
> *And ten thousand fleets are said to have swept over it in vain.*

Sometimes, I am tempted to leave oceanography to Ogden, but thoughtful questions posed about the significance of the sea to human survival and well-being demand deeper probing. To do so, I touch familiar pillars of my understanding of how things are, and peer over the edge of many great unknowns. Common sense forces me to consider first the incredible sweep of time that preceded this moment and the ocean's great age, relative to the infinitesimally small fragment of time enjoyed thus far by humankind.

Life began in the sea more than 3.5 billion years ago and has prospered on land less than half that long. Our distant primate ancestors entered the scene about 65 million years ago, while our history as humans can be traced back only about 5 million years. The roots of modern human civilization are much more recent, starting as the last great Ice Age gave way to more temperate times about ten thousand years ago.

In this ten thousand-year span our species has been successful, slowly increasing in number from thousands to millions. It took the entire history of humankind to produce, by 1800, a billion people. By 1930 there were 2 billion, and within fifty years that number doubled to 4 billion. Now nearing 6 billion we are well along the way to doubling that. At the present rate, this will occur early in the twenty-first century. Our soaring population may suggest success as a species, but the environmental price of modern civilization is high, and our prosperity may be short-lived.

In the rush to "develop" and use the legacy 4.6 billion years in the making, we have struck the earth like a slow-motion comet, wielding powerful new forces of change, rivaling and compounding the impact of natural storms, volcanoes, earthquakes, disease, fires—even, it now seems, nudging the grand and gradual planetary processes that cause ice ages to come and go.

At night, astronauts witness our cities glowing with otherworldly light created by the squandering of millions of years of fossil wealth in a geological moment. Highways crisscross the land, simultaneously forming pathways and barriers. Complex, resilient, and naturally productive ecosystems are disappearing in favor of geometrically regular plots bearing vulnerable, single-species crops. Cement and asphalt cover an increasingly large percentage of life-giving land and water. Other changes are less obvious: the removal of billions of tons of living creatures from the ocean in the past century and the addition of billions of tons of toxic substances.

For a species that prides itself on being "intelligent" and capable of "anticipating the future," and of making conscious plans that are in our best interest, what we are doing to the earth is a stark contradiction. Our cavalier attitude about the state of the sea is especially enigmatic, given the ocean's vital significance. Curiously, no one really knows what the consequences will be of ocean dumping, overfishing, oil spills, whale kills, and thousands of other thoughtless actions that chip and gouge away at the healthy functioning of ocean systems. This much *is* certain: We have the power to damage the sea, but no sure way to heal the harm.

How can we find out what is happening to the ocean? Through observations from ships, satellites, and manned spacecraft, the character of the sea surface is increasingly well known, but what of that enormously important life-filled space below? Despite the swift advances in knowledge about the earth in recent years, the fact remains that below the depth where most divers can venture, about 150 feet, little of the ocean has been seen, let alone explored. Instruments can be lowered to obtain selected samples, a few small submersibles can travel, briefly, to great depths, and once, in 1960, two men spent half an hour at the deepest part of the ocean, 35,800 feet down. Suppose the rest of the planet were known from techniques now used to explore the sea. Suppose knowledge of deserts and forests were based on brief excursions limited by the air supply carried around on the backs of the observers. Or suppose information about trees, foxes, squirrels, bears, ferns, and other terrestrial life had to be gleaned from crushed and mangled samples obtained by blindly dragging nets from somewhere high above. Imagine Jane Goodall's frustration if in her study of chimpanzees she were limited to present oceanographic techniques.

All things considered, it seems so reasonable that people *should* care about the oceans and *should* be driven by a sense of urgency about knowing more. One of the great unsolved mysteries of the sea is why they *don't*. An aquatic atmosphere covers most of the planet's surface, encompasses the continents, and provides a home for most of life on Earth, yet it remains for humankind inaccessible and unknown, by and large ignored, overlooked, or simply taken for granted. How is this possible?

I posed this question during a lunch in 1979 with that wise and worldly woman Clare Booth Luce, a congresswoman, playwright, former ambassador to Italy—and also an avid scuba diver. I was meeting with her to talk about our mutual interest in the fate of a vocal group of Hawaiian winter residents: the singing, spouting—and endangered—humpback whales. While we discussed ways to ensure their survival, it was natural to consider the broader issues of human attitudes toward the ocean.

"All the reasons for justifying going into space can be applied to the ocean, from basic curiosity and the pleasure of being there to scientific, commercial, military, and even lofty philosophical goals," I said. "You know, finding answers to the *big* questions: Where did we come from? Where are we going? How did life come to be? What has kept humankind from exploring the sea?" I asked. "Is it something basically human, such as the fact that we are land-dwelling, air-breathing mammals, blessed with lungs, not gills, and feet, not flippers? Lack of wings has not kept us from flying, but many seem more comfortable with the idea of going to the moon or Mars than descending into the depths of the ocean. Why do you suppose this is so?"

Before answering, she gently lifted a large, pink prawn from her salad (one of those extraordinary creatures whose ancestry precedes ours by nearly half a billion years), paused, and looked out at the blue Pacific beyond the shore of her stately beachside villa.

"Look, it's simple," she said. "I've thought about this a lot and am quite sure that it relates to human culture more than human nature." Casting her eyes toward a radiant bank of fluffy white Hawaiian clouds she said, "After all, *Heaven* is up there . . . and you know what's in the other direction!"

We laughed and began thinking of other everyday attitudes that serve to direct humans' interest skyward.

I said, "Yes, you often hear, Onward and upward! Never . . . Onward and downward!"

Looking pleased, she said, "*Nobody* wishes on a star*fish*. . . ."

It was my turn. "It's okay to feel as though you're flying—but who wants to be in over her head?"

Whatever the real reasons, we agreed that there is a tendency for most people to be complacent about the ocean and ocean life, and to regard the sea and all that it provides as somehow not as relevant to our everyday lives as terrestrial and freshwater environments. Although some are aware of pollution problems, declining stocks of fish, and pressures on populations of dolphins and whales, for most people concern for the state of the oceans is not a high priority.

This apparent indifference also may be related to the widely held view that the ocean is so vast and resilient that there is little reason to worry, either about what is put *in* or about what is taken *out*. Nearly two centuries ago Lord Byron[*] expressed this view when he wrote a few lines often quoted by those of us looking for ways to emphasize the slow pace of change in the sea:

> *Roll on, thou deep and dark blue ocean—roll!*
> *Man marks the earth with ruin—his control*
> *Stops with the shore.*

---

[*] From *Childe Harold's Pilgrimage*: Canto IV, 1817.

No more. Although it is indeed impossible to *control* the ocean any more than it is possible to dictate the course of the world's winds, Byron would be shocked if he could see the ruinous changes that have occurred in the sea since his time, especially in recent decades.

Two centuries ago, there were no forty-mile-long drift nets sweeping through the open sea, indiscriminately taking everything in their path, from targeted tuna to fragile jellyfish, turtles, and even whales. There were no mega-trawlers scraping entire ecosystems into their giant maws, no acoustic fish finders, no factory ships for processing catches at sea, no global marketing and distribution systems that make possible the appearance in Tokyo restaurants of creatures still living, though taken from half a world away. There also were no nuclear wastes, no nerve-gas disposal sites, no oil spills, no plastic debris, no pesticides nor herbicides nor fertilizer runoff from millions of acres of farmland, lawns, and parks, no sewage sludge, and many other substances not known in all of the preceding history of earth.

Since the 1800s, numerous species and entire complex living ecosystems many millions of years in the making have been decimated or significantly altered, from populations of whales and other large mammals to dozens of commercially valued fish species, all marine turtles, many sharks, and numerous small creatures including certain krill, crabs, and shrimp. Worldwide, the living network of microorganisms that shape the basic ingredients of the ocean's "living soup" has been tugged, the system nudged, with unknown consequences. Far too little is known about the earth's living processes to know or predict the specific consequences of our tinkering, but the outcome is not likely to be favorable for humankind.

Curiously, those who claim to believe that the earth and all living things were created by God in fact appear to place greater value on human works and the judgment of mankind. This alarming arrogance was lamented more than a century ago by the English scientist and philosopher Albert R. Wallace, who even then was appalled at the magnitude of extinction of living forms, which, as he said, "the progress of cultivation invariably entails." In 1863, he wrote in the *Journal of the Royal Geographical Society*:

Future ages will charge us with having culpably allowed the destruction of some of those records of creation which we had it in our power to preserve, and while professing to regard all living things as the direct handiwork and best evidence of a Creator, yet, with a strange inconsistency, seeing many of them perish irrecoverably from the earth, uncared for and unknown.

No one can say for sure what such disruptions may mean for human well-being or survival; clearly, however, a global experiment is in progress, and we are in the middle of it, as a part of, not apart from, the rest of life on Earth. Unlike most other participants, though, we have the ability to alter the course of events, and we shall, either through conscious decisions aimed at making a difference, or by default, through inaction or ignorance.

This is a time of pivotal, magnified significance for humankind. The fabric of life and the physical and chemical nature of the planet have been significantly altered through decisions already made by our predecessors and those now living; what happens next depends on what we do, or do not do, individually and collectively, in the next few decades. Depending on choices we make, our species may be able to achieve a viable, sustainable future, or we may continue to so alter the nature of the planet that our kind will perish.

As a child I sometimes longed for a way to cruise through time as described by H. G. Wells in his science-fiction thriller *The Time Machine*. It was exhilarating to imagine traveling into the past or future as easily as taking a trip to the beach. It would be grand, I thought, to stop by places I knew in, say, a hundred or a thousand years; the New Jersey farmhouse where my family lived, the beach at Ocean City where we went for vacations, and maybe even to great cities to see how people lived, what they ate, and whether or not they would like to hear the latest news from the past.

I also dreamed of zooming back in time, to see the earth when dinosaurs abounded—and long before, when trilobites, early relatives of horseshoe crabs, cruised the ocean depths, and great nautiluslike ammonites jetted about in diverse profusion. Skipping closer to the present time, I wished to be able to see for myself what the world had been like when my grandparents—and parents—were young. My father described the Delaware River as a magical place ringing with the laughter of summer swimmers, a place filled with clear water and large fish that watched small boys with apparent curiosity. It did not sound at all like the Delaware River I knew as a child, a place already reeling from upstream pressures that forever changed the nature of that once immensely productive waterway.

My parents, born in 1900 and 1902, grew up on small farms still surrounded by woodland, where flocks of birds darkened the skies, spring and fall. They had twenty kinds of apples, Seckel pears, garden corn, beans, and tomatoes every summer; fresh bread baked at home every day, hand-pumped well water, milk from their own cows, chestnut trees, and snow and rain that fell wondrously pure, without a deadly cargo of exotic hitchhiking chemicals. They witnessed the arrival and impact of domestic electrical service, indoor plumbing, automobiles, paved highways, aircraft, radio, movies, television, supermarkets, world wars, pesticides, herbicides, antibiotics, nuclear power, space travel, satellites, electronics, computers, and thousands of elements of everyday life that my children simply take for granted as a natural part of being a human being.

As alluring as times past and future are, and as much as I wish to be able to rush back to 1800 with urgent messages about the consequences of certain foolish decisions made along the way—or to leap ahead, to see what actually happens—on balance, if I had to choose the most interesting and important time in all of human history to live, it would be now. As never before, and perhaps as never again, the choices made in the near future will determine mankind's success, or lack of it. *These* are the "good old days" sure to be envied by those in the future.

Amazingly, the urgent messages I would take to those who lived long ago are precisely the same as the concerns I wish to make known today. They derive largely from thousands of unusual hours spent in places witnessed by few others, often under circumstances that cannot be repeated. In the chapters that follow, I want to share the exhilaration of discovery, and convey a sense of urgency about the need for all of us to use whatever talents and resources we have to continue to explore and understand the nature of this extraordinary ocean planet. Far and away the greatest threat to the sea and to the future of mankind is ignorance. But with knowing comes caring, and with caring, the hope that maybe we'll find the Holy Grail of understanding, strike a balance with the natural systems that sustain us, and thus achieve an enduring place for humankind on a planet that got along without us for billions of years and no doubt *could* do so again.

Full fathom five thy father lies;
Of his bones are coral made;
Those are pearls that were his eyes;
Nothing of him that doth fade,
But doth suffer a sea-change
Into something rich and strange.

Shakespeare, The Tempest *(1611)*

SEA CHANGE

Part I

SEA OF EDEN

Chapter 1

DEEP TIME

High up in the North in the land called Svithjod, there stands a rock. It is a hundred miles high and a hundred miles wide. Once every thousand years a little bird comes to this rock to sharpen its beak. When the rock has thus been worn away, then a single day of eternity will have gone by.

Hendrik Van Loon, The Story of Mankind

I f the ultimate historians, geologists, were to show the full history of Earth vertically on a scale as long as the depth of the deepest sea, all of human history, about ten thousand years, would fit nicely in the uppermost inch—about the depth of a depression made by a sea gull lightly riding on the surface. It is a staggering concept!

I gaze into the depths and long for a magic submersible to descend through the eons, to get a better feel for the place of humankind in the greater scheme of things and for the significance of swift changes now taking place. Something, possibly the ten-year-old that dances within my decades-old frame, tugs irresistibly and insists that I dive into the ages to glimpse the extraordinary events that have led to and made possible the present moment. Perhaps along the way, some insight may provide answers to the questions posed by mankind: Where did we come from? Where are we going? Do we have some control over the outcome for ourselves as individuals and for our species?

If I could go on such a journey, one of the first places I would explore would be Florida's Gulf Coast a thousand years ago, to a world of deep reefs, shallow sea-grass meadows, and mangrove-bordered shores that I have climbed, swum, dived, and scrambled around during most of the years of my life. I try to imagine gliding into this ocean as it was at the beginning of the millennium—a sea filled with luminous night creatures, with starfish and sea hares, with bright-eyed puffer fish gentle as cows, and rays butterflying by with slow-motion grace. Above, skimming the sea, there surely would be pelicans, wingtips teasing wave tops, and mullet plunging silver into clear blue sky from water as clear, as blue! Below, there would be no glint from beer cans and bottles, no windrows of cigarette butts, no plastic cups, bags, and bits, or PCB's, DDT's, polystyrenes, or other exotic concoctions. There would be plenty of sounds in the sea, from subtle snaps and sizzles of small crustaceans to warbles, grunts, pops, and hundreds of other variations produced by fish and marine mammals—but no throb of engines, no ping of depth sounders, no low rumble of mechanical or electronic subsea thunder.

But what, I would then start to wonder, was Earth like a million years ago, or ten million, or a billion—a time before birds, fish, or large creatures of any sort? A thoughtful venture into deep time might provoke new perspectives on what it has taken to produce the elements that many regard as the property of mankind. Perhaps, after returning from a visit far back into the history of the planet, certain creatures would be treated with more respect—especially those durable ones, like jellyfish and sharks and crustaceans, that have survived, more or less intact, through hundreds of millions of years. Perhaps complex natural systems would be understood for what they are: the distillation of the processes of all preceding time, and the source of the ingredients required for human survival and well-being. It is just possible that a spark of humility could be generated, a flash of awareness about how each species—even *Homo sapiens*—is linked to all others.

The geologist Don Eichler, in his slim volume *Geologic Time*, suggests another way to illustrate the magnitude of time: compressing the whole 4.6 billion–year history of Earth into one year. Considered this way, our generation would be embraced within the last fraction of the last second. Three seconds back, Columbus was setting sail for North America, an instant after the Aztecs built their capital, Tenochtitlan. Zip back ten seconds, and see the end of the Roman empire. Push on for a minute or so, to the close of the Pleistocene Ice Age, when the most recent continental ice sheets began to recede from northern Europe and the Great Lakes area of North America, coinciding with the beginnings of agricultural pursuits by humankind, and the first inklings of modern civilization. To most people, this most recent sixty seconds—if not the most recent millisecond—of Eichler's compressed year is what really matters.

But not to all. Some people—and I am among them—would like to know what preceded this geological moment. When did Earth first become blue with an ocean, green with forests, and blessed with an atmosphere congenial to the likes of us? Why Earth? Why not Mars? What sets this planet apart from the numerous lifeless, oceanless satellites orbiting the sun and similar stars elsewhere in the universe?

Earth scholars who make it their lifetime business to analyze the evidence derived from fossils, from rocks and sediments, and dozens of other sources of subtle clues concerning what has gone before, have some answers. It appears that the oceans formed early on, during the first billion years of Earth time, evidently a by-product of volcanic processes that can still be observed. As hot water or steam escapes from molten rock, clouds of vapor are lofted into the atmosphere, condense in the cool air above, and fall back on the land and sea. Much of the sea existed as ice in the earliest days, when the sun was purportedly one third as bright as it is now. Thawing took place gradually as light intensity increased, and as the earth responded to bombardment and battering from numerous meteorites and other cosmic debris.

The most likely places for life on Earth to begin and prosper may have been deep in the sea, perhaps in hot mineral-laden springs associated with volcanic activity. Another possibility is that life originated in warm, shallow lagoons, or even that life began more than once under varied circumstances. Whatever the actual place or places of origin, it appears that life started in the sea, in the springtime of the compressed year—in May, or perhaps sooner. Gradually,

Earth became a vibrant, and ever-changing living system. Billions of minute, active components drew energy from the sun, directly or indirectly, and through countless transformations, yielded oxygen, carbon dioxide, and water back to that system, slowly developing a protective atmospheric cocoon over land and sea.

It is impossible to comprehend the enormous changes that certainly occurred during the more than two billion years when life on Earth was entirely microscopic. Giants that we are (in terms of size, we are in the top five percent of animal organisms), it is difficult for us to imagine the significance of the microbial life that is all about us *at present*—let alone the chemical and biological workings of deep time and the enormous changes that occurred when much of life on Earth resembled pond scum. It is clear, however, that the habitability of the planet and our success as a species are possible only because we, and the whole living system, stand on the shoulders (if they had any) of microbes.

No words can meaningfully convey the magnitude of time suggested by "April to November," the three-billion-year period during which photosynthesis began and an atmosphere congenial to life as we know it gradually came into being. In early fall of the compressed year something triggered a major change in the way life could be organized, and the presence of a critical volume of oxygen seems to be a key. Within a matter of hours of Eichler time, days at the most, a dazzling array of multicellular creatures developed in the oceans.

It is hard to comprehend that life in the oceans was and still is shaped primarily by ancient ocean processes involving continuous interactions among living organisms, and that even today, the history of life on Earth can be read most completely in living sea creatures, responding still to the rhythms of deep time. Most of the major distinctive categories of animal and plant life that ultimately evolved are still represented in the ocean, including many that are found only there; only about half have ever become established on the land. But despite this explosion of diversity, most of the basic chemical processes that shape the nature of Earth, from the fixing of energy through photosynthesis to the breakdown and decay of organic material, continued then as now through the action of microbes.

The advent of terrestrial creatures is recent in comparison to that of marine dwellers, and many were transients in the symbolic year. The immense swamps that formed coal deposits and have yielded massive fossil impressions of luxuriant forests that spread over much of the earth lasted for about four days in early December. Dinosaurs were abundant in mid-December but disappeared on the twenty-sixth, about the same time that certain hippopotamuslike mammals moved from terrestrial haunts into the sea and gave rise to whale-kind, and small, land-dwelling primates took to the trees. Manlike creatures turned up sometime during the evening of December 31.

One evening within the last second of the last day of December, a great ocean explorer, Hans Hass, spoke. "I feel sorry for my daughter, Meta," he said after a fine dinner, two glasses of wine, and many sea stories. "When I was her age, seventeen, the whole world was unknown. I was the first to take pictures of sharks underwater in the Caribbean, in the Red Sea, in the Galápagos—and the first to dive in many parts of the world. Lotte and I have been so lucky. It's all over now. There's nothing really new left to do."

Within the same second, a famous astrophysicist, speaking in New York in 1992, proclaimed that since Earth has been "fully explored," it is time to focus increasing attention on other planets in the solar system, and beyond. Perhaps the strongly arched eyebrow of the person sitting next to him, the oceanographer-astronaut Kathryn Sullivan, and a squinty-eyed glare from my direction provoked a modest concession: "Well, as Sylvia Earle and others here"—glancing at Kathy—"might point out, the *oceans* are not fully explored. . . ."

Not long before, at the prestigious annual dinner of the Explorers Club in the Grand Ballroom of the Waldorf Astoria in New York City, a distinguished mountain climber told how he had satisfied a lifetime dream by personally reaching the top of the last remaining previously unclimbed peaks on Earth. He added that he felt fortunate to have lived when he did—while there were still unconquered mountains—and lamented that those coming along could no longer have that special thrill of discovery and accomplishment.

A few nanoseconds earlier, during that last second in the compressed year of geological time, I, an aspiring student of oceanography, got the news about the existence of the planet's single most prominent geological formation, a 40,000-mile chain of mid-ocean mountains, unknown until a few decades ago. Included are thousands of peaks, perhaps made even more challenging than their above-water counterparts in that would-be climbers would have to *descend* to get to the tops of most. (A few, such as the Hawaiian Islands, poke through.)

In school, some of my professors had told me not to be fooled by the seductive way that continents seem to fit together when you look at a map of the world, and to laugh at the notion of "continental drift," a concept now vindicated by years of deep-sea research. If entire mountain ranges and a phenomenon as profoundly important as plate tectonics and the resulting movement or "drift" of continents could escape human notice until the latter half of the twentieth century, what else might await discovery in the ocean depths?

This question was put to me in 1982, when I was a member of a panel of advisers to the "Living Seas" pavilion at the Disney Epcot Center in Orlando, Florida. As a scientist notorious for using underwater technology—from subsea laboratories to submersibles and robots—for ocean research, I was among those asked: What new things are likely to be found in the ocean in the next fifty years? What will the oceans be like, and what new technologies will be in use? Will there be underwater cities? Might it be possible to perfect "liquid breathing" by flooding our lungs with hyperoxygenated fluid, or use some version of an artificial "gill"?

Kym Murphy, the project director, wanted to inspire people, get them excited about a bright future and thus help create awareness and greater care—for the present. Disney's clever "Imagineers" had conjured up new exhibits to complement the largest saltwater aquarium then built, nearly six million gallons, and suggested an underwater station outfitted with technology *they* thought might be developed far into the twenty-first century. The question was, what did *we*, the advisers, think? The crystal ball looked murky. Who could tell what the next few decades would bring?

It would be hard, I thought, to convincingly portray what it is *really* like to live fifty feet underwater, gliding like silent birds through liquid sky . . . then, still submerged, to pass vertically through a round door, shimmering silver, into an atmosphere of air peculiarly heavy—dense enough to cause voices to become husky with the effort of talking and turn attempts to whistle into pathetic *whooshing* sounds. Still, I favored the concept of the "Sea Station," thinking that a well-crafted simulation could jar loose the can-do spirit that lurks within us all, and perhaps lead to the creation of more of the real thing. Few people, even experienced divers, have spent weeks, or even days, as residents underwater. Most such diving involves not just compressed air, but exotic gas mixes breathed by highly skilled commercial or military divers who push the limits of what human physiology can withstand while exposed to ambient pressure, sometimes living and working at depths greater than 1,500 feet. To go deeper requires other methods—either wrapped warm and dry inside a one-atmosphere suit or sub-mersible, or vicariously, as the pilot of a remotely operated vehicle equipped with camera eyes and other sensors. These, too, should be "imagineered," I urged, to show what underwater access could be like in a few years—if we can just slip the bonds of complacency and get going!

But, I added, "Even more important than dreaming up new technology is coming up with creative ways for people to see creatures that they might encounter during real dives in the deep sea, perhaps with films or special viewing techniques. There's nothing like a jaw-to-jaw encounter with a friendly shark or whale to get the juices running," I advised. I also strongly urged the creation of exhibits dealing with what has been taken *out* of and is being put *into* the oceans; exhibits that show what can be done to help.

"In our lifetime, we have witnessed what is literally a *sea change*," I said. "Our generation came along at a time when natural ocean systems were still largely intact. In a few decades, our species has squandered assets that have been thousands of millennia in the making—and we're still doing it! Even children have witnessed changes caused by trashing beaches that were pristine a few years ago. What we must do is encourage a *sea change* in attitude, one that acknowledges that we are a part of the living world, not apart from it."

Ideas zinged back and forth. How successful will aquaculture be? Will humans be able to communicate with dolphins? Will there *be* dolphins fifty years from now, given the present worldwide rate of slaughter, reduction in food supply, and pollution? Will there be new sources of power? What new materials will be around that might make ocean access easy? What will become of possible sea-space technology parallels?

William Nierenberg, for many years the director of the Scripps Institution of Oceanography in San Diego, was skeptical regarding predictions, and exhibits based on them. "As a young oceanographer I was on a panel convened to anticipate important new developments for the decades ahead," he said. "I have now lived long enough to see some of the things we missed: going to the moon and the space program generally, the significance of nuclear energy, lasers, the electronics revolution, the impact of computer technology, the revolution in genetics, biotechnology, and the influence of new materials! Plastics, ceramics, carbon fiber, to name a few. So much has happened so fast, and the pace is accelerating. I don't want to make a fool of myself again by pretending to know what might happen in the *next* fifty years!"

Robert Ballard, the famous ocean explorer, then still dreaming of finding the *Titanic*, agreed. "If the exhibits can show that the ocean generally is not just what people see on television—clear, blue, warm water filled with coral reefs and pretty fish—but rather a really immense arena that is mostly cold and dark, filled with the tallest mountains, broadest plains, steepest valleys on Earth . . . if everyone understood what is now known by a handful of ocean scientists, *that* would be an exciting breakthrough. The true nature of the sea, as we now know it—and *don't* know it—would be infinitely more exciting than futuristic fantasy."

As Ballard often says, pacing with an urgency and look of longing that suggests that he is just on his way to do something about the situation, "Less than one tenth of one percent of the deep sea has been explored." Mapped? Yes. Probed with electronic sounding devices? Generally, yes. Observed? Touched? Sampled? Understood? No, not by a long, long way.

School-room maps tend to depict continents in exquisite detail, but show the oceans as big blobs of blue, with no suggestion of the immensely important diversity of features, and creatures, below. I loved the idea of filling in the blanks—and of encouraging all who would look and listen to explore for themselves and help resolve the ocean's mysteries. The biggest problem, we all agreed, is the magnitude of ignorance.

The brainstorming session ended with a sobering but thrilling theme: The greatest era of exploration on Earth, venturing into the vast unknowns of inner space, has barely begun.

Chapter 2

EXPLORING THE SEA

"I love you," I whispered into the ear of the ocean. "Ever since I've known you, I've loved you. I must see all your marvels, know all your beauty. . . ." And the ocean listened and snuggled still closer to me.

Hans Hass, Diving to Adventure

In my dreams, a monstrous wall of green water races my way, hissing, roaring, towering, inescapable, sweeping me into a cascading aquatic mayhem. I am lifted, tumbled, churned, pushed, and fall, gasping, clawing for air. My toes touch sand; a sweet breeze soothes my lungs. I stand, choking, face the next advancing wall, and leap into it, exhilarated!

In reality, when I was three the ocean along the New Jersey shore first got my attention much as it happened in the dream: A great wave knocked me off my feet, I fell in love, and ever after have been irresistibly drawn, first, to the cool, green Atlantic Ocean; later, to the Gulf of Mexico, warm and blue, serving as my backyard and playground through years of discovery; and thereafter to other oceans, to reefs, raging surf, calm embayments, steep dropoffs, and the farthest reaches of the deep sea beyond. The "urge to submerge" came on early and continues, seasoned and made more alluring by thousands of underwater hours, each one heightening the excitement of the last as one discovery leads to another, each new scrap of information triggering awareness of dozens of new unknowns.

The lure of the sea has enticed explorers to probe the mysteries of that vast, sparkling wilderness, probably for as long as there have been human beings. Our origins are there, reflected in the briny solution coursing through our veins and in the underlying chemistry that links us to all other life. We are probably the most versatile of creatures, anatomically gifted with an ability to climb mountains, swing among treetops, leap into the air, race across plains, and briefly enter underwater realms. While we are not naturally equipped with wings to remain aloft or gills to stay submerged for long, we *are* endowed with ingenuity, and have thus been able to respond to another human gift, especially evident in children and those who happily never quite grow up: an irrepressible curiosity. The result has been the creation of an expanding wealth of technology that extends human capability into every place on the surface of the earth, to the deepest parts of the sea, and even into environments inhospitable to any other life form, far beyond Earth's atmosphere.

To some, the word "technology" conveys the specter of an overly mechanized society, a loss of contact with nature, a despoiler of civilization. Yet, without machines to take us into the sky, we would be as earthbound as elephants; without submarines and other specialized diving equipment, our ability to explore the oceans directly would be approximately equivalent to a dolphin's ability to glimpse its above-water realm. And without modern means of discovery and communication, there would be no hope of identifying the critical changes sweeping the planet and of alerting the global community to growing threats to our species. Without technology, the sea would be as fathomless* as the distant stars.

Without special equipment, human beings, like most other air-breathing creatures, cannot live underwater for long. With some practice, I can hold my

---

* A curiously appropriate word, now synonymous with "incomprehensible." The word "fathom" was derived from the Old English *faethm*, meaning "the embracing arms." It was once defined by an act of Parliament as "the length of a man's arms around the object of his affections" and later became a nautical term for six feet. As a verb, "fathom" means to plumb the ocean depths, to probe their mystery.

breath for about a minute—time enough to swim around, touch bottom 30 to 60 feet down, then race back to the surface, to sunlight, to air. Throughout history, humans have been divers in oceans worldwide, many going significantly deeper, and some staying much longer. Over centuries of gathering food from the sea, divers in Japan and Korea—mostly women, known as *ama*—have honed breath-hold diving techniques to perfection, passing traditions from mother to daughter through generations. Even grandmothers dive well into their seventies. Although it is a swiftly fading art, displaced by sophisticated modern fishing techniques and made less rewarding because of sharp declines in desirable shellfish, a few families maintain the old ways. In Japan, the cool air of Hekura Jima Island sometimes resounds with the soft, shrill whistles of the *ama* as they exhale, creating the musical sounds Japanese poets call *iso nageki*, the elegy of the sea. Garbed in long, trailing wraps of white cloth thought to ward off sharks, these women repeatedly inhale and dive into cold water, towing a basket for their catch of abalone, snails, kelp, and sprigs of favored red and green seaweed. Certain South Pacific islanders routinely dive to 100 feet, often remaining submerged for two minutes, and at least five people have successfully dived (that is, they have lived to talk about it) to more than 300 feet, requiring a breathless four minutes. In 1976, Jacques Mayol, the real-life hero who inspired the popular theatrical film production *Big Blue*, was the first to descend to 312 feet (100 meters), a depth once considered to be a physiological impossibility for humans. It was thought that pressure at that depth would cause the thorax surrounding the lungs permanently to collapse, but Mayol—sometimes admiringly referred to as *Homo aquaticus*—not only survived with his lungs and chest intact, but has continued to press the limits in competition with several others who aspire to be the "deepest man alive." How deep *can* free-swimming divers descend? A Cuban-born Italian citizen, Francisco Ferreras Rodriguez, widely known as Pipin (pronounced "pi-PEEN") and thought by some to be more aquatic than terrestrial, is aiming for 500 feet. So far the record stands at 400 feet and is held by the agile underwater athlete Umberto Pelizzari.

Such extraordinary feats notwithstanding, humans are not by nature designed for diving—but many air-breathing creatures are. One group of animals, the reptiles, once were well represented in the sea, many with hauntingly familiar fish- or whalelike shapes. Fossils of graceful plesiosaurs, sleek seal-like sea serpents, and the fish-shaped ichthyosaurs, indicate a long history of reptilian adaptation to aquatic life. Like the dinosaurs, they failed to survive the cataclysmic events that changed the nature of the planet at the end of the Cretaceous period, some 65 million years ago, but one group of reptiles did: the sea turtles, who are magnificent divers and incomparable long-distance swimmers.

Nearly half a ton of turtle may be contained within the massive, flexible body of the leatherback, a species that can measure eight feet across and more than ten feet long when fully grown. Biologists Scott and Karen Eckert have drawn attention to the amazing deep-diving capability of this ancient and distinguished reptile. The Eckerts attached instruments to a 650-pound female who unexpectedly dived deeper and stayed longer than the instruments were designed to record, but the educated estimate is that she went past the maximum on the gauge, 3,300 feet (1,000 meters), to about 4,265 feet. Dives to more than 1,650 feet appear to be routine, but no one knows what the limit may be.

Of all the marine reptiles, the graceful, sculpted shapes of the fifty or so species of modern sea serpents, sea snakes, enchant me the most. Generally these are gentle, curious creatures that live in warm tropical areas of the Pacific and Indian oceans. Some stay submerged for more than an hour, cruising along the sea floor or languishing in the shelter of coral reefs, but diving biologist Ira Rubinoff has discovered that one species, the brilliantly beautiful yellow-bellied sea snake, stays underwater about 87 percent of the time as it wanders the open sea.

The only true marine lizard, the Galápagos marine iguana, swims with agile grace, thrusting its powerful scaly tail back and forth, arms and legs pressed close to its body. I have admired these unusual reptiles in action, munching on seaweed attached to black, volcanic rocks bordering their island home at the surface and down to depths greater than 100 feet. Some contend that these reptiles have made no special adaptations to their aquatic habitat, but their behavioral inclinations clearly set them apart from other iguanas and most other reptiles. Four species of marine crocodiles, all in tropical, coastal waters, and the related and well-known Florida alligator, *Alligator mississipiensis*, move readily back and forth between fresh- and saltwater marshes. All are good swimmers, but none are especially renowned as divers.

In the cool, clear waters around the Galápagos Islands, I have sometimes been disconcerted while cruising along well beneath the surface by swiftly moving black and white missiles that turn out to be penguins seriously pursuing a meal of small fish, or sometimes apparently just larking about. Seventeen kinds of penguins, as well as a fine array of auks, petrels, murres, cormorants, pelicans, grebes, loons, and certain ducks have been at the business of underwater exploration for much longer than humankind, and they have built-in adaptations to show for it. Many have webbed feet, some can withstand prolonged submersion, all are streamlined underwater, and most have a strong preference for the taste of fresh fish.

Three orders of mammals include marine species, including one group that prefers a diet of just plants. The members of the order Sirenia—"mermaids," dugongs, manatees, and the recently extinct Stellar's sea cow—are among the most placid of mammals. Little is known about the diving habits of any of them, but most tend to stay close to the rich, shallow meadows of sea grass and other vegetation in coastal waters.

Not so the order carnivora, of which there are two kinds of ocean-going otters, 18 species of seals, 14 kinds of sea lions and fur seals, and one species of walrus or the wholly pelagic order of cetaceans including 78 variations on the theme of dolphins and whales. Little is known about the behavior of marine mammals underwater, mostly because they are so much at home in the sea, and human observers generally are not.

I have often looked longingly at the speed, agility, and gamboling grace of dolphins, who sometimes fling themselves aloft with deliberate twists and spins that easily surpass the finest human gymnastic displays, all the while keeping pace with a boat speeding alongside at 20 knots, sometimes more, and regularly staying submerged for several minutes with no apparent stress.

Olympic medalist Matt Biondi, much admired as the fastest, although not the deepest-diving, *human* swimmer in the world, enlisted several of his fellow athletes, and Olympic medalists, to join him for a swim-in with a group of wild, friendly spotted dolphins in the Bahamas. The dolphins frequently chose to engage the swimmers, circling them and diving within touching distance. Biondi, enchanted but envious, reported: "The faster I would go to keep up with the dolphins, the faster they would go. They were always faster, always one up on me. . . . Since then I have often tried to imagine what it would feel like to be a dolphin."

In T. H. White's *The Once and Future King*, the young King Arthur, known as Wart, gazes from a bridge into cool, clear water on one especially hot afternoon and impetuously says, "I wish I was a fish." Obligingly, the wizard Merlyn takes off his hat, raises his cane, murmurs the appropriate incantation, and turns the boy into a perch, a long, lithe creature of "a beautiful olive green, with rather scratchy plate armour all over him," delicate pink fins, and a belly of an "attractive whitish color."

Then Wart discovers a wonderful thing. He is no longer earthbound. "He could do what men have always wanted to do, that is, fly. There is practically no difference between flying in water and flying in the air. The best of it was that he did not have to fly in a machine, by pulling levers and sitting still, but could do it with his own body. It was like dreams people have."

For me, such a dream came true in an experience shared with my three children, that in a way combined the wishes of Wart and Matt Biondi: to fly underwater in the company of a wild, free dolphin. Breaking the usual rule of "school comes first," I scooped up my small brood, ages 16, 14, and 8, and enlisted their help for a week of diving and exploring reefs while working on a research project on San Salvador Island in the Bahamas. I have wistfully watched thousands of dolphins during many years spent working in, on, around, and under the sea, often reveling in their exquisite mastery of ocean elements, but I had never encountered one that was willing to stay around for more than a moment in the presence of divers. I was skeptical about the existence of a wild dolphin at San Salvador who would "come right up to you," but my doubts went up in a puff of sea spray when a dark fin appeared in the distance, and a lone spotted dolphin, *Stenella longirostris*, locally known as Sandy, came straight for our boat. We stopped, looked, and leaped in.

My son, Richie, making a polite overture, swam dolphinlike, undulating his whole body, holding his legs tightly together, and thrusting upward with his flippers, which sent Sandy into spirals of apparent delight. The eldest, Elizabeth, blessed with a streaming mane of shining golden-red hair, was an irresistible lure. Approaching close and peppering her with rapid, staccato sounds and soft, high *weeeps*, the dolphin mouthed locks of her hair, then, eyes closed in a look of apparent bliss, gently let strands flow through his teeth, as if trying to guess the nature of this intriguing, silky substance. Gale, an elfin eight-year-old, was the only one of us petite enough to hitch a ride. Looping her small fingers along the leading edge of Sandy's dorsal fin, she allowed herself to be towed in a circle around us, propulsion provided by thrusts of the dolphin's muscular tail. It was a living reenactment of the dolphins and cherubs depicted on ancient Roman coins and Greek mosaics.

Sandy could see clearly underwater as well as above, and so could we, using masks fitted with acrylic windows. Like all dolphins, Sandy inhaled air through a hole conveniently placed at the top of his head, and so did we, via snorkels. We also wore flippers to improve speed and maneuverability underwater. Fully outfitted in the best of modern snorkeling gear, though, we presented a pale, makeshift imitation of Sandy's exquisite design, honed during millions of years of processes that perfected slopes, angles, and surfaces, coupled with finely tuned musculature, energy, and sensory systems. But specialization has a price. We could comfortably enter Sandy's realm for a while, but it was hard to imagine him entering ours—to come on board and go ashore, visit a forest . . . climb a mountain . . . ride a bus.

Until recently, human access to the sea has been limited primarily to the surface. Various craft, from canoes to rafts to sophisticated ships, have transported our kind perhaps for as long as people have lived near the sea, provoked by curiosity, the desire to find food or territory or sometimes to exchange goods with others. The appearance of sails marked the beginning of serious travel across the surface of the ocean during the same era that sponge divers began taking the plunge *sub*sea. Aristotle describes Greek divers breathing air trapped in kettles lowered underwater in the fourth century B.C., but ages passed before anyone devised other ways of working effectively underwater. Legend has it that Alexander the Great used a glass diving bell in the third century B.C., the first

recorded mention of such a system in long series of experimental bells and barrels that preceded modern machines and materials.

In the absence of transparent materials with which to fashion protective goggles for diving, ancient South Pacific islanders captured bubbles and placed them
over their eyes to gain a better view underwater, a technique that persists in
some places today, when modern gear is not handy. Wooden hand paddles were
sketched by Leonardo da Vinci in the 1500s, and an ingenious but unsuccessful
giant snorkel was devised in the early 1600s as part of a hooded leather diving
suit. So it went for many years of trial and error. As recently as a century and a
half ago, exploration of the sea was mostly limited to what could be discovered
by remote means—lowering devices of various clever design from a ship. Even
basic questions could not be answered:

> How deep is the sea?
> Is there life at the bottom?
> What is the nature of ocean currents?
> Where are they, and what drives them?
> Is the sea as salty at the bottom as at the top or the same throughout?

Writing in 1855, Matthew Fontaine Maury, the first superintendent of the
U.S. Naval Observatory, eloquently vented his frustrations concerning the difficulties of, literally, getting to the bottom of things. In *The Physical Geography of
the Sea*, using words equally appropriate today, he noted:

> Astronomers had measured the volumes and weighed the masses of the most distant
> planets, and increased thereby the stock of human knowledge. Was it creditable to the
> age that the depths of the sea should remain in the category of an unsolved problem?
> . . . Indeed, telescopes of huge proportions and of vast space-penetrating powers had
> been erected here and there by the munificence of individuals, and attempts made
> with them to gauge the heavens and sound out the regions of space. Could it be more
> difficult to sound out the sea than to gauge the blue ether and fathom the vault of
> the sky?

Maury was convinced that the deep sea was not the level, monotonous, and life-less plain imagined by some. Rather, he said, more poetically than is allowed in most modern oceanographic tomes:

> The wonders of the sea are as marvelous as the glories of the heavens. . . . Could the waters of the Atlantic be drawn off so as to expose to view this great sea gash . . . it would present a scene most rugged, grand, and imposing. The very ribs of the solid earth, with the foundations of the sea would be brought to light.

Happily, Maury lived at a time when, he said, the "government was liberal and enlightened" and "times seemed propitious." Still, the problem remained of devising the ways and means of resolving the mysteries.

Maury described the earlier, commonly held belief that a chunk of lead sus-pended from a line could be used reliably to gauge depth when the shock of con-tact with the bottom is felt, or when line goes slack. Simple "sounding lines" can be used effectively in shallow water, but not at great depths. The technique was refined with the use of hollow lead cylinders: the degree to which they collapsed under pressure provided a clever but crude gauge of depth. Fine-tuning the tech-nique of dropping lead weights, modified cannonballs, proved most effective overall; thousands of reasonably accurate data points thus obtained led to a tradi-tion of charting the sea floor that continues to the present time. In recent years, the most accurate measurements have been derived by means of a greatly refined version of ingeniously simple acoustic methods used in Maury's time: In calm seas, bells were rung or petards exploded underwater and the depth calculated from the amount of time it took for reverberations from the bottom to be heard.

Sandy, the dolphin, took my daughter's measure, "seeing" with sound as he scanned her body—muscle, bone, soft tissues, and air spaces—with clicks. Humans have adapted this idea. Ships equipped with electronically controlled echo-sounding devices send pulses to the ocean bottom, where they hit and are reflected back to equipment on the ship that translates the pulses into a graphic record. The time taken for sound to return is accurately measured; the water depth calculated to be one half the total travel time (half of a two-way trip), then multiplied by the speed of sound in the water. Many variables can influence the accuracy of such measurements, but new techniques have made it possible to sense the configuration of much of the sea bottom precisely—without ever going there or sending an instrument directly.

Across the ocean from Maury's base in Washington, D.C., other oceangoing scientists in the mid-1800s were building momentum toward an expedition that was to be seen as signaling the beginning of the science of oceanography: the four-year voyage of HMS *Challenger*, undertaken in the service of the British Admiralty. Preceding that historic venture, and absolutely fundamental to its being, were numerous expeditions by British scientists, mostly in the Atlantic Ocean. C. Wyville Thomson, Regis professor of Natural History at the University of Edinburgh and director of the civilian staff of the *Challenger* Exploring Expedition of 1872–76, documented much of what was known prior to 1872 in *The Depths of the Sea* (1873), carefully adopting the "metrical system of measurement" and the "centigrade thermometer scale" throughout, a system still awaiting popular use in the U.S. (Some things change very slowly.) Thomson, like Maury and many modern oceanographers, was frustrated by the magnitude of ignorance concerning the sea, and by the view of the general public and even some supposedly well-informed scientists that the ocean beyond certain depths was a "waste of utter darkness, subjected to such stupendous pressure as to make life of any kind impossible."

Among earlier exploration expeditions launched by the British were the cruises of the *Porcupine* and *Lightning*. On those trips, boxlike dredges built of heavy mesh were towed along the sea floor, and sometimes they yielded bountiful hauls, especially dazzling at night. Living light displays of luminous deep-sea creatures torn from the sea floor can rival the streets of Las Vegas. Such brilliance glistened through Maury's matter-of-fact accounts of a catch snared from 650 feet down in the North Sea. Among long tangles of sea pens with color and substance like moist cotton candy, he found "the round soft bodies of . . . starfishes hung . . . like ripe fruit. The Pavonariae [soft corals] were resplendent with a pale lilac phosphorescence . . . not scintillating like the green light of Ophiacantha [serpent starfish]."

To appreciate the difficulties faced by explorers perched on the deck of a ship rolling through slate-green waters, imagine alien explorers flying a rocking craft slowly over New York City during a storm. They cast nets and hooks into the murky depths and drag them back and forth across Fifth Avenue, along the West Side, down past the Natural History Museum, into Central Park, down to Wall Street, through canyons of stony structures; the instruments bump and scrape, bounce off the tops of high buildings, breaking off chunks as they are towed along, nip some shrubs and blindly snare other objects—perhaps a startled pedestrian walking his dog, a business executive clutching her briefcase, a shopping cart, a cluster of garbage cans . . . What could aliens conclude about the nature of life below from a random trawl sample? They could amass some raw ingredients, given enough tries, but could discern little of the arrangement, and no comprehension of shopping lists and dentist appointments, of gossipy neighbors, purring kittens, parades, speeding tickets, ball games, Barney, the stock market, New Yorker cartoons, pizza, street gangs, or a night at the opera.

Seagoing detectives, otherwise known as oceanographers, have had to face the challenge of trying to imagine from fortuitously grabbed shreds of evidence what the nature of the deep sea is like. At the very best, a fragmentary catalog of some of the slow-moving or stationary residents or structures may be obtained. The wonder is that despite such limitations, when Challenger embarked on her round-the-world cruise in 1872, a significant body of knowledge had been assembled that defined the general lay of the land and the sea. This knowledge also hinted at the magnitude of ignorance about the ocean. During the following four years, new standards were set for ocean exploration, and the science of oceanography was launched.

The instructions from the British Admiralty to Challenger's captain read: "You have been abundantly supplied with all the instruments and apparatus which modern science and practical experience have been able to suggest and devise. . . . You have a wide field and virgin ground before you."

In 1962, the U.S. research vessel Anton Bruun began a four-year cruise that touched some of the same places visited by Challenger scientists: first the Indian Ocean, as part of the International Indian Ocean Expedition, and later the southeastern Pacific and Caribbean Sea. Anton Bruun, formerly the presidential yacht

Williamsburg, often used by Harry and Bess Truman, began her career as a research vessel under the auspices of the National Science Foundation in March 1962, when President John F. Kennedy announced that the ship would be made available for the U.S. Program in Biology. It was quickly converted from a posh yacht to a working platform that served more than three hundred scientists from twenty countries on missions of exploration. By sheer chance, I got to be one of them.

I had heard about preparations for the final cruise of the *Anton Bruun* in the Indian Ocean from a fellow student at Duke University, K. M. S. Aziz, a native of East Pakistan (now Bangladesh). He had been planning to participate for months, but at the last moment he could not go. Our professor, Dr. Harold J. Humm, turned to me and asked, "How about it? Would you like to take his place? They want a marine botanist."

At first, the idea seemed preposterous. For one thing, I was studying hard to pass written examinations that would qualify me for acceptance as a doctoral candidate in botany at Duke, where I had completed a master's degree eight years earlier. When I raised the issue the response was "No problem. Take the tests before you go." Harold Humm always had more confidence in me than I felt I deserved. What made him think I would pass on the first try? There was powerful incentive to succeed, but there were other considerations. A year after completing a master's degree and one year into a doctoral program, my path had taken a sharp turn toward domesticity, and in June 1957, aspiring botanist Sylvia Earle, 21, and aspiring zoologist John Taylor, 22, had married and moved to Florida. There, first in Live Oak, then Gainesville, and finally near my childhood home in Dunedin, I began a great balancing act as active scientist, supportive wife, and, in due course, mother to two small children. I was already scrambling to maintain a household while keeping up my long-term research project on the ecology of marine plants in the Gulf of Mexico, and, later, commuting to Durham, North Carolina, where the Duke campus and my degree beckoned. How could I possibly just "take off" for six weeks?

But my husband and parents were willing to help me on the home front. And the research project could be delayed awhile. So why not go?

I was apprehensive for two other reasons. Aziz had been a perfect choice for the job: hardworking, physically strong, worldly, competent in several languages, an experienced diver, and already an established scientist. I had not yet been out of the United States, not even west of the Mississippi, except along the Gulf Coast, where I had explored for marine plants.

"All the more reason why you should go," said Humm. "It will be good for you. Just do what you've been doing in the Gulf of Mexico." *That* I knew how to do—to explore underwater, where no one had been before, gather samples, identify them, prepare permanent specimens for various museums, catalog them, record everything I could about environmental factors (temperature, salinity, light, depth, current, substrate, tide, associated critters), and ultimately publish the results of whatever was discovered. In the Indian Ocean, in 1964, I would be among the first to see and document the underwater aspects of most of the reefs and islands visited, locations that read like a modern adventure-travel agenda with a few extra twists: Mombasa, the Amirante Islands, St. Joseph's Reef, the Aldabra group of islands, the Seychelles, Somalia, Dar-es-Salaam, Aden. There was little tourism then; how could any ocean scientist worth her salt resist?

But there was a final issue that could be a showstopper. The expedition's co-leader and Chief Scientist, Dr. Edward Chin, diplomatically inquired whether I, or my *husband*, would mind that I would be the only woman on board—and, by the way, some people did *not* think it would be wise for me to go . . . but he personally thought there would be "no problem."

For years, most of the classes and nearly all of the field projects I had participated in had been mostly male, and I had discovered that most potential hassles never materialized if you minded your own business, didn't expect favors or try to horn in on male parties or jokes, were prepared to do twice as much work for half as much credit—and, of paramount importance, kept a well-honed sense of humor. It may sound boring, but it works, even though a price is extracted in terms of bruised self-confidence, stomped-on tendencies toward leadership, and exclusion from being "one of the boys." Usually I did not mind being passed over for things that I knew I could do as well or better than men who were being considered, mostly because I understood the reality: No matter how competent a woman is, sometimes society rules that only a man will do. I could play by those rules—or not play. As for me, I wanted to be where the action was, even if I couldn't be the big boss with a beard. Even so, shipping out with an all-male

crew for six weeks at sea, half a world away from home, was an awesome concept. My four-year-old daughter, Elizabeth, asked me again and again to tell her
the deliciously mysterious-sounding destinations in turn, ending with ". . . and
you'll fly to New York, and be home to us by Christmas!" An incredulous friend
asked, "You're going? You're really going? I thought no women were allowed!"

At the time, women aboard U.S. research vessels were rare. I was told many
times, usually with a teasing smile, "Women at sea bring bad luck." The first evidence that the legend might be true came in Mombasa, our port of departure.
A reporter for the *Mombasa Daily Times* wanted to interview the scientists about
their work; would I mind taking time to describe the research planned? I was
pleased to tell what I could, barely noticing the question "Is it true you will be
the only woman on the ship?" I remember saying yes, then blathering on about
plants and fish and the spirit of exploration. The next day, the headlines, embar-
rassingly large and bold over a photograph of me dripping wet, read:

SYLVIA SAILS AWAY WITH 70 MEN
BUT SHE EXPECTS NO PROBLEMS

None, except with newspaper reporters, I thought, cringing at the ribbing I knew I
would get, and did.

Tucked into my forty-four pounds of baggage were precious pieces of equipment not available to *Challenger* scientists—or anyone else—until recently: a
mask, fins, and snorkel. With these and scuba setups, air tanks, and regulators
already on the ship, we would have access to places never before seen by members of our species. The *Anton Bruun* was not a "dive boat" as such. Like most
oceanographic vessels outfitted for expeditions of long duration over open
ocean, this ship was mostly dedicated to blue-water science, the sort of exploration that takes the ocean's pulse using assorted cleverly designed instruments:
nets, dredges, bottles, meters, gauges, and samplers. Such research does involve
getting wet, mostly from liberal quantities of sea spray washing over the rails
while equipment is launched and recovered. The cruise I was about to join, and
others that would follow, were partial exceptions to the usual rule of deploying
equipment over the side, but keeping oceanographers on the surface.

The misgivings and worries about whether or not I should have come on the trip evaporated with my first glimpse of an Indian Ocean reef underwater, just offshore from Mombasa, an island off the south coast of Kenya. Tridacna clams, the erroneously infamous "man eaters," were everywhere, from teacup-sized juveniles to an occasional giant with a colorful rippling maw large enough to engulf, if not hold, a careless hand or foot. (The clams' reputation as "killers" comes from their tendency to snap shut when a diver with a knife plunges his hand deep inside the clam and tries to cut loose the muscular and tasty hinge tissue. Some say the clam occasionally wins the tug-of-war for vital body parts, and I for one, side with the mollusk.) I focused on the plants, hauntingly similar to those I knew well in the Gulf of Mexico, but with unique ruffles, branches, or twists that set them apart. Like a child turned loose alone at F. A. O. Schwarz, I wanted to be everywhere at once, peering into the great, soft folds of clownfish anemones, poking at giant sea cucumbers, following pairs of yellow butterfly fish, standing on my head to get a better look at a spotted eel in an angled crevice, coaxing a tiny octopus from its lair. . . .

It was easy to forget that I was supposed to be a serious scientist as I careened around, an amazingly mobile "sponge" with flippers, absorbing and savoring every new image in an orgy of discovery made sensuous by the flow of a warm silken sea against my bare skin, a kaleidoscope of blue and green, marked with flashes of silver from fish and fleets of squids flaunting their iridescence.

Excerpts from my notes tell the story of an underwater Eden:

20 November, 0335 A.M. If I don't write right now, events of this extraordinary day will slip by unrecorded. Met the night baker on the way here and he said I seem to get about as much sleep as he does—zilch! Have been working in the lab since 5:30 P.M. after diving on a small slip of an island, mostly reef inhabited by crabs, boobies, and terns—Fungu Kizam Kazi (Elizabeth will like that name!). Found a corroding stack of elephant tusks in about 15 feet of water guarded by an inquisitive grouper. He behaved like the grouper back home: alert, territorial, not in the least intimidated by my bubbling presence.

22 November, 0315 A.M. Up this morning at 6:00 to make sure I did not miss the chance to dive along the outer edge of Grand Comoro Island. In the tidepools of black, volcanic rock, I found slender black lizards skirting the water's edge, feeding on small crustacea; just below, they were mirrored by small black fish—gobies—also dancing along the edge of the water, but from below, and also munching on unwary crustacea. The fish and lizards occupy exactly the same perches, sometimes wet, sometimes dry, as the tide moves up and down the rocks. It is not a good place to be a shrimp.

23 November, 0200 A.M. Was inspected by six huge hump-head wrasse, green giants cruising through the clearest, bluest water encountered so far, bordering the southwest end of Mouniameri Island, between the towns of Pamanzi and Mayotta, Comores. Along the shore, the reef fairly sizzled with the pent-up energy packed in the stinging cells of thousands of sea pens, whip corals, and lush mounds of soft corals—species I have never encountered before. One kind was a startling bright indigo!

30 November, 0130 A.M. Under way, heading for Aldabra. The bos'n, Hank Murranka, just called me to see a magnificent display flying fish, dozens of them skittering out of the way of the bow. Wielding a long-handled dip net, Hank scooped up one as small as a dragonfly—silver, white and blue with fleeting touches of iridescence coloring its translucent "wings."

3 December, 0300 A.M. Was up early again today to dive the central lagoon at Aldabra. A paradise! Lumps of limestone rock are carved by the tides into mushroom shapes that appear poised for toppling with the slightest nudge. Submerged pillars of coral resemble ruins of ancient temples, but the architects for these are millions of small animals, laced together with their own pale skeletons and encrustations of pink algae. The fish seem totally at ease, ambling over from distant parts of the reef as if to politely greet out-of-town guests. They surround us every time we go underwater, and stare, like wondering children. Fortunately, they have not met visitors with spear guns.

Everywhere, we dived, and everywhere, we found fish curious and unafraid, a pristine sea, the distillation of millions of years of history preceding that moment and at that time unmarred by pollution or grotesque overfishing that are now universal ocean plagues. Over the next two years I spent months at sea aboard the *Anton Bruun*, crossing with her through the Panama Canal into the southeastern Pacific, and to islands and currents along Central and South America, several times intersecting the path of *Challenger*. In November 1875, one of the *Challenger* scientists, H. N. Moseley, observed while nearing Más a Tierra, in the Juan Fernández islands, a volcanic rock formation jutting out of the blue Pacific about 300 miles west of Valparaiso, Chile, popularly known as "Robinson Crusoe's Island."

> *It was with the liveliest interest that we approached the scene of Alexander Selkirk's life of seclusion and hardship. . . . The study of Robinson Crusoe certainly first gave me a desire to go to sea. . . . The island is most beautiful in appearance. The dark basaltic cliffs contrast with the bright yellow-green of the abundant verdure; and the island terminates in fantastic peaks which rise to a height of about 3,000 feet. Especially conspicuous is a precipitous mass which backs the view from the anchorage at Cumberland Bay, and which is called from its form "El Yunque" (the anvil).**

In November 1965, *Anton Bruun* anchored in that same bay, and I was among those who gazed at the same angular gray-brown cliffs and yellow-green slopes and eagerly anticipated exploring their craggy underwater extensions. No one had used scuba gear there before, we were told, and thus each dive brought a sense of urgency about observing the smallest details to be able to report back news of a realm unknown.

On the second day in the Juan Fernández, while diving at 100 feet with a small group of ichthyologists, I discovered rocks slippery with a miniature forest of bright pink plants that appeared to have been designed by Dr. Seuss. Clusters of upright stalks were crowned with an explosion of sinuous gelatinous branches that rippled with the slightest current. When still, individual plants

---

* From Moseley's 1892 book, *Notes by a Naturalist, An Account of Observations Made During the Voyage of H. M. S.* Challenger.

resembled pink palm trees or umbrellas turned inside out with the wind. It was unlike anything I had seen before, and subsequent searching proved that it was not just a new species or genus but an entirely new family, perhaps a new order, of red algae.

Finding a kind of plant or animal never before given a name might seem to be cause for great celebration, the culmination of a life's work, a capstone achievement. The honor dims somewhat, however, when the number of creatures yet to be named and described is appreciated, a topic considered in more detail in a later chapter. In short, there are enough new kinds of creatures yet to be discovered in leaf litter, in backyards and forests, and most certainly in the oceans of the world to keep the namers of things fully occupied for centuries. For me as a taxonomist (a classifier and namer of things) myself, most difficult to find is not new species, but the time required to do justice to the scholarly process of preparing a comparative analysis, entering a "legal" description and illustration in the scientific literature, with appropriate specimens deposited in museums dedicated to the business of perpetually keeping libraries of such material in good condition for all time.

Actually, finding the slippery "Seussiform" red alga in the Juan Fernández islands *was* cause for celebration, and I happily set about preparing the nomenclatural formalities. For one thing, it was obviously a common element in the ecosystem surrounding the Juan Fernández islands, yet had not been noticed before. It existed literally right under the feet of numerous explorers attracted to the island but not to life in the surrounding sea. I felt that this plant was of sufficient importance that it would be an honor for Harold Humm if I named it after him. But I also thought it would be useful to choose a descriptive epithet, a name that embodied the plant's distinctive umbrellalike characteristics. One that did both was dreamed up during a brainstorming session among my fellow students at Duke some months later, and *Hummbrella hydra* was born.

When I went ashore at the largest of the three islands of the Juan Fernández, Más a Tierra, I saw a line of ants trooping over a patch of moss near a trickle of freshwater flowing from a tumble of boulders. I collected a few, curious as to how these decidedly nonmarine and flightless insects had made it from the mainland across a substantial expanse of ocean. A year later, the eloquent and insightful Harvard University biologist Edward O. Wilson happily received my insect offering and suggested that, like shipwrecked sailors clinging to debris, a critical number of these small creatures (enough to reproduce and prosper) had probably survived a voyage to the island on a floating log or raft of vegetation some ages ago.

Over time, plants and animals of various sorts had arrived at these islands, some flying or carried by the wind, others arriving by ocean currents. *Challenger* scientists enumerated some of the land plants genetically isolated from their mainland ancestors long enough to have become distinctly different. Included was an endemic (local) palm that, according to Moseley, had almost been exterminated by those seeking to dine on the tree's succulent terminal shoot, described as "excellent to eat . . . quite white, . . . tasting something like a fresh filbert . . . more delicate than that of the shoot of the coconut." A guide then living on the island took Moseley to one of the few remaining trees, hoping to find it in flower. Moseley adds, "As it was not, I cut it down for eating, for the guide was only waiting to let it develop further before felling it for that purpose himself. A few seedling palms grew nearby."

By the time the *Anton Bruun* arrived no palms had been seen for some time; a species of the pleasantly aromatic sandalwood was extinct years before the arrival of *Challenger*. At the time of my visit, I was not aware of the solemn significance of a small piece of sandalwood presented to me as the expedition botanist when we arrived at Cumberland Bay. It was from one of the few remaining roots of plants dead for more than a hundred years and was still fragrant with a spicy essence, a tangible token of something that was wonderful that can never be again. To me, it became a poignant symbol of change.

Chapter 3

ONWARD AND DOWNWARD

How do you get to the great depths? . . . How do you return to the surface of the ocean? And how do you maintain yourselves in the requisite medium? Am I asking too much?

Jules Verne, Twenty Thousand Leagues
Under the Sea

I have often wondered why scientists aboard *Challenger* did not insist on having some provision for diving. Simple things such as masks and flippers were not available, but by the mid-1800s, several scientists had used diving helmets to explore the sea directly. Professor Henri Milne-Edwards and a naturalist friend found the concept irresistible, and were the first known to "dive for science" during their studies in the Straits of Messina off the coast of Sicily in 1844. Years before, in 1819, Augustus Siebe perfected a commercial diving rig with air supplied via a hose to a helmeted diver, that forms the basis of present-day designs—with a few modern twists. An underwater breathing system that enabled divers to swim freely in the sea was developed and used in 1865 by the Frenchmen Benoît Rouquayrol and Lt. Auguste Denayrouze, including use of a container filled with a reservoir of air carried on the diver's back. Three years before *Challenger* set sail, Jules Verne published *Vingt mille lieues sous les mers (Twenty Thousand Leagues Under the Sea)*, the futuristic undersea adventures of Captain Nemo and his mysteriously powered submarine, *Nautilus*. It's a tale thrilling enough to inspire two Hollywood blockbusters many years later, but not enough to arouse the seagoing scientists of the time, about to embark on a four-year, round-the-world *surface* voyage, to pack along dive gear. Whatever the reasons they did not, they were not alone in ignoring the potential benefits of getting under the surface of the sea directly. Many years would pass before there was much serious development or use of underwater technology, except in response to the powerful stimulus of war or salvaging lost treasure.

At the time of the *Challenger* expedition, several submarines had taken people to modest depths. The first was a one-person wooden craft of American design, *Turtle*, used against the British in 1776. It was followed by Robert Fulton's *Nautilus*, built for Napoleon in 1800, and later, several successful European systems were developed for military applications. In 1864, the Confederate sub *Hunley* sank a Union ship in an attack near Charleston, South Carolina, but it is not likely that anyone using the system took serious note of fish or other natural phenomena.

During the next century, access underwater progressed almost in parallel with access to the skies above. Successful ascents were realized during piloted gliding feats in the late 1800s, followed by the Wright Brothers' powered flights in 1903 at Kitty Hawk, North Carolina. Technology designed to master both skies and seas came together in 1911, when Eugene Ely flew a Curtiss Pusher and touched down aboard a cruiser, the USS *Pennsylvania*, in San Francisco: the first landing of an airplane on a ship.

Several decades of unprecedented technological advances critical to the development of manned and robotic devices used in the sea, air, and space followed. Various versions of diving using tanks of compressed air were developed, including the first self-contained system, patented in 1918 by a Japanese inventor, Ohgushi, and a later diving "lung" created by the Frenchman Yves Le Prieur.

In the 1920s helium was tested by the U.S. Navy for use in diving, the first underwater color photographs were taken using artificial light, and Lt. James H. Doolittle made the first transcontinental U.S. flight in a single day (Pablo Beach, Florida, to San Diego, California). The "Kitty Hawk" of rocketry occurred in 1926 when Robert Goddard demonstrated the first successful operation of a liquid-fuel rocket. The following year, Charles A. Lindbergh successfully flew solo from New York to Paris, at about the same time that British inventor Joseph Peress developed the first successful armored diving suit, later known as *Jim*.

The first successful helicopters, the first jet flight, and first passenger plane with a pressurized cabin, all occurred before 1940. It was a heady, exciting era in aviation. Charles Lindbergh and his wife, Anne Morrow Lindbergh, made major flights in 1931 and 1933 to survey possible airline routes in the early days of international air travel. More than half of their "north to the Orient" flight, the first east-to-west flight over the North Pole, was over territory where no airplane had flown before. Something of the spirit of exploration was expressed by Anne Lindbergh in response to a reporter who, prior to the first over-the-Arctic expedition, prodded her to comment on the perils ahead.

"Can't you even say you think it is an especially dangerous trip?" he asked.

She responded: "I'm sorry. I really haven't anything to say. (After all, we want to go. What good does it do to talk about the danger?)"

But danger is the silent partner of exploration, and with gains also come losses. Many paid for new knowledge with their lives, fatally crushed in experimental submersibles or lost in crashed aircraft. The longing to fly faster, farther, higher, and deeper, longer, and ever more distant, is irrepressible, especially when spiced with risk.

Balloonists A. W. Stevens, W. E. Kepner, and O. A. Anderson set a new altitude record, 60,613 feet, aboard the *Explorer I*, in 1934, the same year that zoologist William Beebe and engineer Otis Barton set a depth record, 3,028 feet, while diving a short distance offshore from Bermuda, in a round, steel-hulled submersible, the *Bathysphere*. In his book *Half Mile Down*, Beebe describes discussions that he had with President Theodore Roosevelt about deep-sea diving. Concerning designs for new submersibles he wrote that "there remains only a smudged bit of paper with a cylinder drawn by myself and a sphere outlined by

Colonel Roosevelt, as representing our respective preferences." Several plans for cylinders were considered, but in the end it was clear that there is nothing like a ball for the even distribution of pressure, and "the idea of a perfectly round chamber took form and grew." Unlike aircraft that take into account decreasing pressure with increasing height, submersibles must be designed to withstand an additional 14.7 pounds per square inch for each additional 33 feet of depth. Half a mile down, the pressure upon each square inch is more than half a ton.

The cost of building the two-man *Bathysphere*, $11,000, was uncannily close to the cost, $10,580, of *The Spirit of St. Louis*, the plane Lindbergh used to cross the Atlantic. In both cases, private sponsorship, not government grants or contracts, paid for the machines, and both Beebe and Lindbergh were driven primarily by the desire to use technology to advantage—to test the limits of human endeavor. In the case of flying, though, aircraft such as the one used by Lindbergh for his historic flight already existed, whereas the concept and design had to precede the construction and ultimate use of *Bathysphere*. Otis Barton, *Bathysphere*'s principal designer and funder, planned three windows—cylinders of fused quartz eight inches in diameter and three inches thick—and a "door"—a round, four-hundred-pound lid that had to be lifted on and off by a block and tackle, then fastened down with ten large bolts. Fortunately, both Beebe and Barton were slim enough to slither through the 14-inch-diameter opening, and neither was susceptible to acute claustrophobia. Inside, a tray of chemicals absorbed moisture and carbon dioxide, and oxygen was added as required from a cylinder with automatic valves.

The entire craft was lowered into the sea by a single steel cable nearly an inch thick, wrapped around a cylindrical winch. Beebe puts circumstances of the time in perspective in *Half Mile Down*.

A certain day and hour and second are approaching rapidly when a human face will peer out through a tiny window and signals will be passed . . . to breathlessly waiting hosts on earth, with such sentences as:

"*We are above the level of Everest.*"
"*Can now see the whole Atlantic coastline.*"

"Clouds blot out the earth."
"Can see the whole circumference of Earth."
"The moon now appears ten times its usual size. . . ."

. . . dangling in a hollow pea on a swaying cobweb a quarter of a mile below the deck of a ship rolling in mid-ocean. . . . We were able to adumbrate the above imaginary news items . . . by the following actual messages sent from the Bathysphere. . . .

"We have just splashed below the surface."	
"We are at our deepest helmet dive."	60 feet
"The Lusitania is resting at this level."	285 feet
"This is the greatest depth reached in a regulation suit by Navy divers."	306 feet
"We are passing the deepest submarine record."	383 feet
"A diver in an armored suit descended this far . . . the deepest point which a live human has ever reached."	525 feet
"Only dead men have sunk below this."	600 feet
"We are still alive and one-quarter of a mile down."	1426 feet

Beebe later notes other sea-space parallels:

I knew that I should never again look upon the stars without remembering their active, living counterparts swimming about in that terrific pressure . . . this strange world. . . . When once it has been seen, it will remain forever the most vivid memory in life, solely because of its cosmic chill and isolation, the eternal and absolute darkness and the indescribable beauty of its inhabitants.

I first read about Beebe and devoured his underwater adventures when I was a child, an aspiring marine biologist getting my feet wet along beaches in New Jersey and Florida. His descriptions of life in the sea, told in several entrancing volumes, convinced me at an early age that it was important to know how to swim. My older brother, Lewis (named for my father but always known as Skip), has a graceful, swift, free-style crawl that I always envied but could never duplicate, despite much vigorous kicking and splashing. Underwater, though, I seemed to know just what to do as I held my breath, arms at my sides, and glided around like the tadpole I was said to resemble.

Many years would pass before I could climb into a spherical submersible and emulate Beebe's descents into the never-never realm of deep-sea creatures, but my parents treated all of us to our first flight when I was five. A pilot in a yellow piper cub came to town and offered to take one person at a time aloft for a nominal charge. I do not remember any hesitation about stepping into the seat behind the pilot and being swept skyward for a few turns along the outskirts of town. I *do* remember the engine noise, the exhilarating rush of air sweeping across my face and whipping my hair, and then looking down at what appeared to be doll houses and a scattering of miniature people—Mother, Dad, Skip, my year-old brother, Evan, and a small crowd of watchers, below.

That was 1940, the start of a decade marked by the breaking of the "sound barrier" by Captain Charles E. Yaeger, and the shattering of another kind of barrier, underwater—that of ready, easy access using an aqualung. In 1943, Jacques-Yves Cousteau and Emile Gagnan made a historic breakthrough in the design of an underwater breathing system that virtually anyone can use, and Cousteau proved the point with daring dives to 210 feet in the Mediterranean. The balloonist Auguste Piccard turned his attention to the oceans and in 1948, with Max Cosyno, tested his subsea "dirigible," the bathyscaphe *FNRS2*.

In *The Sea Around Us*, published in 1950, Rachel Carson brought together knowledge of the oceans in a compelling form. It was the beginning of an era marked by records of depth (13,287 feet, in the bathyscaphe *FNRS3*), distance (the U.S. nuclear submarine *Nautilus* traveled from the Pacific to the Atlantic under the North Pole), and aircraft speed (Mach 2 by A. Scott Crossfield; Mach 3 by Captain Milburn Apt). It was the dawn of the space age—and also the era when the first remotely operated vehicle was launched into the sea.

For me, 1952 was a turning point—downward. A school friend borrowed his father's copper diving helmet, compressor, and pump, and introduced several of his pals, including Skip and me, to the fine art of breathing underwater. It was fortuitous for me, a would-be marine scientist, that my parents had chosen to move from New Jersey to Dunedin, Florida, a coastal community near Tampa Bay. Years before, nearby Tarpon Springs had attracted fishermen and sponge divers from Greece. More recently, other fishermen, including my friend and his father, acquired the diving skills and equipment needed for gathering sponges to sell at auctions in Tarpon Springs. How else, in 1952, would it be possible for a 16-year-old girl to have access to a sponge diver's helmet in a wild Florida river?

When I was 12, a U.S. Navy commander, Edward Ellsberg, had transported me underwater, vicariously, with gripping stories about salvage operations reported in his classic volume, *On the Bottom*. Now I had a chance to try diving myself, a venture I looked forward to with pleasure spiced with fear. After all, Ellsberg had cautioned, "Nothing that the ingenuity of man has permitted him to do is more unnatural than working as a diver."

For most divers, little had changed since the 1920s, when, as Ellsberg pointed out, the usual diving dress consisted of a copper helmet and breastplate secured watertight to a flexible, canvas-covered rubber suit. The helmet was essential to the provision of compressed air, but a diver could do without the suit in shallow or warm water. With or without the suit, the laws of nature had the same effect on those who presumed to "slip the surly bonds of earth," and dive. Ellsberg explained:

Water is heavy; as the diver descends he is compressed by the weight of the column of water over him. Over the surface of his body, for each foot he descends, an added load of almost half a ton presses on him. At one hundred and thirty feet, the total load is nearly sixty tons. To prevent the diver from being crushed to jelly by this weight, it is necessary for him to breathe air under pressure slightly exceeding that of the water; this internal air pressure is transmitted by his lungs to his blood and enables him to balance the external water pressure. . . . The diver inflated with compressed air, stands the weight of the sea pressing on him; but if through any accident, he loses the air pressure in his helmet, like a trip hammer down comes the weight of the sea and crushes him . . . flat. . . . As he goes deeper, a diver must increase the air pressure . . . to correspond; it is therefore most dangerous . . . to fall off suddenly

into deeper water and thereby be subjected to greater pressure. If, under such circum-
stances, he cannot simultaneously raise his air pressure, he is crushed by the water
into his helmet, and many men have died from such a "squeeze."

When my turn came to see what it was like to dive with the borrowed helmet, I eagerly slid into the icy, ultraclear water of the Weekiwatchee River, barely noticing the weight of the ponderous "hard hat" biting into my bare shoulders. Ellsberg's gruesome image "crushed to jelly" tiptoed across my mind as air surged around my face, and the river's brisk current kept sweeping my feet out from under me, making it difficult to stay upright. Pain stabbed my ears, reminding me to swallow repeatedly—a small exercise that helps equalize the pressure of air trapped in ears and sinuses at the surface with the compressed air breathed while descending. But as soon as my toes touched bottom, 30 feet down, I stayed put, held in place by the hefty helmet, and then it suddenly sank in: *I am under-water and breathing!*

A few feet away, lurking among stalks of river rushes, the alert eyes of a huge, silver-green alligator gar met mine; then, in a perfect barracuda imitation, this legendary river monster opened and closed its mouth several times, revealing saber-sharp dentition before the creature ever so slowly turned and slid out of sight. Fascinated, I stepped toward the sweep of its speckled tail and was nearly toppled by the rushing current. *Next time, ankle weights*, I thought, very carefully edging my way from the river's main stream to the quieter shore and a scatter-ing of small, golden-brown fish. From the fishes' standpoint, I was a noisy appa-rition of rushing bubbles, hose, and huge helmet with legs, but I willed myself to be inconspicuous and, as stealthily as I could, made my way toward them. Then, something totally unexpected happened. First one, then several, and finally all of the small fish I had been stalking turned and swam in my direction. I was sup-posed to be the watcher, but found myself the *watchee*, the center of attention for a bunch of curious fish, apparently mesmerized by the strange bubbling being that had just fallen through their watery roof. For twenty blissful minutes, I became one with the river and its residents, bending with the current, blending in—and breathing!

Gradually, though, the glistening fish and silvery bubbles blurred and I felt a dizziness that seemed to be more than the consequence of first-dive excitement. I signaled with a tug on the hose that I wanted to go up just as someone free-dived down and pointed for me to ascend. It turned out that exhaust fumes from the generator were being swept into the compressor's intake pipe, and I had been breathing air laced with a toxic mix of carbon dioxide, carbon monoxide, blue smoke, and uncombusted fuel. It was a memorable lesson concerning clean air: Never take it for granted, on the surface or underwater!

The question I now faced was, How could I possibly manage to obtain a diving helmet and compressor that I could use all the time? That problem was eliminated before it was solved. I had a new dream, one inspired by the development of a new way to breathe underwater, free of any connection to the surface: the aqualung.

In the summer of 1952, Captain Jacques-Yves Cousteau captivated many, including me, with the exuberant delight expressed in his book *The Silent World*, in which he described his invention, the self-contained underwater breathing apparatus known as scuba. He wrote:

I reached the bottom in a state of transport . . . To halt and hang attached to nothing, no lines or air pipes to the surface, was a dream. At night I had often had visions of flying by extending my arms as wings. Now I flew without wings. (Since that first aqualung flight, I have never had a dream of flying.)

I yearned to experience for myself the thrills described, to "swim across miles of country no man has known, free and level, with our flesh feeling what the fish scales know." I was not at all charmed, however, by his vivid descriptions of spearing fish and sometimes whales. Like many underwater pioneers, Cousteau and his colleagues quickly became avid subsea hunters; one, Frédéric Dumas, was renowned for spearing thousands of large fish, once taking 280 pounds during five dives one morning . . . on a bet. Ironically, while we exhibit diplomatic behavior when visiting foreign nations and even consider the question of proper etiquette should we encounter sentient life from another planet, we are

amazingly disrespectful in our intrusions on our earthly co-inhabitants, especially fish! Given a chance to explore realms never before seen by members of
our species, and establish a rapport with benign, obviously curious creatures
that had never before met a human, pioneering scuba divers were often grim
ambassadors. Technology that for the first time provided effective access, the key
to understanding the sea, perversely made possible its accelerated destruction.

Attitudes have changed, however.* Like the legions of birdwatchers who stalk
the land to add a new sighting to their treasured "life list," there are growing
numbers of fish-watchers with more than 25,000 living targets to acquire. While
there are far fewer fish to attract spearfishermen now than when diving first
became popular, most experienced divers have lost interest in killing fish, and
those who do tend to take only a few for food. Many former undersea hunters
now use their considerable skills to capture grouper, snapper, morays, and more
on film, or to watch, or even play with, their subjects. Manta rays and whale
sharks have often been slaughtered for sport, but in recent years, divers
have come to value as a most treasured lifetime experience an opportunity to
swim with them, wonder at their magnificence, and sometimes even hitch a
heart-pounding ride on a fin or flipper.

In the early days of scuba diving, as in the early days of aviation, no one had
a license. Those who were so inclined figured out the dos and don'ts for themselves or learned by watching someone else who had done so. Nowadays, a person who wants to try scuba can readily find places to take a course and compress
the years of trial and error by the pioneers into a week or so of training.† No such
courses existed in 1953, however, when I was first given a chance to realize my
dream of diving with an aqualung while taking a class in marine biology at Florida State University's Alligator Harbor Marine Laboratory. My professor, Harold
Humm (who later moved to Duke University), made it clear to his students that

---

* Frédéric Dumas later deplored the use of scuba for spearfishing and, according to a recent letter to me from his
friend Sir John Rawlins, "In 1963, Dumas showed me with pride his 'secret garden' off the Ile de Porquerolles where
every grouper, eel and octopus appeared to welcome him as an old friend, probably the first attempt at creating a fish
sanctuary."

† The National Association of Underwater Instructors (NAUI), the Professional Association of Diving Instructors
(PADI), Scuba Schools International (SSI) and the YMCA are among the reputable organizations that approve courses for
certification.

he thought the best way to study marine creatures was to take ourselves where they live—underwater. He provided us with several ways to view the action: glass-bottomed buckets, face masks and flippers, a Desco mask—a full-face system with about 100 feet of air hose connected to a surface air compressor—and, best of all, two gleaming new air tanks equipped with double-hose regulators.

My first scuba dive was preceded by two important words of instruction: "Breathe naturally." By not holding my breath, I avoided the potentially deadly problem of an embolism, the painful consequence of compressed air rapidly expanding as a diver ascends. I knew nothing about decompression tables and half-believed the nonchalant rule of thumb that it was not possible to get into trouble diving with a single tank of air (a lethal misconception). I did know something about the behavior of gases under pressure, though, having devoured Ellsberg's somber litany of diving perils in *On the Bottom*. His words slithered into my thoughts as I prepared to take the plunge:

The inert component of air, nitrogen, . . . is the cause of "the bends." Nitrogen, which forms four-fifths of our atmosphere, is ordinarily breathed in and out, having no effect except to dilute the oxygen. However, when air is much compressed, the nitrogen entering the lungs, instead of being all exhaled again, dissolves in the blood, and the heavier the pressure, the greater the quantity of nitrogen dissolved. While the diver remains under pressure . . . he notices nothing. But . . . when the pressure is released on . . . coming to the surface, the nitrogen dissolved in his blood bubbles out and forms a froth. . . . These bubbles clog the arteries, impeding circulation, and causing convulsions or "the bends." In many cases, the bubbles gather in the spinal column, where they affect the nerves, causing paralysis. In less aggravated cases, a favorite place for bubbles is in the joints, resulting in great pain.

To get the bends you would have to spend a long time in shallow water or just a short time in deep water—in either case, enough time for more nitrogen to become dissolved in tissues than can be expelled safely while ascending. I planned to spend a *short* time in *shallow* water; the greatest danger I faced at that moment was possible terminal enthusiasm. Without hesitation I strapped on the heavy steel tank, adjusted my face mask, bit down on the mouthpiece of the regulator that was draped over my head, and dropped into the sea five miles offshore, 15 feet down, into a meadow of green-brown sea grass. With a gentle kick I glided to a small clump of sponges and found a feisty three-inch-long damselfish who was not pleased by my intrusion into its territory. Balancing myself with one finger, I found it easy literally to do a headstand so that I could peer into the dark crevices of the fish's lair. Often, I had seen such fish and glimpsed vignettes of their behavior during brief, breath-holding dives, but now I could stay, waiting and watching, until the fish became relaxed and went about its business, apparently no longer concerned about the passive bubbling hulk on its doorstep.

Before yielding the tank to the next lucky aquanaut, I spent a few minutes exploring the potential of weightlessness, just as astronaut Buzz Aldrin would years later while training underwater for the first moon landing. "Eventually," he said, "I mastered the intricate ballet of weightlessness." This was vital preparation for spacewalking when, as Aldrin puts it, "flexing your pinkie would send you ass over teakettle." Easing away from the damselfish, I hovered, blending with the sea like a jellyfish and then, inspired by the athletic grace of every other creature in sight, tried something I imagined to be impossible: a midwater slow-motion backflip. No problem! Forward, then—soaring, rolling, swimming upside down, then a spin. The fish were almost certainly perplexed by the large mass thrashing about in their midst, but in my mind I was dolphinlike—but with an edge on the dolphins. They have to surface to breathe every few minutes, but with the air tank I could stay submerged for an hour.

Reluctantly I returned to the surface convinced, more than ever, that I really needed gills, or at least an aqualung! I was not alone. Thousands, then millions, responded to the wonder of being able to breathe underwater. For marine scientists, scuba was a breakthrough comparable in some ways to the development of the first microscopes. In both cases, technology provided a means of access, a way to see things otherwise not visible, and gradually, scientists effectively used such tools to better understand the nature of the world. Nevertheless, many scientists viewed scuba diving for scientific research as nothing more than a wispy disguise for having a good time, implying that what you're doing can't be very serious if you're having fun! Since then, several decades of underwater observation using aqualungs and other diving techniques have permanently transformed human perspective, forcing those who have tried it to look at the earth—and themselves—with fresh eyes.

New perspectives were rapidly coming from other directions during the 1950s, most notably from advances in space exploration. In this decade, *Sputnik I*, the first man-made earth satellite, was placed in orbit by the Soviet Union, the U.S. launched its response—the satellite *Explorer I*, and the Soviet Union landed the first man-made object on the moon, *Luna 1*, and photographed the far side of the moon for the first time, using *Luna 2*. There was considerable talk of looking for life on other planets. Curiously, few took note of new insights about life on this one, when in 1951 scientists aboard the Danish research vessel *Galathea*, using towing gear, captured and brought to the surface animals from 33,400 feet deep in the Philippine Trench. This was not quite the deepest part of the ocean, but was the greatest depth at which life had been found since a fish, some brittle stars, and a few other creatures had been captured in a trawl at a depth of 20,000 feet by Prince Albert of Monaco half a century earlier. It is notable that Sir Edmund Hillary and Tenzing Norgay observed *no* sign of life at 29,028 feet, the summit of Mount Everest, during their historic first ascent to the highest point of land on Earth in 1953.

In the sixties, space exploration accelerated. The first weather satellite was launched, and in 1961 Major Yuri Gagarin became the first man to view Earth from space. Later that same year, Navy Lieutenant Commander Alan Shepard, Jr., became the first U.S. astronaut to go into space, and in 1962 Marine Lieutenant Colonel John Glenn, Jr., orbited Earth aboard the Mercury spacecraft, *Friendship 7*. *Mariner 2* became the first spacecraft to conduct a flyby of another planet, Venus.

Underwater, a significant milestone was achieved in 1960 when U.S. Navy Lieutenant Don Walsh and the Swiss engineer Jacques Piccard descended in the bathyscaphe *Trieste* to the deepest part of the ocean, the Challenger Deep at the very bottom of the Marianas Trench: 35,800 feet down. (Located about 190 miles southwest of Guam, the Marianas Trench is one of the deep trenches that rim the Pacific like a necklace.) More important, they returned, safely. Walsh, who piloted *Trieste*, knowingly points out, "One-way trips don't count."

It is worth pausing here to reflect on the magnitude of that accomplishment. In the eyes of many, myself included, it was a feat comparable to placing men on the moon—but at a fraction of the cost, and almost a decade earlier.

The distance does not seem great when compared to the lofty reaches of air and space: 35,800 feet, 11,000 meters, seven miles. People fly at that altitude in comfortable aircraft watching movies, eating lunch, taking naps, drinking fine wine. But there are some obvious differences, starting with pressure. Someone standing at sea level is under an "ocean" of air that weighs 14.7 pounds per square inch (psi). At the summit of Mount Everest, the pressure is only 4.5 psi, and when balloonist Auguste Piccard ascended 10 miles, he experienced a mere one-pound of pressure on each square inch. Descending into the sea, twice atmospheric pressure, 29.4 psi, is reached at only 33 feet (10 meters). At about half a mile (1,000 meters), the depth first attained by Beebe and Barton in a bathysphere in 1934, the pressure is 1,360 psi. Some mammals, sperm whales and Weddell seals among them, dive to more than twice as deep simply holding their breath and staying submerged for as much as an hour. At the bottom of the Marianas Trench the pressure exerted by the weight of seven miles of water on each square inch of a fish, sponge, submarine, or the sea floor itself is 16,000 pounds, or eight tons.

Most trenches are narrow gashes in the sea floor close to continental masses. At their bottom are subduction zones, areas where the edge of the earth's crust collides with a continent and is drawn beneath it. New crust is continuously being spewed forth from the volcanoes at the center of the midocean ridges that run the length of the major ocean basins like giant backbones. Despite its edges' being swallowed under continents, the Pacific Ocean grows in overall width at a rate of about 3 inches (8 centimeters) a year as its sea floor spreads. This doesn't seem like much, until the immense time factor is taken into account. Walter Sullivan puts the process into perspective in *Continents in Motion*:

> *Can one believe that 7,000 kilometers (4,300 miles) of ocean floors have slipped under the rim of North America? Or that an equal amount has vanished beneath the Pacific coast of Asia? Or that an ocean possibly as large as the Atlantic once lay between Russia and Siberia? Such concepts . . . have emerged from efforts to apply*

the new theory of crustal plates to all mountain systems of the world, including those that now lie far inland, such as the Urals and Rockies.

In 1960, when Walsh and Piccard embarked on their epic voyage to the bottom of the sea, knowledge of "crustal plates" was just beginning to unfold, and the idea of "continental drift" was still scoffed at by many. At that time, years of deep-sea drilling and hundreds of submersible dives that would help resolve the mysteries of Earth's ongoing rearrangements were still in the future. But the importance of physical human access to the deep trenches was appreciated as a means to explore part of the puzzle.

Another mystery concerned verification of the existence of life at the ocean's greatest depth. In *Seven Miles Down*, Piccard describes how this issue was conclusively resolved:

As we were settling this final fathom, I saw a wonderful thing. Lying on the bottom just beneath us was some type of flatfish, resembling a sole, about 1 foot long and 6 inches across. Even as I saw him, his two round eyes on top of his head spied us—a monster of steel—invading his silent realm. Eyes? Why should he have eyes? Merely to see phosphorescence? The floodlight that bathed him was the first real light ever to enter this hadal realm. Here, in an instant, was the answer that biologists had asked for decades. Could life exist in the greatest depths of the ocean? It could! And not only that, here, apparently, was a true, bony teleost fish. . . . Yes, a highly evolved vertebrate, in time's arrow very close to man himself.

The discovery of such an organism makes it possible to make certain inferences about the deepest sea, and start to answer previously unresolved questions. We can presume the presence of free oxygen, needed for the flounderlike fish to survive, and thus the likelihood that deep-sea currents transport gases and other materials throughout the oceans. The simple fact that vertebrate life can withstand 16,000 pounds of pressure per square inch, coupled with continuous near-freezing cold and darkness, should be cause for news accounts roughly equivalent to, say, finding life on Mars. But like in the mid-1800s, the mid-1900s were characterized by a mind-set that was distinctly aimed skyward.

In the foreword to *Seven Miles Down*, Robert Dietz comments, with more than a hint of bitter wonderment:

> *Over the past decade, the two most powerful nations on earth have spent billions of dollars for rocketry, hoping eventually to send a man to the moon for direct observation. In contrast, two citizens of landlocked Switzerland with only private assistance succeeded in building a vehicle to take man to the deepest hole in the ocean.*

Eventually more funding and encouragement for ocean technology, research, and exploration began to become available than had been the case for many decades before. President John F. Kennedy, champion of space exploration, also valued the importance of the sea, observing in 1961, "Knowledge of the oceans is more than a matter of curiosity. Our very survival may depend upon it." Fine words were followed with action, and the results of unprecedented support for ocean exploration, research, and the technology needed to implement them were swift and dramatic.

In the heady sixties, it seemed that interest and commitment concerning exploration and understanding and the oceans, while still lagging far behind the special human passion for the skies above and space beyond, might actually be able to "catch up," and win acceptance on a comparable basis. Among the most memorable accomplishments were the creation of programs such as the U.S. Navy's Sealab series, the Smithsonian-Link Man-in-Sea program, and Cousteau's Conshelf project, leading to the establishment of more than fifty underwater habitats for humans in six countries. The world fleet of manned deep submersibles grew from three in 1960 to nearly fifty by the end of the decade; one of these was the Woods Hole Oceanographic Institution's famous *Alvin*, the durable workhorse of research submersibles. At its base in San Diego, the U.S. Navy began to design the first practical Remotely Operated Vehicle, or ROV. All of these technologies got a boost from the rapidly expanding offshore oil and gas industry, which both spurred and funded significant innovations and provided rich opportunities for practical use. Furthermore, a growing audience tuned in to the seas with Jacques Cousteau's compelling films about the beauty, excitement, and importance of the oceans.

In 1962, the year John Glenn orbited Earth in the *Friendship 7*, a new depth record of 1,000 feet was set in California by a young Swiss mathematician, Hannes Keller, and his companion, Peter Small.*While Glenn breathed an atmosphere of oxygen within the protective shell of his spacecraft, the divers inhaled a special, changing mix of oxygen, nitrogen, and helium in proportions calculated by Keller to overcome some of the problems brought about by the unusual behavior of gases under pressure.

Each gas in a mix exerts independent pressure, called partial pressure, on the lungs. The partial pressure of oxygen is nearly doubled with every atmosphere of weight added (each increase of 33 feet adds one atmosphere, 14.7 psi). The same is true of gases such as helium and nitrogen, each gas causing separate physiological responses that limit how deep a diver can go. Pure oxygen, a generally "friendly" vital substance, can be a dangerous foe when breathed at depths greater than about 30 feet. Cousteau quite suddenly discovered the gas's dark side when using a pure oxygen rebreather at 45 feet, as recounted in *Silent World*:

> *My lips began to tremble uncontrollably. . . . My spine was bent backward like a bow. With a violent gesture, I tore off my weight belt, and lost consciousness.*

Even mixed with nitrogen, oxygen can be toxic. Compressed air, with 20 percent oxygen, can cause convulsions at depths greater than about 250 feet. However, long before this depth is reached, problems with nitrogen are likely to discourage most from continuing downward. Nitrogen under pressure causes a peculiar euphoric effect much like the dreamy state induced by laughing gas, nitrous oxide. At about 100 feet and deeper, divers get "high," often experiencing a tranquil, giddy "buzz." Some divers happily hallucinate, become forgetful or confused about which way is up, or decide that the regulator is a nuisance and offer it to passing fish.

The cause of the dangerously delightful "rapture of the deep" was pinpointed in 1935 by a lean, lanky, and totally engaging U.S. Navy doctor, Al Behnke, who was also my neighbor in California. Forty years after his pioneering work with

---

* Sadly, Small died during the experiment, and Keller was nearly lost.

nitrogen, Behnke charmed me in his book-lined living room in San Francisco with thrilling, white-knuckle accounts of the learning-by-doing days of diving.

"The sinking of the U.S. submarine *S-4* in 1927, after colliding with USCGC *Paulding*, was a turning point in the history of submarine rescue," he told me. "The worst of it was that the crew survived the crash and sank, safely, to one hundred two feet, then slowly suffocated while waiting for a rescue that never came. What those men went through . . ." He paused, both of us thinking about their tortured death, the air inexorably becoming more foul with each breath by each man, all listening to tapping and scraping on the sub's hull, but no one on the outside able to get in, and no way for those inside to escape.

That catastrophe provoked an all-out effort to come up with submarine rescue plans including a movable, cylindrical chamber that could be fastened onto a sub's hatch to transfer people in and out. Another milestone came some years later when a freak accident sank the submarine *Squalus*, killing twenty-six men immediately and trapping thirty-three survivors 243 feet down off the New England coast. Using the new chamber, those living were quickly rescued, but recovering the bodies of the others and eventually the submarine itself required hundreds of hours of diving. Driven by the necessity of finding an alternative to nitrogen, Behnke helped pioneer the concept of using a mixture of helium with oxygen, leaving out nitrogen, to achieve a non-toxic, nonnarcotic mix of gases to breathe for deep diving. Helium, the lighter-than-air substance used to inflate party balloons as well as divers' lungs, is an odorless, tasteless, invisible element. It is one seventh as light as nitrogen, the lightest substance known next to hydrogen but without hydrogen's explosive properties. Its use as a diving gas—to serve as a "mixer" for the life-sustaining oxygen—had been proposed but not tried in the ocean.

Helium's extreme lightness has two disconcerting side effects: Breathing helium causes the body to lose heat much faster than breathing normal air of the same temperature, and vocal cords vibrate more quickly, turning even diver bassos into trilling falsettos. Nonetheless, the value of helium was proven during the dives on the remains of *Squalus*. "We decided it was time to make it happen," Behnke recalled, "and there wasn't time to practice." The techniques developed through the subsequent salvage eventually became standard for commercial and military divers worldwide.

In the 1960s I read and reread the diving adventures of Beebe, Cousteau, Ellsberg, and others, and was inspired to scrounge whatever equipment I could to explore my own blue backyard, the Gulf of Mexico. Often I went alone, diving repeatedly at the same sites at different times of the year to look at seasonal changes underwater. Shrimp fishermen who worked just offshore from where I lived in Dunedin, sometimes let me go along and paw through the mountains of "trash fish" and "gumbo" that came in with their desired catch. Navy divers at Panama City, Florida, taught me much of what I needed to know about the serious, not-to-be-ignored diving protocols, especially important in the absence of a formal diving course. And Eugenie Clark, the celebrated "shark lady," tucked me under her flipper and allowed me to tag along on ventures into the Gulf in exchange for helping her assemble a nascent plant collection for the young and growing Cape Haze Marine Laboratory that she started at Placida, Florida.

Diving with Genie was like diving with *a genie*. She magically turned sober senior scientists into wide-eyed little kids by leading them into the sea and introducing them to creatures she had discovered along the south Florida shore; and she made me, the wide-eyed youngster, believe I might someday turn into a serious scientist. Also, as a happily married wife, mother of four, world-class explorer, author, speaker, and scientist, she made it look easy to juggle several simultaneous lives. In 1965, when she asked if I would take on the job of resident director of the lab, then based in Sarasota, while she went to live in New York City, it did not occur to me that I wouldn't be able to manage, even though it would entail a daily two-hour commute each way from my home in Dunedin. Somehow it worked, but only because my parents lived next door and helped.

As resident director, I continued Genie's shark research program; this involved "long-lining" for sharks—using a heavy-gauge line with a series of baited hooks—releasing or bringing back to special holding pens those that were alive, and measuring and dissecting those that were not. Rarely did I see a shark while diving along that coast, but every day that we set the lines, several kinds showed up: lemon, tiger, hammerhead, bull, and now and then a great white.

Shark biology seemed only distantly related to my ongoing research on marine plants, but in fact, dealing with the big predators day to day forced me to keep an "ecosystem" perspective, and deliberately look for the connections between the photosynthesizers at the bottom of the food webs in the sea, and the great toothy consumers at the top. An ecologist at heart, I could see the sense to try to take into account animals, as well as the plants and the physical and chemical characteristics of the places I observed, but underwater, it wasn't always easy.

I envied my colleagues who conducted ecological research in deserts, mountains, and rainforests, using elaborate instrumentation that would instantly sputter and die in saltwater, and making detailed observations while camped for weeks on-site. I dreamed of being able to do such things—underwater. Dr. Dwight Billings, the pioneering desert and alpine ecologist at Duke and a member of my doctoral committee, was sympathetic. "I have this picture of you in my mind: holding your breath, a thermometer clutched in your teeth, a bagful of sample bottles, a fistful of stakes, a hammer under one arm, and a coil of rope, setting out to establish a transect line underwater. It probably takes you ten hours underwater to do something that would take ten minutes on land."

He exaggerated a bit, but some things—everyday things—*are* more difficult to accomplish underwater: for example, sitting on a rock for a few hours writing up notes; changing film in your camera; eating lunch; talking; marking a place so you can return to it reliably the next day or week or year. But there are compensations. I explained to Billings how I could "fly"—like Superman!—over the undersea forests of my choice, taking conspicuous advantage of three dimensions, gliding down whenever I chose and landing precisely where I wanted to (as long as it was not more than about 150 feet deep). Finally, the most important advantage for me as a biologist was that diving made it possible to go where most of the living action on Earth is concentrated: underwater.

Throughout the 1960s I went to sea as often as a busy life with two small children and my scientist husband would allow. Many times, four small hands helped me arrange plant specimens on stiff, white sheets of herbarium paper and place them carefully between sheets of cardboard and blotting paper for drying. I made thousands of records this way of the marine algae occurring along hundreds of miles of Gulf of Mexico coastline from the tip of the Florida Keys to the broad marshes bordering New Orleans. This was also my decade of personal exploration in waters far away from home, including five expeditions aboard the U.S. research vessel *Anton Bruun*. And, I had a fateful encounter with Ed and Marion Link aboard their research vessel, *Sea Diver*, as they explored the waters bordering the Tongue of the Ocean in the Bahamas, as the only *woman* scientist participating in the Smithsonian-Link *Man*-in-Sea program. In due course, the meeting led to my involvement as leader of an all-women scientific and engineering team for carrying out an experiment in underwater living, the Tektite project.

Ed Link, a pilot and friend of Charles Lindbergh, had pioneered the use of an on-the-ground flight simulator for training pilots, a technique still famous as the "Link Trainer." Perhaps the most fully and productively occupied individual I've ever had the pleasure of meeting, Link frequently said, with a broad smile and certain jut of the chin that left no room for argument, "I've never worked a day in my life." "Retired" at fifty-eight and lured to the sea by a love for sailing and a passion for archaeology, Link turned his considerable inventive talents to revolutionizing access to the sea. With two others, Captain Jacques-Yves Cousteau and the U.S. Navy captain George Bond, Link now helped pioneer the development of living and working underwater.

The goal was to extend both duration and depth for divers at ambient pressure. Bond, a medical doctor as well as an officer, affectionately known as "Papa Topside" by his friends, recognized that once a diver's body was saturated with compressed gas and an equilibrium between pressure inside the body and in the surrounding water was reached, the decompression time would be the same no matter how long the diver stayed submerged at that depth—hours, days, even months. Decompression time depended on the depth and the gases breathed.

Starting in 1954, Bond tested the concept, eventually conducting an experiment with volunteers who successfully lived for fourteen days in a pressurized chamber that simulated pressure conditions at 200 feet. To live underwater, it would be necessary to bring the pressure of the atmosphere inside a chamber—the living quarters—into equilibrium with the water pressure outside. Then, people could swim in and out of their underwater dwelling as freely as they could walk in and out of their dry-land homes. Well, almost. It would still be necessary to travel in the water with air tanks strapped on; on land, even in smoggy Los Angeles, such measures are not yet necessary.

In its wonderfully methodical way the U.S. Navy planned to try the concept in the ocean in about five years. Cousteau and Link could move much faster, and did. Aquanauts with white fur coats and pink ears and tails—mice—participated in experimental saturation dives conducted by Link and a medical doctor with a passion for the ocean, diving, and exploration, Dr. Joan Membery. The mice were placed in special chambers containing a breathing mixture of helium and oxygen and pressurized to simulate depths greater than 1,000 feet. When the mice were brought back intact from their expedition into the unknown realms of physiology, Link decided to try the technique on himself, but at a lesser depth. In 1962 he stayed for fourteen hours (including six hours of decompression) at a depth of 60 feet in the first undersea station. A month later, Robert Stenuit used the same station longer (twenty-four hours) and deeper (200 feet), breathing a mixture of helium and oxygen. The time required to allow compressed gases to escape from his tissues took sixty-six hours, but once "saturated," the time would have been the same whether he stayed for a day or a month. Greater depth demands longer decompression, e.g., twenty-four hours at 1,000 feet may take ten *days* of slow return to surface pressure.

Shortly after Stenuit's dive in 1962, Cousteau finished development of the first of several of his experimental undersea dwellings, Starfish House, a bright yellow structure with a large central room and four radiating arms to accommodate apartment-style living for five people, anchored about 400 feet down in the Red Sea. Conshelf II, in June 1963, used Starfish House as the main settlement but also had a second habitat, Deep Cabin, anchored some distance away in 90 feet of water and occupied by two men who made excursions downward to as much as 165 feet. A domed sea-floor hangar housed Cousteau's small submersible, the diving saucer.

In an ongoing bout of "one-downsman-ship," it was Link's turn next. Link said of his next venture in undersea stations, in a 1963 *National Geographic* account: "Call it the deepest long dive. Call it the longest deep dive. Both definitions describe our goal beneath the bright water of the Bahamas. . . . We had come to put two men a long way down for a long time—more than 400 feet for more than 48 hours."

Robert Stenuit was again chosen to push the limits of going deeper and staying longer, but this time he would be joined by marine biologist Jon Lindbergh. Both would breathe a mix of helium and oxygen and have exhaled carbon dioxide scrubbed out chemically using a new system that Link had designed. Their domain was distinctly different—a bright yellow and black "submersible, portable, inflatable dwelling"—SPID. At 400 feet, they were very much on their own. A submersible could approach, but not free-swimming scuba divers.

Stenuit accepted the risks in good spirit, saying, "I knew what Ed had in mind . . . to live, eat, sleep, and work in depths so far unreachable by free divers, and in so doing, to take a long step toward the conquest of the continental shelf. To me it was the most extraordinary adventure of which a diver might dream."

Upon his return, Stenuit said:

I have returned from a strange journey in an alien world. . . . Living in the depths, I have become in certain ways a creature of those depths, adapted to their pressures. Now the human environment is temporarily intolerable to me. I need pressure. . . . I must wait inside this lifesaving prison of a decompression tank until I have been slowly weaned . . . and made once more fit to live on Earth.

After a two-day underwater stay in SPID, four days of slow pressure reduction were required for Stenuit and Lindbergh to "return to Earth," but as before, had they remained a month, or a year, the return time from saturation at that depth would in theory still be four days.

In July 1964, the U.S. Navy staged its first open-sea saturation dive, Sealab I, near Bermuda. Under Papa Topside's watchful eyes, four men remained at 193 feet for eleven days. The following year, Sealab II was organized, a project involving twenty-eight men in three teams who stayed underwater offshore from La Jolla, California, for fifteen days each. From time to time, the aquanauts were joined by Tuffy, a trained dolphin, a diver designed by nature with a warm subcutaneous "wet suit" of insulating tissue and flippers, and whose adaptations enabled him to make repeated excursions to the station 200 feet down and back with no need of decompression.

Meanwhile, Cousteau continued experimenting in the Mediterranean Sea with extended diving techniques. In 1965 he launched Conshelf III, designed for operation in 328 feet of water, near the coast of France. This time, the demonstration was intended to do more than simply show that people could survive in a pressurized underwater chamber, swimming in and out to make observations in the sea beyond. Conshelf III would also demonstrate that subsea stations manned by divers could be effectively used to support offshore oil and gas exploration and recovery.

Six thousand miles away, in California, M. Scott Carpenter, leader of the first two Sealab II teams submerged 205 feet underwater, was able to converse by a special telephone link with Conshelf III's team leader, Andre Laban. In a final symbolic gesture of exploratory camaraderie, astronaut Gordon Cooper sent greetings to Sealab via a special radio channel from his perch within the *Gemini 5* spacecraft then orbiting 150 miles overhead. The stage was set for an underwater equivalent of the *Skylab* and perhaps, someday, a sea station to parallel the space station, with the U.S. Navy taking the lead.

Ed Link continued to push ahead with another variation on the theme of saturation diving, in collaboration with the talented engineer John Perry. Perry had already developed successful deep-diving systems for commercial and military uses and, like Link, was motivated by a desire to develop technology vital for exploration and gaining new insight into the nature of the oceans. Perry would now design and build a new kind of system.

The idea was to develop a sub that would serve as an underwater taxi for divers, a mobile decompression chamber to transport them, to allow them to swim in and out at depth and be moved from place to place while under pressure, to begin decompression aboard the submersible or be brought back, still under pressure, to a decompression chamber on the surface. The pilot of the sub stayed in a separate compartment maintained at one atmosphere—the same as surface pressure.

In 1967, the Perry-Link *Deep Diver* was launched and with it, a new and valuable concept comparable to the World War II Royal Navy *X-craft* midget submarines, soon adopted and widely used by offshore industry to support commercial deep-diving operations. In a *National Geographic* account (1968) Link described *Deep Diver* as "more than a sub. It's a system. It includes a special hydraulic crane that can launch and retrieve her even in rough weather. . . . We park her on the bottom and build up the gas pressure in the divers' compartment until it equals the pressure on the outside; the hatch drops open and out they go. . . . When they get back in, they simply close the hatch to lock the pressure in with them, and come on up. We hoist her aboard and sail away."

A chance to use this little sub as a scientist in the Man-in-Sea Project in the Bahamas brought me aboard the vessel *Sea Diver* in 1968. I had never before been inside a submarine, let alone had a chance to swim *out* of one underwater. From *Sea Diver*'s deck I climbed into the dive chamber with a seasoned submariner, Denny Breese, my notebook and camera, and a passenger not on the manifest—my daughter-to-be, Gale, then five months under way. Nowadays, most physicians advise women not to dive while pregnant, but the doctors I consulted before the project thought there would be no problem either for me or the baby, and there were none. Now, my daughter explains her love for the ocean as something she can't help. "After all," she says, with an impish grin, "*I* was diving before I was born."

At last the hatch closed and the sub was picked up by the crane and gently deposited in the sea. Through tiny ports in the dive chamber I glimpsed aquamarine merging into deep open-sea blue as the sub glided down a gradual slope. With a gentle bump we touched down on an open sandy area peppered with small coral-crowned rocks and miniature forests of the plants I had come to see. Denny signaled the pilot that we were ready for action, and compressed air hissed into our chamber. Immediately my ears felt squeezed, and I held my nose and swallowed to relieve the stress on aching cranial canals and sinuses. About a minute later, pressure inside the sub reached that of the surrounding sea and Denny pounded at the pegs holding the hatch until, with a slow, metallic sigh, they yielded and the door swung down. Despite my knowledge of the laws of physics, it seemed improbable to me that pressure inside the sub could keep water from rushing in, but there I sat, warm and dry, staring into a shimmering, well-behaved circle of blue, rimmed with silver. I poked my finger into the pool, then my toes, then let go, plunging feet first onto soft sand three feet under the sub, 125 feet beneath the ocean's surface. For a moment I stood in waist-deep water, my top half still dry, then waved at Denny and ducked out, under the sub and away to the reef, reveling in the chance to have an hour and a half to look around—more than four times longer than I had ever enjoyed in a continuous stretch at this depth. To impress nondivers with what this means, I ask them to imagine trekking anywhere else on Earth—up mountainsides, into grocery stores, libraries, or parks—with life-or-death limits on how far (125 feet) and long (20 minutes) you can go.

Even my extended passport expired too soon; reluctantly I scrambled back through the mirrored underside of *Deep Diver*'s open hatch, and helped Denny dog down the hatch. Wistfully I watched the reef through the sub's port as we lifted off and eased our way slowly back to the ship. En route, the pressure gradient was reversed, gradually returning to one atmosphere over more than an hour of decompression. Time enough for me to seriously dream about going deeper, staying longer . . .

I could not imagine how soon, nor how spectacularly, I would have a chance to do both.

Chapter 4

BLUE INFINITY

There is no way I can adequately describe the astounding events of the last five days. The language does not have adjectives superlative enough to capture what we have seen and experienced. . . . Heaven may pale by comparison beside the wonders we have witnessed.

Gentry Lee, The Garden of Rama

Moonlight and the brilliant sparkle of Venus flickered through fifty feet of ethereal ocean, faintly illuminating my dive partner, Ann Hurley. Our underwater flashlights were off, our senses fine-tuned to the pale shadows of rocks and reefs, as we made our way toward a distant glow that marked our underwater home. There, fifty feet under a warm Caribbean Sea, awaited four rooms filled with air, a hot shower, dry clothes, cooked meals, a nicely equipped laboratory, and clean sheets on a soft bunk. With three other women, Ann and I were participating in an experiment in living and working in an underwater laboratory called Tektite. It was a fine way to spend the summer of 1970.

The previous year, the U.S. Navy was inspired to continue the highly successful and largely trouble-free Sealab series, and prepared a new undersea station to accommodate aquanauts at a depth of 620 feet, with plans for excursions to a maximum of 1,000 feet. Originally, fifty divers had been trained and in 1969 the new station had been built and placed in the sea near San Clemente Island, California. At that time the Navy was embarking on a program that would mirror the growing commitment by the National Aeronautics and Space Administration (NASA) to make it possible for people to live and work in space. But it was not to be. A tragic accident with one of the diving units resulted in the death of a brilliant young aquanaut, Barry Cannon, and concerns about safety caused Sealab III to be canceled before it really got started. The Navy terminated its seafloor habitat program, but continued to help sponsor another saturation diving project—Tektite I—in collaboration with NASA, the Department of the Interior, and other governmental and nongovernmental agencies.

NASA officials believed that much could be learned from undersea experiments that would have useful carry-over for space. Monitoring aquanauts and doing psychological tests was thought to provide insight on how people work in isolation, in hostile environments, and on characteristics that might be useful to know about when selecting future astronauts—and aquanauts. Even the name of the program, Tektite, conveyed the sea-space theme, tektites being small, glassy nodules from space most often found in the sea.

Five hundred million people watched Buzz Aldrin and Neil Armstrong take their historic first steps on the moon in July 1969, but only curious fish and round-the-clock television monitors and psychologists observed the first team of Tektite aquanauts 50 feet underwater just offshore from Lameshur Island at St. John, in the U.S. Virgin Islands. Four men stayed submerged for two full months, living and working in a four-room habitat, two large vertical cylinders joined by a short tunnel. Built by the General Electric Company, the system seemed inspired by GE's popular assortment of large, white kitchen appliances, complete with plug-in cord. Cables extended from the habitat to a shore-based source of air, water, and power, about 600 feet away.

At the time, I was enjoying the third year of my first professional role as "Dr. Earle," simultaneously holding an appointment as a Radcliffe College Scholar and Harvard University Research Fellow, concentrating on the less glamorous but vital side of exploratory field research: carefully examining specimens, researching existing records, contemplating what it all means, then writing up results. Life had never been fuller, professionally or personally. My marriage to John Taylor had come to an amicable end, and I had remarried and taken the current positions to be with my new husband, Dr. Giles Mead, Curator of Fishes at Harvard's Museum of Comparative Zoology.

Like other young scientists, I was reveling in the breakthrough insights and discoveries that tumbled forth from my research and I longed to spend more time at sea, in the lab, and in my basement office at home on Beacon Hill in Boston, where I managed to do a lot of after-hours writing while watching ankles and boots pass by on the sidewalk above. But unlike most of my scientific colleagues, I also longed to be a good wife, chef, hostess, homemaker, and occasionally glamorous companion for my scientist husband, and good mom to my three young children, with time to do more than just drive them to and from school with ice cream stops en route. Maintaining a semicivilized home in our row house just around the corner from Boston's historic Louisburg Square was in itself a full-time challenge.

Thus, it was extremely unlikely that a 1969 announcement from the Smithsonian Institution that appeared on the bulletin board at the Museum of Comparative Zoology could or would affect my life in any way. It was Giles who insisted that I should check it out as something too important to pass by. Phase II of Project Tektite was getting under way, the flier said, and as many as fifty scientists and engineers were invited to conduct independent research projects starting in the spring of 1970. If accepted as one of them, I might arrange to go during the summer when the children were enjoying a visit with their grandparents in Florida.

I was intrigued with the idea, especially after experiencing hundreds of hours of stopwatch diving, and the extended time possible using *Deep Diver*. I thought, *Two weeks! Not just in and out for half an hour or so at a time, three or four times a day, but all day every day and all night. Underwater for fourteen days straight! What a concept!*

It wouldn't be exactly like camping. Some people referred to the habitat as the "Tektite Hilton" because of its temperature and humidity control, carpeted floors, multicolored interior, soft bunks, taped music, television, range, sink, refrigerator, freezer filled with Stouffer's prefrozen meals (such as those provided to astronauts during quarantine after reentry), microscopes and other research instruments, a private toilet, a freshwater *hot* shower—and the best swimming pool on the planet.

The amount of time spent outside the habitat would be up to the aquanauts, much as it would be for scientists working on the land. The physical and physiological constraints on diving time and depth are inflexible, but saturation diving using the Tektite habitat approached the rules in a different way, providing unlimited time at the depth of the laboratory at 50 feet, with generous excursion time downward to at least 100 feet and upward to within 20 feet of the surface. At the end of the mission, aquanauts would swim to a diving bell, a pressurized capsule open at the bottom. Once the divers were safely aboard, the hatch would be closed and the bell would be lifted to the surface while maintaining pressure. A hatch on the side would allow the bell to be fastened to a large chamber with pressure equal to that inside the bell. Once connected, the doors could be opened and the divers transferred to the chamber for 21 hours of decompression.

It did not seem like a bad trade-off: 21 hours parked in a metal cylinder decompressing in order to have 336 hours of continuous underwater time. I had traveled much longer crunched into the confines of other metal cylinders—airplanes—to have far fewer hours of diving on distant reefs. Without much hope that I would be selected, I drafted a proposal for work that I wanted to do, given a chance to stay submerged on a reef for two weeks. For a long time I had been intrigued with the impact grazing fish have on the number and kind of plants in an area, and this would be an ideal opportunity to have a close look at the issue. I learned that several ichthyologists—fish scientists—wanted to team up for a dive, and discussed with them the mutual advantages of combining their interest in fish with my work with plants. But the most important consideration with respect to my participation had nothing to do with science or diving experience.

"We didn't expect women to apply" I heard a voice saying on the telephone. It was Dr. James Miller, project manager. "What would you think about being part of a team of just women? There are concerns about having men and women living together in the habitat."

I could hear, and almost feel, the embarrassment in his voice, tangible right through the telephone line. Applications submitted by would-be women aquanauts had turned out to include a number of first-rate projects that were impossible to disqualify on the basis of merit or the scientists' experience, but the concept of having "mixed teams" was simply too hot to handle in 1969. Women astronauts had been discussed, but no serious actions taken to include them in space programs, no matter how impressive their performance or credentials. One way to deal with the possibility of women *aquanauts* co-habitating with male colleagues, clearly, was to sidestep the issue and select a team composed entirely of women.

I shrugged and said, "Why not?"

Given a choice, I would have preferred going with people I knew and had worked with before, but I did not mind the concept of being part of a team of women. In fact, it sounded as though it could be fun. Miller asked whether I, as the applicant (male or female) with the most documented dive time coupled with strong scientific credentials, would be willing to be the team leader. Again I shrugged and said, "Sure." (Later I learned that when he had been asked whether he thought women should be involved with the program, *he* shrugged and said, "Sure, *why not?*")

No one anticipated the reaction to the announcement that a team of women would be conducting research projects while perched on a reef 50 feet below the surface of the ocean in the Virgin Islands. The first clue that my life was about to take a radical turn came via a headline on the front page of the *Boston Globe*, accompanied by a photograph someone had taken of me—once again, soaking wet:

BEACON HILL HOUSEWIFE TO LEAD
TEAM OF FEMALE AQUANAUTS

All of us—Dr. Renate True, an oceanographer from Tulane University; Margaret Lucas, an ocean engineer with her master's degree from the University of Delaware; and Dr. Alina Szmant and Dr. Ann Hurley, both from Scripps

Institution of Oceanography—were dead serious about wanting to explore the use of new equipment and accomplish the research objectives we had set. None of us expected to be the subject of scrutiny, other than as participants in NASA's ongoing psychological research studies. We were prepared for the circumstances described to us: We would be watching the fish, who in turn would watch us; meanwhile, the team of psychologists would watch us watching the fish, and using a special two-way television monitoring system, we could watch ourselves watch the watchers while they watched us watching them—and the fish. That was okay, part of the program. Being watched by various media representatives was something else.

The behavioral studies required that all missions be conducted in isolation, that is, there could be no face-to-face contact with anyone other than team members for the duration, much like conditions during a flight in space. We were on our own, even though people were not far away. Two-way communication was maintained, however, especially with the watch directors, who were the conduit for official messages both ways. Through them we learned that the women's team could expect to be blitzed with questions when we returned to the surface, and we were. In fact, I did not mind this. It was frustrating, trying to fire public interest in an ocean that was rapidly changing. *If people could see what we see underwater*, I thought, *they will be motivated to take care of the place. But they can't know if nobody tells them.*

Fifty aquanauts were selected to participate in a series of ten missions over a period of seven months, but as it turned out, most of the attention focused on the women's team, referred to variously as "aquanettes," "aquabelles," "aquababes," and even "aquanaughties." Such treatment was more cause for amusement than consternation, and on balance was an indication that at least someone was paying some attention to something that was happening in the ocean, even if for the wrong reasons. We even had fun rewriting some of the headlines giving the words a fresh twist, such as "Beacon Hill Husband Leads Team of Male Aquahunks."

The scientific side of the project focused on several objectives. Individual research remained a high priority, as did the psychological evaluations conducted by NASA. But there was an underlying aim to learn as much as possible about the practicality—and limitations—of saturation diving. Already, commercial applications for the technique were growing, and other countries were taking a strong interest in using this and other approaches to diving. If Tektite II proved successful, there could well be a boost for ocean awareness generally, and especially for the direct approach—*in situ* research—to ocean exploration. The method enabled marine scientists to use field facilities in the sea in ways that their terrestrial counterparts have always taken for granted.

To help take advantage of unlimited excursion time a new tool was made available, a rebreathing system, comparable to the life-support units worn by astronauts, that could supply air in depths to as much as 1,000 feet for twelve continuous hours. Excursions from Tektite mostly were in less than 100 feet, but for my purposes, rebreathers were perfect—no noise, no bubbles, and lots of time. The secret to their success is simple. The diver rebreathes air after oxygen has been automatically added as required from a cylinder carried in a special backpack. Exhaled air passes through a container in the backpack, where carbon dioxide is chemically scrubbed out and the clean air is recirculated. Using rebreathers, it is possible to hear subtle sounds obscured by scuba—the crunch of parrotfish teeth on coral, the sizzle and pop of snapping shrimp, the grunts of groupers, the chattering staccato of squirrelfish. The "silent undersea world" hums, trills, drums, and crackles with a symphony of vital signals sent and received. I was thrilled to be able to tune in, and to be accepted as a big, silent "fish" rather than an intrusive, rowdy, bubble maker.

It is easy to look at the published records to determine the magnitude of the work accomplished during thousands of hours of observation and research on the coral reefs of Lameshur Bay; it is less easy to quantify the change that took place in those who for the first time were able to see the ocean not just as passing-through visitors, but as day-to-day residents. It is comparable to taking a walk in the woods for half an hour versus camping there, day and night, for a couple of weeks. Patterns come into focus, shy creatures reveal themselves, concepts emerge from the shadowy places around the edges of your mind when you have the luxury of uncommitted time. Stay underwater long enough and individual fish become recognizable, not just as gray angelfish or filefish or barracuda, but as very specific characters whose habits become as familiar as those of neighbors—which they are, when you yourself are living underwater.

One of the great joys of being a dweller on a reef is the ease of being up and about early, while it is still full dark, slipping into quiet water, and gliding without lights to the reef's edge where many dawn risers tend to congregate. Five gray angelfish were among the first to emerge from their nighttime crevices and every morning, same time, same place, begin their slow waltz around the mounds of coral, pausing now and then to nibble a bit of sponge or nose at a lump of algal debris. There were also several species of parrotfish, all fitted with fused chompers that much resemble the bill of the birds for which they are named. As colorful as macaws, the parrotfish, like many birds, are active by day and sleep at night, but unlike any parrot or other bird I have ever heard about, these extraordinary fish often spin a clear gelatinous cocoon around themselves and slumber in a trancelike state robed in diaphanous goo. Among the first to tuck in at sunset, they are also among the first to appear by dawn's faint light, just as squirrelfish, cardinal fish, and other night-active creatures move in to occupy for their daylight slumber the places just vacated by the day-active kinds. A clear case of "hot bunking," I reckoned, watching a triggerfish emerge early one morning from a comfortable-looking niche where it had been resting on its side (lightly covered by a blanket of white sand, no pillow), just as a large, lone squirrelfish sidled in.

Another advantage of being a resident is getting to know the individual places claimed by specific creatures. Even the large, feathery basket starfishes proved to have strong homing instincts. Open like enormous lace handkerchiefs by night, at some point before dawn (I could never determine exactly when), they return to their niche—always the same place—where they collapse into their dishrag mode and spend the day tightly curled up.

The Tektite habitat itself soon became populated with settlers: colonies of algae, sponges, polychaete worms, and many other reef creatures, who set up housekeeping on the outside while the aquanauts did the same within. Curious fish often gathered around the entryway and looked up through the strange surface into the illuminated room above. On full-moon nights, silhouettes of large, sleek predators—amberjack and tarpon—arched and turned, silver on silver, attracted to the clouds of small fish who in turn were attracted to the hordes of minute crustacea drawn to the lights of the habitat. Sometimes scientific detachment was put aside in favor of joining in the exuberant rush of fins and scales, not as a predator, but as an active witness and benign participant in the ebb and flow of life.

Late one afternoon, returning to the habitat along a familiar part of the reef, I felt so at ease that it seemed my wish for gills had been granted; I merged with the water, allowed myself to hang, motionless, neither sinking nor rising, just there, like a sleepy barracuda or a single molecule. Other times, three or four of us would romp along, punctuating the short journey from reef to open plain now and then with somersaults—definitely not a part of the scientific program, but instructive, especially at night. A kick, a swift flick of a hand or arm, or a full body flip left a fiery trail of luminous sparkles, living light generated by numerous small planktonic species barely visible by day but dazzling when touched at night.

Most days and many nights we stopped by the habitat only to recharge air tanks, grab something to eat, write up a few notes, change film, and be off, putting in ten to twelve hours in the water. Within two weeks I had observed the day and night behavior of 35 kinds of plant-eating fish in 14 families, and 154 kinds of plants, including 26 species not previously found in the Virgin Islands. It is a modest inventory, but not likely one that I could have made given two *months* of diving from the surface.

A curiously fateful change in policy that influenced the future commitment of the United States to oceanic and atmospheric matters occurred during the time we were submerged. When we entered the habitat on July 6, 1970, the Department of the Interior was managing the program and most other ocean issues on behalf of the nation, but by the time we emerged, on July 20, a dramatic change had been announced. A new agency, the National Oceanic and Atmospheric Administration (NOAA) would embrace elements ranging from the U.S. Weather Service and National Marine Fisheries Service to the U.S. Coast and Geodetic Survey, a national network of research laboratories, and the budding U.S. environmental satellite program. The original idea was to create a strong, independent agency, a counterpart to the space program (and consequently referred to by some as a "wet NASA"). Instead, NOAA was placed under the Department of Commerce, where it has gradually languished, without power or authority commensurate with the great and growing scope of environmental responsibilities. I had no inkling that exactly twenty years later I would be appointed by President George Bush and confirmed by the U.S. Senate to be the Chief Scientist of NOAA, one of four people making up that agency's management team. I was an often lonely voice in support of what remained of the concept for a vigorous national underwater research program.

As predicted, the return of the women's team was greeted with much fanfare. We spent more in-water time than any of the others, and it was said that we were the "best housekeepers," and got along together "amazingly well." (The idea of women cooperating, enjoying one another's company—actually working together as a team—was thought by some to be impossible. "We thought you'd tear each other apart," one of the psychologists later told me!) In the end, all of the ten teams accomplished what they set out to do, and then some—but only we were brought bouquets of roses as we emerged from decompression, and only we were whisked off to Chicago for a tickertape parade with Mayor Daly down State Street, riding in a fur-lined limousine built for an earlier visit by the Pope, all followed by a reception at the Shedd Aquarium, where Mahalia Jackson sang to us. And only we were asked to address Congress and accept Conservation Service medals presented by the Secretary of the Interior, Wally Hickel, and to join Pat Nixon (not Richard) for a luncheon at the White House. Some regarded this as a perverse form of reverse discrimination, while others were stung by the underlying message of age-old stereotypes. I tried to smooth the rough edges when asked to speak in response to receiving the awards—accepting on behalf of all fifty aquanauts—but, understandably, it did not keep some from grousing.

For me, an unexpected consequence of participation in the Tektite program was a change in my attitude toward communications and the media, a consequence of repeatedly facing a bristling array of microphones and thousands, sometimes—via radio and television—millions of people, apparently eager to listen to what I might say. *I* was eager to hear what I might say, too. I had no idea what would come out when formidable interviewers—Barbara Walters, Hugh Downs, Dick Cavett, and many others—asked questions ranging from "Did you see any sharks?" (No! Not one!) to "Did you eat any seafood while you were down there?" (Yes! Frozen fish sticks and lobster Newburg).

It quickly became apparent to me that it might be possible to bounce from the light and entertaining to more serious topics. More often than not, it seemed to me that this is what really did interest the interviewers, and maybe the audiences, but many were not sure what questions to ask to get to the heart of meaningful issues—just as I was not sure how to respond.

In the year following, I gave lectures or talks somewhere at least once a week, and gradually began to warm up to the concept of trying to articulate for everyone, not just my learned ivory tower colleagues, the results of scientific inquiry in the sea. Still, when William Graves, an editor from *National Geographic*, called and asked me to write a story about the experiences of the women's team, my first response was, "No way"—despite my great respect for that venerable institution. I had grown up with the fine academic tradition of scientific communication, a form that requires precise, often arcane expressions, a detached arms-length approach, with personal interpretations avoided unless carefully couched in the most cautious terminology. A price is exacted from scientists who are perceived as "popularizers," and I already had felt the heat from my colleagues because of the publicity generated by the Tektite dive. With some soul-searching, though, I had to admit that part of my reluctance in accepting the invitation was that I already had found it extremely challenging to talk to general audiences without falling back on technical words that I had trouble explaining in straightforward terms. (As my daughter Elizabeth admonished me from time to time, "No gobbledygook, Mom!")

In the end, I not only agreed to accept the opportunity provided by the *National Geographic* to reach more than 10 million readers, but also, softened up as I was, I took a real leap over the edge and wrote a short piece for *Redbook*, a magazine more noted for its steamy stories than for scientific reporting. Certainly, it had a different audience than the *Journal of Phycology* or *Occasional Papers of the Farlow Herbarium of Cryptogamic Botany*. I worked especially hard on crafting the *National Geographic* piece, but one glaring problem emerged: There were no photographs to illustrate my description of living and working on a reef for two weeks.

Of special interest was the incident I related concerning what happened when a sand-clogged regulator and faulty reserve valve left me without an air supply more than 1,000 feet from the habitat. Sunlight and air were only 71 feet away—overhead—but I was separated from them by a physiological barrier as effective as a brick wall. Racing for the surface for air with my tissues saturated with compressed gas would spell swift disaster: bubbles in my blood, pain, possible paralysis, even death. My options were limited to a long swim back to the habitat sharing air with my buddy, the engineer Peggy Lucas, or activating an emergency pinger to alert shore-based safety divers to try to find us and drop spare tanks.

The greatest threat to survival underwater, and perhaps elsewhere, is not the bends, sharks, poisonous jellies, or other imagined terrors of the deep—it's panic. Fortunately, my buddy had a cool head and calmly handed me her mouthpiece when I appeared before her and gave her the traditional signal for "out of air," a slicing motion across my neck with my index finger. I took two breaths, then handed it back. "Buddy breathing" is one of the basic maneuvers taught in most diving courses, but this was the first time either of us had had to seriously put our practice into *practice*. We made it back "home," swapping the precious supply of air back and forth, with no further problems, even pausing for a moment to observe fish grazing on the grassflats along the way. To illustrate this event and other elements of the article, the *National Geographic* enlisted Pierre Mion, an artist well known for his hauntingly effective images of space and undersea technology. Pierre called and asked if he could come to my home to take photographs that would help him visualize scenes for the proposed paintings. It meant a trek for him to Los Angeles where my husband had recently been appointed director of the Natural History Museum. A logical place to simulate deep-sea drama was nearby—a backyard swimming pool. He enlisted a diving friend, Dr. Joan Membery, to stand in as a model for my "buddy." When I discovered that his friend was *the* Dr. Membery who, with Ed Link, had pioneered studies of diving with breathing mixes of helium and oxygen years before, the backyard venture was transformed into a symposium on diving physiology—and the beginning of an enduring friendship.

Nothing relating to Project Tektite was really dangerous except posing for Pierre's paintings. To create an image of me entering the habitat after a dive, I propped a ladder against a large sycamore tree in the backyard and, in full scuba

regalia—wetsuit, tank, mask, flippers, the works—climbed into the leafy canopy, with Pierre snapping away from various strategic angles. I risked not only
life and limb (mine and the sycamore's) for art, but also my reputation for sanity
among my new neighbors, who watched, silently, from a safe distance. It did not
help that my extended family, now including six children, "Fiddi" Angermeyer,
a young man from the Galápagos Islands who lived with us, and our domesticated Australian dingo dog Fang—all stood observing the backyard antics with
supreme indifference, coolly conveying the impression that the Meads—or
Mom, at least—*always* climbed trees in full scuba gear.

After that, I thought I was shock-proof with respect to strange outcomes of
the Tektite project, but I had not reckoned with the title chosen by the *National
Geographic* for my maiden plunge into popular-science writing—one that truly
set me up for much ribbing from my loftier-than-thou scientific colleagues. Following an article by John Van der Walker, "Science's Window on the Sea" in the
August 1971 issue, my piece was announced in bold letters as:

ALL-GIRL TEAM TESTS THE HABITAT

At first I cringed. I was a 34-year-old "girl" at the time; the youngest participant was the engineer Peggy Lucas, 23. Later, I philosophically reminded myself
that the reason I had been asked to do the article was not, after all, primarily my
scientific expertise but was rather a response to the wave of curiosity generated
by the unexpected participation of women—that is, *girl*—scientists. If I could
turn some minds toward ocean issues along the way, so much the better.

To many at the time, the idea of females of any sort participating in oceanography, especially as divers, was unusual. Just how unusual was conveyed to
me personally by Papa Topside, Dr. George Bond, the revered "father of saturation diving." I had been invited to participate with him in a scientific diving
project offshore from Panama City, Florida, where I had years before acquired
some of the basics of underwater survival by tagging along with tolerant Navy
divers. After one dive, as Bond and I sat drying in the afternoon sun, he began
talking about Sealab, and the moment seemed right for me to tell him how much
I admired his pioneering work—and that I thoroughly enjoyed being a Tektite II

aquanaut. This formidable, notoriously gruff Navy captain looked at me carefully, then with a small shrug said, "You know, I opposed having you go."

"You mean you didn't like the idea of a women's team?" I asked. I was prepared to agree with him. All things considered, I really would have preferred my original scientific program of working with several male colleagues.

"No," he said. "I mean I opposed having *you* go. You were the only one who is a *mother*. A lot of what you did was really experimental, from using rebreathers to going through a decompression schedule that hadn't been tried before. You have a family to think about, you know."

"But weren't there *fathers* participating?" I asked, not really believing how serious he was.

"Yes," he said. "But that's different. Anyway, I was overruled, and I want to tell you that you did a great job. But if it were my decision alone, I would never have let you go."

A slight crinkle that might have been sheepishness played around the edge of Bond's grin. I laughed, my chin jutting forward with an end-of-conversation tilt, and we got ready to go diving again.

While the Tektite II project was in full swing, in June 1970, progress continued in the development of new approaches for underwater living. In Hawaii, the bright yellow habitat Aegir, named for the Norse sea god, provided a safe haven for six men in 520 feet of water—nearly 100 feet deeper than any previous habitat. Thereafter, numerous projects blossomed. By 1975 there were about fifty underwater habitats in existence around the world, ranging from very simple structures for use as "camping" facilities in shallow water to highly sophisticated versions such as West Germany's Helgoland habitat and the Soviet Union's Chernomor, both used for scientific research in cold-water areas. Most, like the Tektite habitat, were retired after a year or two of operation, but one, the Hydro-Lab, demonstrated remarkable staying power. Built by Perry Oceanographics and launched in Florida in 1968, Hydro-Lab was moved in 1971 to a location one mile offshore from Freeport, Grand Bahama, and later to the U.S. Virgin Islands. For years, Hydro-Lab was regarded as an underwater Mecca by several hundred scientists—343 in all, including 37 women. I managed to be one of them during five missions, each lasting from a few days to a full week, including one brief stay shared with Senator Lowell Weicker and two members of his staff. While in the Bahamas, Hydro-Lab residents maintained constant radio contact with the shore-based management team headed by Robert Wicklund, a biologist and the director of the Caribbean Marine Research Center, who envisioned and helped implement a long-range research program on the ecology of the reefs near Freeport.

One of the special benefits of having an underwater laboratory stay in one place for a long time, with many people bringing varied talents to bear, is the gradual accumulation of knowledge that provides something more than a superficial inventory of what lives where. Details come into focus, relationships among the reef residents gradually become known, the subtleties that make a system really work become evident. The few such places where reliable long-term baseline information exists, on land or in the sea, are rare treasures, ongoing reference points against which change can be assessed with more confidence than can be provided by recollections.

In 1975, some new tools were brought to Freeport that helped extend the baseline downward. Years before, at the very outset of the Man-in-Sea programs, Link had entertained a vision of one day achieving regular access to the continental-shelf areas of the world, and felt certain that underwater dwellings coupled with a fleet of submersibles would be vital. In 1963, he observed, "If man could find a way to work there in safety and relative comfort, he would at once possess the key to more than ten million square miles of sea bed. He could tap the scientific secrets and mineral, animal, and vegetable wealth of these immense submerged plains. . . ."

Link developed a partnership with Seward Johnson, an industrialist, philanthropist, and avid sailor who shared his desire to develop and apply technology to gain better access to the sea. Together, they formed the Harbor Branch Foundation (now the Harbor Branch Oceanographic Institution), based at Link Port in Fort Pierce, Florida, and set about implementing an ambitious program of engineering and science.

In 1969 and 1970, while the Tektite project was under way, Harbor Branch launched two new diver lock-out submarines, *Johnson-Sea-Link I* and *Johnson-Sea-Link II*, affectionately called bubble subs because each featured a spectacularly clear acrylic sphere at the front end. Inside, a pilot and an observer enjoyed unprecedented visibility for dives in depths to 2,000 feet. As in the subs' predecessor, the Perry-Link *Deep Diver*, divers traveling in the aft "lock-out" compartment could be pressurized to match the pressure outside the sub and swim out and return, using the sub as a decompression chamber, while the pilot stayed in a one-atmosphere chamber. Now, Ed Link's plan to combine the use of an underwater habitat and his new diver lock-out submersibles was funded by the small Manned Undersea Science and Technology division of NOAA; the special program was dubbed "Scientific Cooperative Operational Research Expedition," better known as the "SCORE Project."

The SCORE Project would for the first time validate the concept of using an underwater habitat housing scientists and a diver lock-out sub, one of the *Johnson-Sea-Link*'s, as a taxi to and from deep water. Five teams of four people each would try the system, not only demonstrating the usefulness of saturation diving and lock-out subs, but also adding significantly to the growing documentation about the Freeport reefs.

As one of the team leaders, in April 1975 I once again crouched in the pressurized chamber of a lock-out submersible, only this time when the hatch swung open, I could not touch bottom. Beyond the open port beckoned more than 1,000 feet of blue infinity! Instead of parking on a flat surface, the sub was 250 feet underwater suspended from a chain imbedded in the face of a steep dropoff, and I was free to swim out and for nearly an hour cruise along the face of the cliff like a subsea hawk.

Dives made at such depth using compressed air must be done very, very, carefully. Nitrogen fogs the brain with euphoria, and oxygen-induced convulsions can strike without warning. For an excursion of twenty minutes, two hours of decompression are required, and each additional minute at depth carries a magnified decompression penalty. In the submarine, my partner, Steve Nelson, and I were not starting at surface pressure, however. We were already saturated at 50 feet, and would return to 50 feet. An hour at 250 feet would require relatively short decompression—but the problems with nitrogen and oxygen were still there.

Two hundred fifty feet is not a great distance on dry land, but underwater, until recently that distance down was mostly the stuff of dreams and science fiction. Many times I had lingered briefly at the top of the cliff where I now soared and had peered over at the vast blue space below, wishing to glide deeper and see what creatures lived in the indigo wilderness just beyond my fingertips—but I resisted, obeying the rules, knowing I could not venture deeper or stay longer unless I chose to stay there forever.

Steve, who would remain inside to tend the air and communication lines, helped adjust the straps on the gleaming new Kirby-Morgan dive helmet feeding air past my face in a soft, steady stream. I launched myself into clear blue space and flippered my way in slow motion toward the cliff. As I looked back, the sub was an apparition of bright lights, an Oz-like bubble with two humans inside wearing T-shirts and blue jeans, magically suspended in midwater. Balancing myself with one finger along the vertical rock wall, I explored a narrow shelf, lightly covered with bits of fine, shelly sand, and discovered there a miniature forest of green plants, each one a perfect parasol.* I selected a few to examine on the surface, photographed what I could to document this newly explored realm, and earnestly wished for gills. More time! Greater depth! Each new level of access increased the sense of urgency for more.

Others were driven by similar desires, and significant progress was soon forthcoming. The U.S. Navy, although no longer engaged in developing undersea stations, was actively researching deep-diving techniques. In 1975, Navy divers made the deepest plunge into the open sea up to that time, 1,148 feet, using a tethered, pressurized sphere from which they emerged and returned for transport back to the surface for days of decompression.

Commercial applications of saturation diving progressed rapidly, driven by the needs of an expanding offshore oil and gas industry. The limits of depth and

---

* My original hunch, that the plants had not previously been discovered or named, was later confirmed, and so I had the pleasure, in collaboration with the Harbor Branch scientist Nat Eiseman, of honoring Seward Johnson, Ed Link, and the submarine by calling the little green alga *Johnson-sea-linkia profunda.* These little plants were later discovered in large numbers elsewhere in the Bahamas in depths as great as 660 feet, but never in less than 250. So far, discovery of a site where the plants grow has always been from one of the *Johnson-Sea-Link* subs.

endurance were pushed in the cold waters of the North Sea, with saturation dives progressing to depths greater than 1,000 feet and divers sometimes remaining pressurized for six weeks or more. Dives to 1,000 feet are never routine, but during the 1970s, such depths were often attained by many unsung heroes who tested the limits of human physiology with their own blood, muscle, and bone.

High rewards drive high risks, and many were taken in the 1970s, leading to the recognition of limits beyond which it was difficult to go. For many years, a team of scientists and physicians headed by Dr. Peter Bennett at the Duke University Medical Center methodically probed the nature of those limits. A maze of giant pressure chambers still serves as a special laboratory where volunteers can experience simulated dives to great depths while being carefully monitored and observed. When asked recently what the maximum depth might be for free-swimming humans, Bennett gave a qualified answer. "It depends on what you want to do," he said. "Using controlled mixes of helium, oxygen, and a little nitrogen, regular working dives to a thousand feet are possible, but there are still many unknowns. Experimental dives to two thousand feet are promising, but not yet practical."

In 1976, the Manned Underwater Science and Technology office of NOAA proposed the development of OceanLab, a large, mobile underwater laboratory that might be self-contained for prolonged underwater operations. The idea most often discussed involved a submarine about 100 feet long with a special built-in saturation-diving facility that could be pressurized to allow six divers to swim out in depths as great as 1,000 feet and then return to OceanLab for transport under pressure to swim out at other locations, or remain inside, ultimately decompressing to join their one-atmosphere companions. William Busch, a tall, lanky scientist-engineer from NOAA, came to my office in San Francisco to discuss my possible participation in the project.

At the time I was getting settled in a new research position as Curator in the Department of Botany at the California Academy of Sciences in San Francisco's Golden Gate Park. In my personal life, I was on my own at this point, my second marriage a casualty of the pressures of two impossibly full lives. I emerged emotionally bruised but intact, with three children and loving parents able and willing to help me care for them. The thought of undertaking a project such as the one Busch described was irresistible, despite concerns such as those candidly delivered by George Bond regarding the risks associated with the Tektite project. I had no intention of undertaking anything I regarded as unsafe, let alone life-threatening. But diving from a submarine in 1,000 feet of water seemed to me to offer far better odds for enjoying a long life than the daily drive in freeway traffic between my home in Oakland and my office in San Francisco, eating pesticide-laced vegetables from local supermarkets, combating viruses encountered in crowded public places, or enduring an impossible personal relationship.

I diligently set about devising a research project that would take advantage of the proposed new facility, and awaited word of progress on funding, but the timing for a U.S. proposal was inauspicious. In the late 1970s, federal funding for ocean science and technology, never very substantial, was cut in favor of other federal priorities, and OceanLab was among the casualties. Many scientists opposed the concept anyway, believing that appropriating $50 million for OceanLab would mean reducing *their* budgets; others, convinced that the initial cost estimate was low, objected on the basis that once begun, the project would become a monster—again, at the expense of what they regarded as other, more urgent science and technology. A few, such as I, were greatly disappointed, and concerned that important momentum would be lost, with no clear substitute that could provide a strong focal point for "pushing the envelope" of saturation diving and related undersea technology. In fact, as it turned out, the OceanLab idea was sound, and in due course, a similar system was developed in France.*

The demise of OceanLab and the unexpected changes at home meant that I could concentrate on an idea that had begun to take form years before, but for which the circumstances had never been quite right. I had always wanted to observe the planet's master divers, whales, in action on their own terms. A conversation with the whale biologist Roger Payne inspired me to stop waiting for an opportunity to come along and instead take the initiative. *After all*, I thought, the "why not?" approach worked once before. It just might work again.

---

* Worldwide, interest in using saturation diving techniques has waned, although NOAA has maintained modest support through the years and now sponsors a six-person underwater laboratory, Aquarius, in 60 feet of water near Key Largo, Florida.

Chapter 5

DIVING WITH
THE EXPERTS

If We Do Nothing,
In A Few Years
The Whales Will Live Only
As A Legend,
A Marine Daidara-Bochi Who,
It Was Said,
Sang Unspeakably Beautiful Songs.
                    Scott McVay, in Osaka, August 18, 1970,
                    "The Alternative?"

Looming from deep water, well below where I floated but fast approaching was the largest creature I had ever seen, a plump, pregnant humpback whale, now less than 30 feet away . . . then 15 . . . then a mere arm's length from collision. It was too late to worry whether or not the whale had brakes or to attempt a swift sideways slither to get out of her way. My fate depended entirely on the agility, perception, and attitude of a creature more than 800 times my weight. I was acutely aware at that moment that my species has a long history of actively hunting and killing members of hers in all of the world's oceans.*With that track record, there were plenty of reasons to expect hostility. There was no precedent for what I was doing, no "guidebook" with thoughtful descriptions of proper etiquette for greeting humpbacks underwater or surefire protocols to let the whales know that my intentions were entirely benign.

My two companions, the expert photographers Al Giddings and Chuck Nicklin, and I wanted desperately to be inconspicuous and unobtrusive, but there are no bushes or rocks to hide behind in the open sea. Concerned that scuba noise and bubbles might send unfriendly signals, we had hovered near the surface, free-diving, and did not attempt to pair up as "buddies." Even had they been close enough to notice that I was about to make colossal contact with one of the 40+ ton giants we had come to observe, there was absolutely nothing Chuck or Al could do to keep me from becoming 110 pounds of history.

One moment, all I could see was a massive black head, 15-foot-long flippers, and a grapefruit-sized eye, then, in a mountainous blur of motion, she swept by, miraculously avoiding contact by inches. With a small gasp, I realized that I had been holding my breath . . . and so had the whale. Her great back lifted slightly out of the water as she emitted a great *whoosh* of air and salt spray, inhaled sharply, then arched back down, propelled by a modest flip of her huge tail-fluke. As supple as an otter—not at all like the stiff images in most books—the whale moved with stately grace, tilting her great eye slightly, enough to suggest that her approach had been intentional. We had come to Hawaii to observe the whales, and in the first of numerous encounters, I discovered that the term "whale watching" has a double meaning.

Until 1977, no one had attempted to get to know whales on their own terms, to see whales as whales see whales: in the vast blue ocean amphitheater that they call home. Part of the reason, of course, is that human beings are newcomers to the business of diving and until very recently have not had the means to be able to spend much time under the sea. It is therefore not surprising that prior to the

---

* A few humpbacks are still killed by whalers from the island nations Bequia and Tonga, and by pirate whalers.

1970s, most discoveries about humpbacks as well as most other whales had been derived from whaling records, from dead bodies and pickled parts, or from fortuitous encounters with live whales at the sea surface. Likewise, whales had little opportunity to get to know us except as conveyers of death and destruction.

My interest in diving with whales has nothing to do with Indiana Jones—style heroics; as a would-be marine mammal myself, I have high regard, great admiration, and a deep empathy for creatures who have a 65 million—year evolutionary edge on the business of living in the sea. It seems only logical to go to where they live and observe directly what they do, rather than sitting on the surface, wistfully wondering.

Humpbacks occur throughout the world, migrating annually from cold-water feeding grounds to the tropics for courtship and calving. Planetary giants sometimes exceeding 60 feet in length, they are larger than any of the dinosaurs and the gray and minke whales, but less than half the size of blue whales and smaller than fin, sperm, bowhead, right, and Bryde's whales. These are the nine so-called "great whales." By the 1970s commercial whaling had reduced most of the great-whale populations to a fraction of their former numbers, and while much had been learned about the inner structure of whale anatomy, little insight was gained about how whales live or what they do. Family bonds were inferred by the reaction of a whale to an injured mate, or a mother to a harpooned calf, but little was known about whale societies on peaceful terms, and virtually nothing was known about what they do underwater. With rare exceptions, I have discovered that animals underwater are *not* aggressively dangerous, and whales generally do not attack people, except when injured or otherwise provoked. Descriptions of humpbacks observed at the surface have suggested a dolphinlike "good nature." Herman Melville, for example, in his fictional classic *Moby Dick*, called them "gamesome and lighthearted," a tribute to their apparent playfulness and conspicuous above-water acrobatics—leaping high and smashing the surface with their flukes and winglike fins.

By 1977, the most detailed studies on the day-to-day activities of any of the dozen large cetacean species had been pioneered by Roger and Katy Payne during months of observing southern right whales that came close to shore in a protected embayment in Argentina. A chance conversation with Roger about his adventures in South America—and the ten-year study he and Katy had made of humpback whale songs using underwater microphones—led to discussions about the possibility of actually observing humpbacks singing and doing whatever else it is whales do—day and night, at the surface, and deep below. Neither Roger nor Katy was inclined to dive, but their considerable expertise concerning the intricacies of whale song and surface behavior would dovetail well with on-the-spot observations that I could make underwater, reinforced by film and still photographs of what the whales were doing. The more we talked, the more exciting and right the concept seemed, but we had to face the usual realities: time and money. Roger and I set about patching together financial support and managed, during six months of letters, proposals, and countless calls, to assemble the necessary wherewithal needed. Al Giddings obtained additional financial support for the project from Survival Anglia Television and the *National Geographic* magazine.

Roger and Katy thought they could arrange for field work in Hawaii in the winter of 1977, although it would take some doing to organize care for their household of four youngsters, assorted critters, and an active ongoing research program. When possible, I tried to do what the Paynes often did: enlist offspring as part of the research team. My three children, Elizabeth, Richie, and Gale, were able to join the expedition some of the time, as did the Paynes' daughter, Holly. Roger's graduate students Peter Tyack and Jim Darling signed on for the duration, as did Al's partner in many underwater adventures, Chuck Nicklin, and two versatile assistants.

It was also necessary to have a permit to get close to the whales. Although killed by whalers in U.S. waters as recently as the 1960s, humpbacks have been protected since 1972 by provisions of the U.S. Endangered Species and the Marine Mammal Protection acts. A permit from the National Marine Fisheries Service (NMFS), a branch of NOAA, is required to "take" a whale, a euphemism interpreted variously as kill, relocate to an aquarium, approach closer than 165 feet (50 meters) (later increased to 330 feet), or deliberately interfere with their "natural" behavior. Fishermen receive special consideration; within limits (i.e., up to 20,000 a year), they are allowed to kill dolphins and some-times whales without a special permit and without penalty when the animals are taken "incidentally," as a result of fishing techniques. A permit to "take" even members of endangered species such as humpback, gray, and bowhead whales may be awarded to those engaged in offshore oil and gas operations or activities involving the underwater use of explosives that might damage or kill marine mammals. Our intentions were wholly benign, but so was the increasing atten-tion of whale-watching enterprises, researchers, writers, artists, photographers, and the general public who, altogether, posed pressures sufficient to warrant a "limited-access" approach. Roger was granted an NMFS research permit that covered our expedition, and starting in February, we began research on hun-dreds of individual whales who would become familiar participants in studies that in time would expand to encompass much of the eastern Pacific, from Hawaii to Alaska and south to islands offshore from Mexico.

Finally the heart-stopping moment marking the true beginning of our project came: I was perched on the side of a small rubber boat several miles offshore from Maui, and I had to convince *myself*—not some funding organization—that it was a good idea to step into the sea with four actively curious humpbacks. The whales had been moving on a straight course in the deep, gem-clear channel between the islands of Maui and Lanai, when suddenly, they changed direction and headed straight for us. By the time the engine on our small rubber boat had stopped, the whales were underneath us, upside down, gliding and turning, looking up with apparent curiosity. For a moment, I hesitated . . .

Humpback whales have a reputation for gentleness, but I was unsure what forty tons of gentleness might feel like. Would the whales take notice of us? Would we be able to watch them without disturbing their natural behav-ior? Could we stay with them long enough to make meaningful observations, or would we succeed only in snatching passing glimpses? At the time, unless we encountered a distinctly rotund, pregnant female or nursing mother, or had a particularly clear view of the whale's belly, we did not even know how to distin-guish males from females.

There was only one way to find answers. I plunged in. At first I saw only clear, blue water, but four swiftly moving shapes soon came into focus as high-speed humpbacks. My first encounter, described earlier, was over in a moment as a large female sped by within touching distance; then, the same whale swam toward Al, only this time, a collision between Al's head and one of the whale's winglike flippers appeared certain. Al was oblivious to the impending disaster, as he was completely absorbed in filming the image of a more distant whale. Chuck hooted a warning, not wanting to witness a decapitation on our first day—but, at the last possible moment, with a deft twist, the whale artfully lifted her flipper over Al's head, barely avoiding contact. I stopped worrying then, casting my fate to the whales. Underwater, there was little choice. The whales could—and often did—outmaneuver, outswim, and outdistance us easily.

The very next day we once again watched a group of whales change course and head our way. Like enormous swallows in formation, five sleek black forms glided under the boat, but two, exhibiting most unbirdlike behavior, slid by upside down, their white underbellies glistening like snowbanks as they passed.

This time, with no hesitation I breathed deeply, jumped in, and dived to meet them. Two whales flexed swiftly toward me, gliding like ice dancers, an arm's length away; had I stretched, I might have touched the smooth white underside of the largest, the same colossal beauty who had inspected me the day before. It was easy to identify her by the distinctive pattern of markings on her face, tail, flippers, and belly. Collections of whale-tail portraits had been started by various researchers as a means of cataloging numbers of whales, but underwater we could see the unique, individual markings of entire animals, not just their flukes. Those watching from the surface could sometimes see what we now saw, a whole whale, when one leaped out of the water, as humpbacks often do. They propel their great bodies skyward, then thunder back, smacking the surface with a booming *thwack* that can be heard for miles—underwater, as well as above. Typically, however, humpbacks show to surface observers a puff of exhaled vapor, a fin, a flipper, and just before diving, an upright tail with markings as distinctive as a fingerprint.

Awash in a vast, saltwater Jacuzzi (the wake of two thrusting tails), I surfaced for air. With envy I waited for the whales to do the same. I could stay submerged for less than a minute. Nine minutes passed before one of the whales, then the other four, expelled misty fountains of foggy breath, inhaled, and dived back under. Often, they remained underwater for twenty minutes, sometimes longer.

For nearly an hour the whales stayed near us. Repeatedly I thought, *Well, now they're gone*, and repeatedly I caught my breath as two or three or all five came streaming by again. And again. Occasionally the whales produced loud but pleasant sounds, not the melodious cascading sequences called whale song but rather, short *wheeeeps* and grunts that seemed to be a form of socializing, like the yips and barks of wolves.

Quite abruptly it was over. Five tails disappeared from view and the sea was silent. An ocean without whales suddenly seemed empty, devoid of joy. I touched the side of the boat, ready to lift myself out, when Al's assistant, who had been watching from the surface, shouted, "Look! They're back!" One whale nearly grazed my foot as all five swept under the boat.

I had resisted using a scuba tank because of concern that the noise and bubbles produced might cause alarm or otherwise disturb their natural behavior. But I decided to take a chance that they would stay, and now entered the whale's subsurface realm with an hourlong passport of air, occasionally descending to as much as 150 feet. In the midst of five whales, I flowed with them, gliding down when they went down, flippering up as they did so. They moved with such power and grace! I felt as if I were swimming in slow motion next to them, and it was not an illusion.

The large, curious female who had first come my way remained the most attentive. Several times she moved slowly by, turning her great eyes, registering awareness, perhaps locking away in her large brain impressions for future reference. In my considerably smaller brain, the encounters with this whale are deeply engraved and readily retrievable, sometimes bringing an unexpected tonic of pleasure during long, boring meetings. That graceful animal with her dolphinlike look of good humor and apparent exchange of awareness will haunt my sea thoughts forever.

She and her four companions stayed near us for almost three hours, mostly minding their own whale business, but sometimes taking notice, evidenced by close passes and breathtaking interactions—all without the slightest indication of intent to harm. Ideally, from the standpoint of research concerning whale behavior, we wanted to observe the action without causing a reaction, but it was difficult not to be moved on the occasions when the whales made the overtures, engaging *us*.

Nearly ten years before, Roger and a brilliantly creative colleague, Scott McVay, had analyzed the sounds of humpbacks recorded offshore from Bermuda and had begun using the term "songs" to describe their regular, intricate sequences of repeated sounds such as those made by crickets, birds, and frogs. They also discovered that all humpbacks in a given area sing the *same* song, with an occasional individualistic twist.

Katy Payne, a trained biologist and musician, compared recordings over a period of years and came to the astonishing conclusion that, while all the songs in any one year were the same, *no two years of song were alike!* That meant that the whales do not just mechanically repeat preprogrammed sounds, but, rather, compose as they go along, deleting a few elements and occasionally adding new segments. The songs of two consecutive years are more alike than those more distantly separated, indicating evolutionary rather than random change. Five years can produce striking dissimilarities, however. Roger notes that two years of Bermuda recordings, 1964 and 1969, are "as different as Beethoven from the Beatles." Almost anyone listening to a recording of humpback whales can identify them as humpback whales—but experts can pinpoint *where* a song has been sung, and even *when*.

In Hawaii, those of us diving could hear directly what surface listeners with hydrophones recorded on tape: a constant background chorus, day and night, of soloists, each belting out the same song over and over, no one synchronous with any other, but together creating a curious harmony. No one had witnessed, let alone filmed, a singing whale underwater before, and there was still speculation as to whether they do or do not emit bubbles through their blowhole when the sound is produced. Some thought the large double-nostril blowhole at the top of the whale's massive head might be used as some ingenious, colossal wind instrument.

Hearing humpbacks was easy; seeing a singer in action was not. Al Giddings was the first among us to do it, and also the first to record it on film. Chuck and I went with him and dived down near where we saw a singer submerge. At close range, the sound of a singing whale is so intense that it is almost unbearable—like being in the front row during a Wagnerian opera. Air spaces in our heads and bodies vibrated, as the whale went through its repertoire of eerie *wheeps*, arcing trills, and low rumbling groans and sighs, ranging from ultralow bass through bubbling, rippling sequences, hee-haws, then to high, violinlike squeals. It seemed that the whale was equipped simultaneously with an orchestra—and a barnyard.

Above us, shafts of cathedral-like light beamed through a violet sea, as we settled down to wait. Chuck and I elected to stay put, certain that at any moment we would see the singer, while Al, following an uncanny sixth or seventh sense, moved westward. We surfaced fifteen minutes later, and so did the whale—600 feet away! Al *was* able to see and film the virtuoso in action, however, and was the first to describe what many later observations confirmed: no bubbles emitted while singing. We still did not know how they produce their awesome sounds, but at least we now knew how they did *not*. We were also impressed with an enhanced awareness of the importance of sound and hearing to the whales in their mostly dark, three-dimensional world. Sensing with sound is as vital to whales in their realm as sight is to us in ours. Hearing is so important to cetaceans, in fact, that it is hard to imagine that a deaf or hearing-impaired whale or dolphin could survive for long.

Three months of looking produced numerous observations that helped provide new insight into the life and times of humpbacks. As a side project, I kept track of evidence of swim-along fish, dolphins, and hitchhiking barnacles and other organisms associated with humpbacks, which are like moving islands in the sea. One tenacious hanger-on is the beautifully structured acorn barnacle, *Coronula*, a creature so intimately associated with the whale's hide that living skin is laced into the barnacle's stony framework. These in turn provide footing for another species, clusters of soft pink and brown goose barnacles. Masses of these species, sometimes as much as half a ton aboard a single whale, tend to cluster about where the chin would be if humpbacks had one; others border the edges of the flukes and flippers like brass knuckles.

It is generally thought that while in Hawaii humpbacks do not feed at all, but live on reserves of blubber developed during months spent in distant, cold-water areas where they gorge on dense populations of schooling fish and small crustaceans, mostly krill. I stayed on a constant lookout for evidence that some might be sneaking a snack and concluded, after several sightings of whales lunging, mouth open, and occasional observations of fecal plumes, that some intake might be occurring, but certainly not enough to contribute much to the energy they were expending.

I also paid special attention to interactions between calves and their mothers and others in groups that sometimes assembled. Identifying mothers and newborns, then tracking them through the years, seems an obvious thing to do to evaluate population health, reproductive success, and potential for the future, yet little information existed in 1977. That year marked the beginning of a classic ongoing study of this vital issue by Deborah Glockner-Ferrari and her research partner and husband, Mark Ferrari. It also was the year that pieces of the puzzling question of humpback migrations and population interactions throughout the eastern Pacific began to lock into place, and connections were made between disparate observations.

It had been noted that every spring, humpbacks leave Hawaiian waters, and every summer, humpbacks arrive in the Aleutian Islands and the quiet, plankton-rich waters of southeastern Alaska. From time to time whales were also seen in the coastal and offshore island areas of California and Mexico. As to how many humpbacks there were in all, or how the populations were related, no one knew. In the years that followed, answers would start to emerge from the process of matching the summer and winter "rogue's gallery" of tail prints from distant places and comparison of songs provided other clues.

In mid-1977, Al Giddings and I joined Charles Jurasz, who with his wife had for years been observing whales in southeastern Alaska, mostly around Glacier Bay. Whales migrate here in the summer, and it seemed logical that we should, too. Jurasz enticed us further with the promise of being able to witness whales engaging in "bubble-net" feeding, a technique whereby one or more whales produce a fizz of fine bubbles that encircle and concentrate masses of krill or sometimes schools of small fish such as herring or capelin. When Jurasz first reported it at a meeting of whale scientists in 1976, few took seriously his completely accurate description of this extraordinary use of bubbles to snare food. We were not disappointed; bubble-net feeding was clearly and often evident at the surface, although Glacier Bay's rich, murky waters reduced diving observations to a game of blindman's bluff.

About twenty-five whales entered the bay that summer, and about forty more occupied nearby waters. Several years of records kept by the Jurasz—a "fluke file" accompanied by colorful and memorable names—made it easy to start getting acquainted. Among those encountered most often were Spot, a presumed male notable for a mark the size and shape of a baseball on the underside of his fluke; Notchfin, named for obvious reasons, but also known for a blotch on her tail that bore a striking resemblance to a map of South America; and White-eyes, so called because of two round white patches that appeared like glowing orbs when the whale's tail was lifted high. Spot and Notchfin were often near each other, and we suspected amorous intentions, but they, like all the other whales in Glacier Bay, were mostly preoccupied either with eating or with sleeping. What seemed special was that they often ate and slept *together*.

News of one significant discovery about Glacier Bay's whales came to me two years later, in the spring of 1979. I was in the midst of a very serious meeting in Geneva, seated at a long table at the World Wildlife Fund's international headquarters with about thirty other representatives to the International Union for Conservation of Nature (IUCN), listening to discouraging news about the global loss of wildlife habitat, when a secretary cracked open the door and urgently motioned for me to come. She handed me the distinctive yellow envelope of a cable. Years of intelligent scientific spying were summarized in a six-word message: SPOT AND NOTCHFIN SIGHTED IN HAWAII.

Jurasz had called my daughter Elizabeth, who immediately recognized the significance of the news and fired off the cable. Confirmation had at last come about that at least some Hawaiian humpbacks feed in Glacier Bay. Or were they Glacier Bay humpbacks, traveling to Hawaii for calving? Either way, it was a breakthrough discovery, made even more intriguing when Spot turned out to be female, not male. She and Notchfin both produced calves that year.

The ability to make matches between whales from distant locations was possible because in 1978, a team of dedicated young researchers, including Roger Payne's student Jim Darling, began to organize and catalog the photographs of

Hawaiian humpbacks, as Jurasz had done for more than a decade in Alaska. By the spring of 1979, descriptions of 264 individual whales had been assembled, and seven Glacier Bay whales, including Spot and Notchfin, were among them.

The whale-song connection was another matter. Whales vocalize constantly, producing a wide range of socially relevant grunts, squeals, and warbles wherever they are, but they tend to sing only during the winter months, not while they're in their summer feeding areas. In Alaska, we joked about the reason: Maybe it wasn't considered polite to sing with a mouthful of krill. The real reasons were deciphered through hundreds of careful observations by Peter Tyack and other researchers who demonstrated clearly what others had suspected, that singing is related to courtship, a favored activity in tropical waters.

An unexpected dividend of the whale-song research was the discovery that whales recorded offshore from Baja California, Mexico, were singing the same song as the Hawaiian whales 3,000 miles away, and thus were not a distinct group, but part of one interacting population. Matches of fluke prints from Mexico and Jim Darling's Hawaii file later clinched the relationship.

Bit by bit, such tantalizing morsels of information gradually provide increasingly sophisticated insight into the nature of whales and dolphins, creatures who share not only our mammalian heritage, but also a parallel history. At about the same time our ancestral primates made an appearance and as the era of dinosaurs faded, the first whalelike creatures descended into the sea. The oldest fossil whales had short legs and resembled elongated pigs. From this unpromising beginning, nine families of elegantly streamlined creatures have developed, including seventy-eight modern species of dolphins, whales, and porpoises, all superbly adapted to aquatic life.

Nonetheless, cetaceans, like other mammals, must stay within convenient range of breathable air. Most dolphins do not hold their breath for more than a few minutes and tend to remain near the surface. Humpback whales seem to favor depths above 600 feet, often submerging for more than twenty minutes. Sperm whales routinely dive half a mile down, perhaps more than twice as deep, and they can—and often do—stay underwater for more than an hour.

The average depth of the ocean, 13,000 feet, is well beyond the range thought possible for a round-trip journey, even by the most exquisitely adapted aquatic air breathers. Among creatures with lungs, only humans have ventured to such depths—and beyond—by using technology to overcome the constraints of nature.

My "year of the whales," 1977, took me to the edge of where I could go as a free-swimming diver, and allowed me to peer into the distance beyond, vicariously glimpsing deeper realms through the eyes of the whales that we watched and got to know personally. It was frustrating to descend with a diving whale to the limit of my range and then be forced to stop—or risk slipping deeper, for a last blissful excursion into the blue. I always managed to pause at that invisible wall, where my depth gauge indicated I had reached my limit, midwater, and watch dark flukes pass beyond, into the fluid darkness below. I wanted to go too! But how? Once, I had been deeper than humpback whales are thought to dive, a little more than 800 feet, using the three-man whale-shaped sub *Deep Diver*, but like most submersibles, this sub was not really designed for tracking whales into the depths and no attempts had ever been made to do so.

The marine mammal expert Ken Norris succeeded in following and observing dolphins swimming close to the surface in a vehicle designed just for that purpose. The cigar-shaped system, fondly known as the SSSM, for "semisubmersible seasick machine," was attached by a cable to a powerboat speeding along at eight or nine knots, and featured a cylindrical subsea viewing area with Plexiglas ports and a cushioned swivel chair where Norris sat. Spectacular observations were made of dolphins and other high-speed creatures, but the SSSP was a creature of near-surface waters, unable to descend more than a few yards.

To explore the whale's environment from top to bottom requires deep-diving submersibles such as the two Russian submersibles *Mir I* and *Mir II*. Using these subs during many hours in the deep sea, Joe MacInnis—one third physician, one third fish, one third poet, and 100 percent Canadian—has gained some insight into places familiar to many cetaceans, but largely unknown to primates. He observes:

The secret pit of the ocean holds a universe of tangled infinities: perpetual currents, enduring pressures, and a darkness measured in hundreds of millions of years. Within its hidden depths are mountains higher than Everest and crustal slashes deeper than the Grand Canyon of the Colorado. Because of its inaccessibility, it is a world with the dew still on it. . . . Because it is so colossal, covering more than half the surface of the planet, the deep ocean is earth's last great untouched place. A mile or two beneath the sea, the twentieth century seems like a rumor.

I dreamed of being able to step into the ocean in something small and light, a diving suit that would behave more like a second skin protecting my vulnerable body while I traveled with the whales at will, without concern for decompression or stiff constraints on depth and time. Lacking gills, I might settle for a small submersible. I tucked the idea away in the back of my mind for future reference, and two years later, when a chance came to try a miniature submarine—a one-atmosphere personal diving suit—I was eager to explore the possibilities.

Chapter 6

TO THE EDGE OF DARKNESS

I could ask, "Why risk it?" as I have been asked since, and I could answer, "Each to his element."

Beryl Markham, West with the Night

Like celestial champagne, bubbles sparkled over my head in the clear blue water six miles offshore from Oahu, Hawaii, near Makapuu Point. It was the nineteenth of October, 1979, and I was returning from the most amazing dive of my life: walking solo for two and a half hours 1,250 feet underwater in a forest of bamboo coral, touching the long, spiraling whiskerlike forms and watching rings of blue light pulse magically up and down from slender tip to solidly anchored base.

A science-fiction thriller could not have been more bizarre than my recent encounters with the blue-flashing coral whiskers and some of their associates: huge deep-sea rays, hovering like enormous butterflies a few feet from the sea floor, a full-grown 18-inch-long shark with glowing green eyes, dozens of long-clawed red crabs, clinging to a lush shrub of pink coral, luminous jelly-creatures, and a slender silver-black lanternfish spangled with a lateral row of blue lights—the first I had ever seen alive, in its own realm. Thousands of people in boats cross right over this wondrous place every year, oblivious to the beautiful and complex deep-sea communities below. I had been oblivious, too—and would likely have remained so, except for *Jim*.

My first face-to-face meeting with *Jim*, at the Commercial Diving Training Center near Los Angeles, was not at all what I expected. From what I had heard, *Jim* was rugged and reliable, the strong, silent type with a no-nonsense reputation for doing tough jobs in difficult circumstances underwater. I had seen photographs of *Jim* emerging without a shiver from a hole cut through several feet of ice covering the Arctic Ocean in northern Canada. I remembered images of him moving with surefooted confidence along a narrow walkway at the base of an oil rig, several hundred feet underwater, wielding a large shackle and, with apparent ease, twisting a huge connecting bolt, then ascending effortlessly from deep water as a tether brought him back to the surface with no concern about the effects of changing pressure. But here at the Training Center was *Jim*, in pieces, both legs separated from the body, and one arm completely dissected.

My friend Phil Nuytten, who had arranged for this meeting, saw my look of dismay and explained. "*Jim* needs to be checked out before you two can start working. We're going to try to fix things for a better fit. Don't worry, it won't take long!" Phil was right. I soon stood in front of the fully assembled articulated metal suit, *Jim*, which would provide flexible protection for my vulnerable body against the pressure of the deep sea. Named in honor of Jim Jarrett, an intrepid diver who was the first to test the prototype system in the late 1920s, the modern version I now faced looked similar to astronaut attire: white, with silver trim, formfitting arms and legs, and helmetlike head gear with clear ports through which to view a realm with an atmosphere hostile to humans.

The resemblance to a space suit is more than superficial. Both suits are designed to protect the human inside while allowing him or her to retain reasonable comfort and working capability. Both are supplied with self-contained life-support, including a rebreather system to provide a recycled air supply that lasts for many hours (carbon dioxide is scrubbed out chemically and new oxygen is automatically added as required). A critical difference between the two types of suit concerns the pressure environment. Space suits, designed for use in low- or zero-pressure environments, are soft and pliable; one-atmosphere diving suits such as *Jim* must be made of hard materials—metal, ceramic, or composites—to resist the pressure imposed by the weight of the water above. Whereas a fully outfitted astronaut's suit may weigh more than 100 pounds in air, *Jim* weighs about 1,000 pounds. Yet in their respective mediums, both are "weightless." *Jim*, in fact, must be additionally weighted with lumps of lead hung outside at the waist to offset buoyancy caused by air inside the suit. The suit's in-water weight of about 60 pounds makes it possible to maneuver the system without bobbing about. The added lead ballast can be released by turning a lever from within *Jim* to restore buoyancy and cause the system to shoot to the surface, presumably with its human cargo intact.

Before I could start training with *Jim*, some adjustments were necessary. Using discreet engineering terms, Phil pointed out that I was not optimally pro- portioned to operate the muscle-powered system. Unlike astronaut suits that are custom-fitted to the individuals designated to wear them, this one came in one size—large. *Jim* is scaled for husky commercial divers who tend to be six feet or so tall and weigh in at 200 pounds plus. To accommodate my smaller dimensions (5 feet/3 inches, 110 pounds), one section of the legs was removed and lifts were installed inside *Jim*'s metal boots, but I still had to stand on my toes a bit to touch bottom. Nothing could be done to shorten the "sleeves," so to reach the end of each arm and thus operate the mechanical claw at the tip, I had to stretch one way or the other, depending on which hand I wanted to use. The only benefit of this was that unlike most operators, I had no difficulty pulling my arms back in to take notes or scratch my nose or ear.

Superficially, I may have resembled a sky-walking astronaut, but when maneuvering underwater, I felt more closely akin to a drunken crab or a walk- ing refrigerator. I would certainly have been rejected as a commercial *Jim* diver, but some concessions were allowed to me as a scientist. I was about to become the first person trained to use the system for research dives, a radical depar- ture from *Jim*'s usual salvage and oil-rig assignments. I was also the only one ever to dive *Jim* with enough space to be able to withdraw both arms from the "sleeves," extract my feet from the "feet," and turn completely around while inside the suit. With room to spare, I was able to tuck inside a Nikon with extra lenses, a notebook, and my trusty leather Coach shoulder bag stuffed with a few essentials—an apple, Snickers bar, spare pens and film, and a rag to wipe condensation from the ports. As a thoughtful last-minute addition for the deep dives, Phil thoughtfully tucked in a hammer for me to use to strike the release mechanism on the weights in an emergency, if my muscles alone were not up to the job.

After several days of intensive training in a test tank, I learned to master the fine art of walking by rolling first one way, then the other, to achieve a bearlike gait; of throwing my weight forward to cause *Jim* to fall facedown so I could get a close-up view of the sea floor; and then of thrusting back to shift the center of gravity enough to make the suit stand upright again. *Jim* does not bend in the middle but, rather, behaves something like a weighted doll that when knocked down tilts back to a standing position. With a satisfied smile, Phil explained that *Jim*'s no-frills design was endearing to proponents of KISS ("Keep it simple, stupid") engineering.

About fifteen such suits were in active commercial use worldwide, but this one, number 9, had been temporarily diverted for our experiment. Al Giddings first proposed the idea. We were collaborating on a book on the history of underwater exploration for the National Geographic Society, and a two-hour ABC television special on the same topic. Al was convinced that by using an atmospheric diving suit, we could dramatically illustrate many of the problems inherent in human access to the sea—and their solutions. I was skeptical about the practicality of using *Jim* for effective scientific research, but excited about the prospect of zipping down to 1,250 feet and back with *no decompression!*

The more I thought about it, the more intriguing the concept seemed. I could envision enormous advantages to being able to drop in on the ocean of my choice for a few minutes or hours, just as scientists go into a forest or desert—or into a laboratory—with *Jim* as an exotic but necessary "lab coat."

It is possible to dive to more than 1,000 feet without a pressurized suit or submersible, but it's stressful and requires using a carefully balanced breathing mixture of oxygen, helium, and nitrogen. In tightly controlled experiments, a

few men have gone to slightly more than 2,000 feet, but there are physiological constraints that make such ventures difficult and dangerous. A high level of
physical fitness is a prerequisite for such diving; moreover, the pressure exerted
at such depth requires a long decompression, which imposes additional wear and
tear even on young, healthy bodies in top condition. Few scientists have shown an
interest in subjecting themselves to such rigors to explore the ocean—perhaps
one of the reasons that so little is known about the nature of the sea below the
first few feet. In *Jim*, surface pressure is maintained throughout the dive, and at
the end of it, the pilot simply steps out, moving from a one-atmosphere diving
suit to the one-atmosphere pressure all of us experience at sea level. Training is
needed to use *Jim*, but virtually anybody can do it.

At least, that was the theory. It seemed reasonable to suggest to Phil Nuytten,
an executive with Oceaneering International, the owner and operator of all of
the *Jim* suits, that if a pint-sized marine biologist could successfully use such a system, others might be inspired to do so. This rationale may have helped influence
Phil's decision to get behind the project, but I suspect that he was also drawn by
the irresistible appeal of using *Jim* for scientific exploration and of experimenting with a new approach: deploying the suit from a small submarine, rather
than using a tether back to the surface, which some might regard as dancing too
close to the edge of safe and sane diving. Phil and Al both seemed uncommonly
attracted to wild and crazy ideas (often their own), and both had an uncanny
knack for emerging with wild and clearly not-so-crazy successes.

I earnestly hoped this expedition would be one such success, but I listened
carefully to the concerned murmurings of a British engineering expert, Graham
Hawkes, whom Phil had invited to come along as a technical adviser. We took
Graham's judgment seriously. He had helped reconfigure the original *Jim* design
for greater depth and dexterity, and had also designed and developed a related
system called *Wasp*, a spiffy yellow and black tube-shaped, one-man, one-
atmosphere suit that literally flies through the water, powered by several strate-
gically placed thrusters. If I got into trouble while using *Jim*, Graham would be
ready to "fly" to my rescue using a *Wasp*.

With thoroughly British politeness and delicacy, Graham questioned my sanity, and that of Phil and Al, when he heard the specific details of what we were

proposing to do. It wasn't just that I was a woman, or a scientist—although he admitted to entertaining some reservations on both counts. Mostly, he was concerned about the mode of deployment. Typically, the suits are lowered over the side on a cable, much like a fish at the end of a line, and are reeled in and out using a winch that is firmly welded to the deck of a stable surface platform, an oil rig or large ship anchored with strong, four-point mooring. The support ship for our operation had no such refinements, but Al and Phil had worked out an innovative alternative for launch and recovery. "Using the cable for launch and recovery isn't practical, especially if *Jim* and *Wasp* have to be in the water at the same time," Al said.

Al planned to observe and film *Jim* (with me inside) in action from a nearby underwater vantage point, crouched inside the two-person submersible *Star II*. The sub had welded to its front end a sturdy metal platform for recovery of deep-sea specimens, mostly coral. "What do you think of standing *Jim* on that platform, strapping him to the sub, and using the sub as a taxi?" he asked. "Once we're on the bottom, the suit can be released by a lever operated from inside the sub, allowing you to walk freely, without the usual tether connection back to the surface." A short, light line leading from *Star II* to the top of Jim would make voice communication possible. When it was time to return to the surface, the sub would lift off first, and *Jim* would follow, towed by the line. After the *Jim*-towing sub arrived at the surface, the support ship would recover *Star II* and *Jim* using a special submersible launch and recovery platform.

There were lots of questions about what might go wrong. What if the line between the sub and *Jim* breaks? What if the suit leaks when it is jerked off the bottom at the end of the dive? Suppose there isn't enough air ballast in the sub for the suit and sub to lift off together? What will happen if the support ship can't spot the sub and *Jim* when they get back to the surface? What if we haven't thought of some perfectly obvious stupid detail that could cause the whole plan to fail?

Dr. John Craven, Director of the University of Hawaii's Marine Science Institute, the person responsible for the operation of *Star II*, minced no words when he heard the dive plan. "No. Impossible. The risks are too great," he said. Craven knew quite a lot about risks. As Chief Scientist for the Polaris missile project and longtime colleague of the irascible Admiral Hyman Rickover, Craven had been right in the thick of things through perhaps the most colorful era of underwater operations in the history of the U.S. Navy, including the successful deployment of the *Trieste* to the deepest part of the ocean, the search for the lost nuclear sub *Thresher*, and the secret recovery of a Soviet submarine by the giant research and undercover "spy ship" *Glomar Explorer*. A thoroughly reasonable man, he was interested in seeing our project succeed, but did not want to encourage what appeared to him to be a catastrophe in the making.

In the end, despite his own obvious concerns, it was probably Graham, using a vest-pocket calculator, yellow-pad sketches, and cool logic, who convinced Craven and the rest of us that the proposed plan would work. (Much later, Graham confided, "I was asked, *could* this be done. I showed that, technically, it could. No one thought to ask whether or not I believed it *should* be done!")

Several shallow-water launch and recovery maneuvers helped us all gain confidence that a real working dive in deep water could be accomplished. It was important to get the procedures worked out, because no one had tried deploying *Jim* or any other such diving system without a cable connecting it firmly to a surface platform for secure launch and recovery. This would be the deepest solo dive ever made without a tether to the surface.

Twice, the sub was launched with *Jim* on its bow and descended several hundred feet into clear, deep water. And twice, it was necessary to return to the surface before touching bottom because a tiny copper wire crucial to voice communication between the sub and *Jim* snapped. Each time, it was necessary to delay the operation for a day or two, long enough for a complete checkout of all systems. The small breaks in the wire fueled the arguments of the skeptics.

I did worry about those small, stupid things that can go wrong, but had long ago balanced my involvement with what some regard as perilous exploratory ventures at sea against the risks that most people unblinkingly accept in everyday life, whether tripping over a doorstep, slipping on a shoelace, or succumbing to a virus. I felt much safer piloting *Jim* solo in a dark sea than I have often felt walking alone on dark city streets.

Finally, after an all-night series of checks and last-minute grooming of the suit and sub, preparation was complete for the third try for a successful round-trip journey to the sea floor. And then it was happening. I stepped into *Jim*, attached to the platform at the front of the sub. *Jim*'s hinged headgear swung closed, the strap connecting the suit to the front of *Star II* was tightened, and descent on the launch and recovery platform began. Seventy feet below the surface, the sub separated from the recovery platform and began falling through prism-blue water that seemed to glow with internal light. Shafts of gold from the midday sun penetrated the upper few feet, but gradually the colors mellowed into blue infinity, deep indigo shading to blue black, the blue-edged darkness of deepest twilight or earliest dawn. And there were stars, or so it seemed: small biolu-minescent creatures sparkled with brilliant silver-blue flashes as they brushed against the pane through which I viewed Earth's "inner" space. These mysterious, living constellations are as unknown and nearly as remote from human experi-ence as distant galaxies.

As *Star II* gently touched down on the gradually sloping sea floor, 1,150 feet beneath the surface, I abandoned for a moment all thoughts of the careful sci-entific evaluation I had planned. Rather, I simply enjoyed the sensation of being there, knowing that whatever might happen would be unlike anything I had ever experienced before—unlike anything *anyone* had ever experienced before. *Star II*'s lights created a clear pool in the deep-sea darkness, and I could see sev-eral strange shapes outlined by the glow—a white sponge, delicate corals, and several translucent red shrimp. A smooth gray eel undulated at the edge of the light, then disappeared into the blackness beyond. I wanted to hurry after it, but leaning forward, I was stopped by the suit's hard shell, still strapped to the front of the sub.

Al's voice interrupted over the intercom.

"Sylvia, how are *Jim*'s joints? Are you okay?"

I wanted to yell, "Let's go!" but tried to muster a properly dignified response, mindful of Al's concern.

"I'm . . . fine," I said carefully. "I think it would be good to edge a little deeper and find more coral or a patch of rocks. *Jim*'s left arm is a little stiff, but I'll keep working it." At this depth, the pressure on the suit's oil-filled joints caused them to lose mobility. Frequent movements are required to maintain lubrication and keep the "elbows" and "knees" from going rigid. Standing on the front of *Star II*, held by the midsection strap, I flexed and stepped in place, like the Michelin Man doing a slow-motion jig.

Al asked Bohdan Bartko, the sub's pilot, who was a veteran of thousands of hours of underwater cruising, whether we could go deeper. Bo suggested that, given the air and battery power available, he could allow about half an hour to look around for a more suitable place for the proposed deep-sea walk. With the driver in back and I, the passenger, chauffeured along in style on the front of the limousine, we cruised slowly, pausing now and then to consider the options. A field of stately bamboo coral amid clumps of sponges and branching gorgonians came into view, and we stopped. I thought it was a fine place to test the suit's capability as a useful tool for exploratory observations.

The local residents greeted us with equanimity. None rushed over to attack, and none seemed panicked. The sub's lights soon attracted a number of small crustaceans, and may have enticed several fish—a small shark, several rays, a silvery lanternfish—to investigate. I mused about the reception members of my species might give to a giant yellow creature of unknown origin with flashing lights whirling down from above, clutching a wiggling, crablike thing. But I left that thought for future review and got down to the business of discovering what I could about the nature of the extraordinary community before us. I told Al I was ready, and he pushed a lever inside the sub that unfastened the strap around *Jim*'s waist, and I was free to move around on my own.

There was, literally, one small catch. As I stepped forward, I stubbed my metal "toe" and could not disengage one of *Jim*'s boots from the sub's platform. I thought—again—of how some seemingly insignificant detail can cause otherwise perfect plans to collapse in a heap. Ironic, too, to get this far and be unable to free myself for the final phase of the experiment. I had a vision of the sub surfacing with *Jim*, boot still stuck, toppling headfirst off the platform for a most undignified finale.

Al and Bo discussed what to do. Bo thought a full reverse followed by full-speed-ahead and a second reverse might shake *Jim* loose. With one swift, smooth backward motion of the sub, the snagged boot dislodged. At last I was free to do what I have so often done as a diver, and as an observer in fields, in forests, in deserts, and on rocky hillsides: I stood quietly, then moved about slowly, pausing to watch the action of the red crabs and small white crustaceans moving in and out of the illuminated area, taking notes and experimentally touching whatever I could reach—a crisp sponge, the tough fabric of a soft coral, the firm texture of rock. *Jim*'s pincerlike claws, fine for turning bolts and putting a pin in a large shackle, were frustratingly inept in making contact with living organisms—no match at all for my own precise, sensory arms and hands that I had always, before, taken very much for granted. But despite the limitations, I could gently nudge the flowerlike coral polyps, and was able to grasp and hold one small piece for later close viewing. I looked—enviously!—at a large, pale, well-coordinated crab, gracefully tiptoeing along the sea floor while holding high a bouquet of gorgonian coral in each of its two small, hindmost pincers. Several kinds of deep-sea crabs have this habit, and while the reasons for such behavior are completely mysterious, I thought with a wry smile that a perceptive crab might be equally puzzled about the interest the strange upright creature awkwardly roaming in its territory might have in the wisp of coral *it* held.

I waited until Al documented the nature of the place with all lights blazing, then asked for the sub's lights to be turned off so that I might observe bioluminescent activity more effectively. Al must have been low on film, because with uncharacteristic forbearance, he agreed to sit in the dark for a while, unable to expose any film (his version of Hell on Earth, especially underwater), while I explored, looking for creatures equipped with living light.

A fleeting thought touched my consciousness as the lights went out: *I wonder whether Mother and Dad—and the kids—know what I'm really doing here in Hawaii.* Probably not. There were no directions to tell us how to do this, let alone enable us to tell anybody else. But if they had known I would be standing next to a little submarine with its lights out on the bottom of the ocean six miles offshore from Makapuu Point, looking for small creatures with blue lights, they would probably have said, "If you think it's all right, dear (or Mom), it's all right with us. Just be careful." They are the best parents and offspring an oceanographer with incurable curiosity could hope for.

I waited for my eyes to adjust to what I expected to be total, absolute darkness. But no! Looking upward through *Jim*'s topmost port, I could see an immense amphitheater of blue-blackness arching over deep gray-blackness. Midday sunlight was apparent even to my eyes, and would be magnified many times over for creatures whose senses are tuned for living at the edge of light. I could make out irregular shapes—sponges and clumps of low, branching coral, and the rounded hulk of the sub, 20 feet away. I was at the edge of darkness, that elusive, almost-dark-almost-light realm, the twilight zone.

Chapter 7

THE TWILIGHT ZONE

At 1,000 feet . . . I tried to name the water; blackish-blue, dark gray-blue. The last hint of blue tapers into a nameless gray, and this finally into black. . . . At 1,900 feet, to my surprise, there was still the faintest hint of dead gray light, 200 feet deeper than usual, attesting to the almost complete calm of the surface and extreme brilliance of the day far overhead. At 2,000 feet the world was forever black.

William Beebe, Bermuda, 1934,
Half Mile Down

On the surface of a calm, clear sea, *Jim*'s support crew wore wide-brimmed hats and slathered on sun block for protection from the fierce midday sun. More than 1,000 feet below, lights out, I stood in near darkness, suddenly mindful of the precarious nature of my existence. It was the absence of light that prompted the reflective mood, not my dependence on the good working condition of the special dive gear I was wearing.

It was one thing to know, intellectually, that the sun's energy, coupled with water and oxygen, are essential for life on Earth. It was another to feel in my bones the miraculous combination of circumstances that make life possible—my life, of course, but most incredibly, all life, through all time. I was sobered by the concept, looking upward through the dark sea, acutely tuned to the reality of sunlight sparking processes in the legions of microscopic plants that absorb carbon dioxide and water, churn out oxygen—and produce the basic simple sugar that in turn is transformed into other compounds munched upon by the countless creatures that have lived, died, and been reconfigured into other ingredients that make up the planet's current organic chemistry. Today is different from all other preceding days because of these ongoing processes, and tomorrow and all days that follow will be slightly changed as a result of whatever happens now, and the next day, and so on. It is about the most fantastic concept I can imagine, the interacting processes that somehow work and produce, for now at least, a planet hospitable for humankind.

To understand what makes Earth different from any other place, and why life can thrive here, it would be logical to focus on the part of the biosphere that supports most of the action—that is, the sea, of course. Since I am by nature an air-breathing sun-loving mammal, it has taken some time for the awareness to seep into the cracks of my brain that *most of the biosphere is ocean*. All of life on Earth lives in the dark at least half of the time, and much of it lives in dark all of the time—in the depths of the sea.

Unfortunately, human beings do not always behave in a logical manner, either as individuals or as a species. Instead of applying 95 percent of our scientific efforts to 95 percent of the biosphere, most of the attention has been and still is focused on the brightly illuminated surface portion in which humankind normally functions—a dangerously cavalier bias, given the utter dependence of the five percent on the rest. One of the reasons for this curious imbalance is the surprisingly wide-spread belief that the sea is a place apart *from*, not fundamental *to*, the basic Earth processes that make human life possible.

Another is the continued lack of access that has until recently inhibited generations of explorers. Even in 1979, there were few diving suits or submersibles in existence to enable people to go 1,250 feet into the sea—and never before had anyone had a chance to venture solo into that nearly dark, nearly light realm with a mandate simply to explore, unencumbered with heavy work assignments or preconceived experiments that had to be accomplished during the limited time available. During my momentous dive with *Jim*, I was given license to let my curiosity take me where it would—to prowl around like a cat in a new house, whiskers twitching, alert to the slightest movement, sensitive to subtle nuances of shape, light, and sound. I was free to do what explorers elsewhere on Earth take for granted they can do—to walk around, touch strange objects, look for familiar patterns and be sensitive to new arrangements, and sometimes, simply to be quiet and reflect on the significance of that special environment and the nature of the creatures that live there.

Standing at the upper edge of the great, sparkling cold darkness, I looked up to a full spectrum of light dancing from the waves overhead, each photon newly arrived, most having traveled from the sun through 93 million miles of space, a far lesser amount arriving from distant stars. When light reaches Earth's blue atmospheric "cocoon," it encounters its first significant obstacles. Some is reflected back into space or is absorbed by gases, clouds, water vapor, dust, pollen, and other ingredients that make up "air." But the barriers to free passage of light through the atmosphere are slight compared to the swift changes wrought the moment light enters the sea.

First, the quantity of light drops exponentially, so effectively absorbed by water that only one percent travels to a depth of 325 feet (100 meters), even in the clearest freshwater lakes or most transparent expanse of open ocean. The intensity diminishes tenfold with every 227 feet, yielding at midday a moonlightlike ambiance where I stood in *Jim*. At 1,950 feet (600 meters), illumination is equivalent to starlight; at 2,275 feet, the intensity is approximately one ten billionth of that at the surface, and at 2,925 feet and beyond—there is total blackness. It is not simple to determine the exact point in the sea where light fades and darkness begins, because there are numerous complicating factors: the place, the time of day, season of the year, the weather. Where the air and sea meet, light is further modified by reflection and refraction, both affected by surface roughness—waves and ripples—or the presence of ice and, in modern times, perhaps floating debris or a slick of oil. Water itself appears clear when viewed in a glass, but every drop contains organic and inorganic bits, including living creatures that scatter, diffuse, reflect, and absorb light, each in its small way altering the amount and kind of light that passes downward.

Plankton, silt, and the undefinable mix of substances that flow into the sea from most urban areas cause light in many coastal waters to diminish rapidly a few feet under the surface. One sunny October afternoon, I jumped overboard in San Francisco Bay to retrieve an outboard motor that had fallen off a dock next to a boat slip. I found the motor 13 feet down, by touch—not sight—half imbedded in silty mud. Only the slightest suggestion of a dark greenish-brown glow gave me a clue about which way was up as I turned to ascend. In harbors at San Francisco, Boston, Hong Kong, Bombay, and Tokyo—to name but a few—less than half of the light striking the surface makes it to as much as 6.5 feet; by 25 feet, more than 90 percent is gone.

Photosynthesis is light driven. In the sea, this process, fundamental to oxygen generation and food production, and thus to the vital functioning of the biosphere, occurs in the illuminated "photic zone," mostly in the uppermost 60 to 100 feet, but extending downward to the gray limits of where enough light can be effectively absorbed by plants to spark photosynthesis.

There is a level of illumination in the ocean where food production precisely balances the amount of energy required for cell maintenance and growth, a pivotal place in the sea known to oceanographers as the "compensation point." In less light (at greater depths), photosynthesis may still go on, but not fast enough to keep pace with energy needed for organisms to stay alive. For many years, textbooks suggested that one percent of surface illumination, or, in ideally clear water, a maximum of 325 feet (100 meters) framed the limits where photosynthesis could not only maintain a kind of cellular status quo, but also yield a dividend in terms of growth. At greater depths—anyplace with less than one percent of surface illumination—the need for energy for respiration and other activities would outstrip food production and the plants would not long survive.

Alas for the authors of the texts, a number of deep-dwelling algae did not get the news. Some plants, including various remarkable bottom-dwelling algae, prosper in depths greater than 650 feet, where the quantity of light is but a fraction of one percent of surface illumination. In murky coastal waters, where the one percent level of surface illumination may be reached about 30 feet underwater, and in polar seas, under several feet of ice, the photic zone is considerably more narrow. In the Arctic and Antarctic, light is also seasonal, with darkness prevailing for four months of the year and a state of twilight for another four months. Nonetheless, plants abound in polar seas, from jewel-like microscopic plankton and ice-hugging diatoms to large, slippery brown kelp fronds many feet long.

How deep *do* plants grow in the sea? The deepest-growing plant species thus far discovered* is a kind of crustose red alga found growing in profusion on a sea mount in waters 871 feet deep, where the light is calculated to be about .0005 percent of full surface sunlight. I am confident that plants will someday be found growing just a little bit deeper than the most recently published pronouncement of the absolute maximum depth where they can occur. Below the depth where plants can grow, plant eaters live a catch-as-catch-can existence, relying on an occasional drifting banquet followed by a long period of fasting or, if omnivorous, munching on their fellow citizens while awaiting the next salad.

While the most dramatic impact water has on light is in reducing its quantity, it also strikingly modifies light's quality: the portions of the spectrum it permits to pass. Light filtering through a dense canopy of rainforest plants may be of low intensity at ground level, but the full range of light visible to human eyes is all there in gloriously colorful hues of red, yellow, blue, and many shades between. At the moment light enters water, however, changes are swiftly wrought in the character and color transmitted downward. The long red and yellow wavelengths are quickly absorbed in the first few feet; shorter wavelengths—green and blue—penetrate the deepest. The effect is startling. In the prevailing blue atmosphere, brightly flowered swim trunks undergo a strange metamorphosis to conservative shades of Wall Street gray.

The all-encompassing blueness beyond 81 feet (25 meters) transforms even brilliantly colored creatures into monotonous tones of blue-gray or black, much as blue lights of a theater marquee can cause rosy cheeks and red lips to appear sickly bluish gray. I was horrified the first time I noticed wisps of a greenish-black substance oozing from a nick on my knee while diving, even though I knew from reading Jacques Cousteau's adventures that blood is supposed to look

---

* By Mark Littler and Diane Littler, Department of Botany, Smithsonian Institution, while diving in the Bahamas in a *Johnson-Sea-Link* submersible.

that way in deep blue light. In this sapphire world, red fish, like scarlet swim trunks, appear dowdy, blending well with the darkness below, and seeming to merge with the surrounding sea, blue on blue. At 162 feet (50 meters), coral and sponge-encrusted reefs appear to be painted from a palette featuring only shades of blue and gray, but their vibrant red, yellow, and orange secrets are readily revealed by the full-spectrum beam of a flashlight. Residents from hundreds of feet below are shown to be brightly colored when dragged to the surface in a net and exposed to sunlight, or when viewed in place using the incandescent lights of deep-diving submersibles.

In the small area I explored using *Jim* I noted dozens of red galatheid crabs clinging to shrubs of pink coral, several bright red shrimp, and a purplish-red jellyfish. There were no obvious plants, other than a few strands of drifting sea lettuce, a kind of bright, papery green algae that normally attaches to rocks in shallow water.

Because light in the deep sea is characteristically blue—whether from sun, moon, or starlight penetrating from the surface or from the predominantly blue light of bioluminescence—it is not such a bad idea for inhabitants of the deep to be red. In general, red pigments absorb and do not reflect blue light and red fish thus appear dark. It probably should not be surprising that many oceanic animals living in the deep sea are scarlet to purplish red, or black, because they appear black-on-black, and thus, in effect, invisible.

Peter Herring, an eloquent expert on bioluminescence, goes further with his explanation for the success of scarlet shrimp in the deep sea. He suggests that most deep-sea animals do not see color; since they have only blue light by which to view their surroundings, many have eyes tuned to see *only* blue light. Since blue light is perfectly absorbed by the scarlet shrimp's red pigment, the shrimp appear black and are simply not visible to predators with the ability to register only blue. A few fish provide a notable exception to the general rule of blue; they are among a small, elite number of creatures capable of producing—and seeing—red bioluminescence. A headlight on these fish emits deep red light, and they also have a red-sensitive visual pigment. According to Herring, it is:

. . . in effect, a private channel of communication. In addition, this personal sniper-scope enables its possessor to observe potential prey without their realizing that they are under surveillance; unlike most other inhabitants of the deep, it will be able to see, for instance, the unwary scarlet shrimp, oblivious to the fatal beam focused on it.

Some fish, sheathed in silver scales or highly reflective skin, employ a quite different slight-of-fin disappearing technique. Their bodies, like mirrors, reflect light coming from the surface—or the searching beam of one of the ingenious light-generating devices of predators.

Other creatures escape notice by being transparent, a fine strategy for survival adopted by various jellies, squids, octopuses, certain fish, and the larval stages of many creatures. A plankton net drawn through an apparently lifeless ocean typically yields a catch resembling minute fragments of broken glass mixed with quivering iridescent lumps of jelly creatures. Some of the glistening bits are minuscule but fully grown plants and animals, but many of the near-invisible dwellers of the open sea are the offspring of lobsters, crabs, eels, urchins, and a wide range of others. All are trying to avoid being eaten while they capture and eat as much as possible to prepare for the next phase of life on a reef or craggy rock crevice hundreds or thousands of miles from where they began life as seafarers on the open ocean.

The transparency trick is useful for many, but a full gullet or stuffed stomach is a glaring giveaway to an otherwise effective disappearing act. Also, for many organisms eyes, hearts, livers, and other organs are hard to conceal, and these creatures have developed other methods to survive in a realm where nearly everybody, even fellow hatchlings, may regard you as a tasty morsel—and there are no hiding places. Countershading, a dark upperside and pale or silvery belly, is a useful approach for avoiding detection, one that is commonly employed even in shallow water. Halfbeaks, needlefish, flying fish, mackerel, and many other fish seem to disappear when viewed from above because their dark topsides blend with the surrounding dark water, while from below, their silvery-white undersides merge with the surface light.

Peering out of one of *Jim*'s acrylic ports, I glimpsed a small, metallic-silver hatchetfish, a creature equipped with a suite of strategies to avoid being munched upon while seeking sustenance for itself. At first glance, hatchetfish, like camels, appear to have been designed by a committee, in this case, a drunken committee with a twisted sense of humor. Large tubular eyes are directed upward, while an impressive array of gleaming light organs, photophores, shine blue light downward. I laughed incredulously when, many years before I saw one of these wondrously puzzling creatures snagged in a midwater trawl and brought into the midst of a dozen eager oceanographers. I asked who could answer the question, "Why would any creature in its right mind point a flashlight in one direction—down—while looking in another—up—meanwhile moving resolutely straight ahead?" There was much speculation, and no one knew for sure, but the best guesses were that the downward-pointing light provided countershading. Some beam-producing predators with a taste for hatchetfish might be fooled by the reflecting light striking the fish's silvery sides, but an attacker looking from below would see the distinctive silhouette of a favorite meal. By turning on its photophores, a hatchetfish can illuminate its body with a glow to match light from above, and thus reduce the risk of its silhouette being seen.

Cleverly controlled laboratory experiments with various deep-sea fish and observations from submersibles have confirmed the validity of this strategy. A similar technique was tested successfully during World War II by torpedo bombers who used lights against a bright sky to avoid detection by subs as they swooped in on their target, surfaced submarines. Many species of fish, squid, and shrimps inhabiting the ocean's upper 2,600 feet have developed an assortment of lights on their underbellies. Some, such as the not-so-crazy hatchetfish, the appropriately-named lanternfishes, and certain squids have an enhanced ability to fine-tune the angle and level of blueness to that of the changing levels of illumination from above.

The ocean shapes the character of Earth, governing climate and weather, regulating temperature, and providing 97 percent of the planet's living space—the biosphere. Elsewhere in the universe there is water without life, but nowhere is there life without water. No water, no life. No blue, no green. PHOTO COURTESY NASA

If all of Earth's history were compressed into a single year, human civilization would be contained in the last few seconds. We are newcomers, appearing as aliens in ancient ecosystems filled with creatures, such as sharks, whose history precedes ours by hundreds of millions of years. Those shown here are swimming in Cuba's Jardines de la Reina protected area. PHOTO COURTESY MICHAEL AW

Of the forty or so known phyla of animals, all are represented in the sea and the majority occur on coral reefs, such as the one shown above in the Swan Islands of Honduras. Only about fifteen phyla of animals occur on the land, even in the richest rainforests. PHOTO COURTESY KIP EVANS

Cold/warm systems can be as colorful as a carnival and as diverse as a coral reef, as shown in this section of South Africa's Algoa reef, dominated by sponges, bryozoans, echinoderms, and anemones. PHOTO COURTESY KIP EVANS

This breeding aggregation of Limulus *horseshoe crabs near Cape May, New Jersey, embraces a history that has persisted for 400 million years, but they may not survive the current century owing to habitat loss and use by fishermen as bait. Four species represent an enormous wedge of unique genetic history; more than 750 thousand kinds of insects represent a comparable category of animal life.* PHOTO COURTESY GIGI BRISSON

Cephalopods—octopuses, nautilus, cuttlefish, and squid—are essentially living fossils, considering their ancient history. I met this one, a so-called "market squid," near Catalina Island, California. I prefer the scientific name, Loligo opalescens. *People consume them by choice, but they are a vital part of the diet of numerous birds, fish, mammals and invertebrates.* PHOTO COURTESY AL GIDDINGS

Sunlight powers photosynthesis in phytoplankton that combine water and carbon dioxide to produce sugar and oxygen, the start of the great ocean food webs. The zooplankton shown here may include more than a dozen phyla of animals, from larval fish and sea stars to fully mature arrow worms, comb jellies, and various crustaceans. PHOTO COURTESY SYLVIA A. EARLE

"Big fish eat little fish," they say, and it is happening here in a marine sanctuary at Cabo Pulmo, Mexico, where large grouper, snapper, and other sizable species are dining on a large gathering of small fish who in turn have consumed significant amounts of plankton. It is the "blue carbon cycle" in action. PHOTO COURTESY SYLVIA A. EARLE

The first of four oceanographic expeditions aboard the National Science Foundation research vessel Anton Bruun *gave me an opportunity to dive, explore, and document previously unknown places in the Indian Ocean using traditional nets and dredges, but also to use scuba to observe behavior and see living plants and animals, not just dead specimens.* PHOTO COURTESY HAROLD VOKES

The headline of a 1964 Mombasa newspaper story reads "Sylvia Sails Away With Seventy Men—But She Expects No Problems." Here, I am shown with some of my shipmates, who shared with me the only real problem we all had: how to explore the mysteries of the vast Indian Ocean, thousands of feet deep, from the deck of a ship. PHOTO COURTESY HAROLD VOKES

I earned my doctorate degree and also my "flippers" as a certified diver in 1966, although I began diving in 1952 with two words of instruction: "Breathe naturally." PHOTO COURTESY EUGENIE CLARK

The broad shelf along Florida's west coast was my laboratory for ten years of research on the ecology and systematics of marine algae at Florida State University, then Duke University, and now throughout the Gulf of Mexico at the Harte Research Institute at Texas A&M in Corpus Christi. PHOTO COURTESY JOHN L. TAYLOR

Long expeditions at sea come at a heavy price: separation from my family. Two of my three children, Elizabeth and Gale, are shown here with some of our geese, Christmas and Thanksgiving, and their two goslings, Fois gras and Polarguard. PHOTO COURTESY SYLVIA A. EARLE

Whenever possible, I share time at, on, or under the sea with my children. Elizabeth is now president of Deep Ocean Exploration and Research (and mother of two boys), Richie has worked for years with California Fish and Game (and is father to two boys), and Gale is a talented singer and also pilots submarines. PHOTO COURTESY SYLVIA A. EARLE

Diving makes it possible to get to know fragile creatures in their own environment, such as these moon jellies, Aurelia, *here forming a gelatinous blizzard in Truk Lagoon. Previously, I had seen them as crushed blobs in a net.* PHOTO COURTESY AL GIDDINGS

With scuba, the ocean becomes a laboratory for scientists and also provides access to a natural gallery of masterpieces, such as this enormous sea fan in the tropical waters of Palau. PHOTO COURTESY KIP EVANS

Scuba diving provides a passport into the sea, but time and depth are limited to brief excursions. In 1970, Project Tektite, sponsored by NASA, the US Navy, and the US Department of the Interior, enabled fifty scientists and engineers to experience saturation diving in the US Virgin Islands, prolonging observation time from hours to weeks underwater. PHOTO COURTESY BATES LITTLEHALES, *NATIONAL GEOGRAPHIC*

Men and women living together underwater was unacceptable in 1970, so a women's team was formed and I was appointed the leader. Every bit as qualified as our male counterparts, we were hailed with enthusiasm by the media as aquababes, aquabelles, and even aquanaughties, but rarely as serious professionals. PHOTO COURTESY FLIP SCHULKE, *BLACK STAR*

Large ports provided a great view, but we spent as much time as we could—eight to twelve hours a day—in the water. PHOTO COURTESY BATES LITTLEHALES, *NATIONAL GEOGRAPHIC*

We dined well with sufficient frozen meals to last for the duration of our stay. The members of the women's team were awarded Conservation Service Awards by the US Department of the Interior and made honorary citizens of Chicago by Mayor Richard Daley after a ticker-tape parade down State Street. According to NASA advisors, our successful performance helped pave the way for women as astronauts more than a decade later. PHOTO COURTESY BATES LITTLEHALES, *NATIONAL GEOGRAPHIC*

By 1975, more than fifty underwater laboratories had been built to enable divers to go deeper and stay longer underwater. Shown here is Hydrolab near Freeport, Grand Bahama, which operated for many years as the durable "workhorse" of underwater laboratories. During one series of dives, the Johnson-Sea-Link *submersible served as a pressurized "taxi" to take saturated divers from the Hydrolab to dive sites over the edge of the nearby deep drop-off for prolonged excursions at 250 feet deep, returning to the sub to start decompression as it traveled back to the Hydrolab.* PHOTO COURTESY SYLVIA A. EARLE

Small submarines enable divers to go deep in the sea without being exposed to increased pressure. A few special submersibles, such as the Johnson-Sea-Link shown here, have a special compartment that can be pressurized to equal that of the surrounding sea and enable a diver to swim out while the sub waits. To take this photograph, I emerged from the sub 250 feet down along a steep wall near the Hydrolab in the Bahamas. PHOTOS COURTESY SYLVIA A. EARLE

The Aquarius underwater laboratory, built 5.6 miles offshore from Key Largo, Florida, in 1986, was a NOAA facility until 2012 when Florida Institute of Technology assumed ownership. Six aquanauts (and sometimes astronauts in training for working in space) use the lab as a home base for operations. PHOTO COURTESY SYLVIA A. EARLE

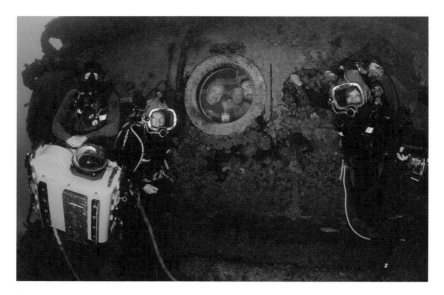

Three Aquarius team members inside, three outside. It is now acceptable for men and women to work and travel together in space, on land, and under the sea. PHOTO COURTESY KIP EVANS

Soon after it was placed on the reef, Aquarius became home for more than aquanauts. Over time, the structure took on the appearance of an extension of the reef, festooned with sponges, corals, and various encrusting organisms. Fish such as these striped grunts took up residence there as well. PHOTO COURTESY SYLVIA A. EARLE

It should have been obvious to me before, but while living underwater, getting to know the local residents, it became clear that fish have distinctive individual behaviors and are recognizable as individuals. Most Nassau grouper, like the one shown here, are curious by nature, but some are more inquisitive than others. PHOTO COURTESY KIP EVANS

There are many kinds of moray eels, but I have yet to meet one with a bad temper. PHOTO COURTESY SYLVIA A. EARLE

Whale sharks, known as "domino fish" in Mexico, can be identified by their distinctive arrangement of spots. A computer program makes it possible for each one to be identified from photographs of their unique speckled patterns. PHOTO COURTESY BRYCE ROARK

These white tip reef sharks at Cocos Island, Costa Rica, have no distinctive spots, but they do have individual markings and behaviors. PHOTO COURTESY SYLVIA A. EARLE

Jim, the "submersible you can wear," enables divers to go deep in the sea without being exposed to increased pressure—a system similar to that used by space-walking astronauts. Al Giddings crouched inside a small submersible to take this photograph of my record dive to 1,250 feet, six miles offshore from Makapuu Point, Oahu, Hawaii, in 1979. My return from the dive in Jim was greeted with a wave of publicity that helped bring the need for ocean exploration and understanding into sharper focus. Even now, access to the deep sea (and thus most of the biosphere) is elusive, limited to brief glimpses made possible by special technology. PHOTO COURTESY AL GIDDINGS

In 1984, I had the pleasure of cracking champagne over the hull of an innovative new submersible, Deep Rover, designed by Graham Hawkes and built by Deep Ocean Engineering, Inc., the company we cofounded to design, manufacture, and operate equipment for undersea exploration. During several days of diving Deep Rover offshore from San Diego, California, Graham Hawkes, Phil Nuytten, and I (and, later, Bruce Robison) established records for diving solo to 1,000 meters. Here, I am just beginning my descent along the Coronado Escarpment in southern California. PHOTO COURTESY DEEP OCEAN ENGINEERING

Deep Rover 1, *shown diving here at Lee Stocking Island in the Caribbean in 1986, has since logged thousands of dives and been deployed from dozens of platforms ranging from a helicopter and wooden raft in Crater Lake to sophisticated ocean-going ships.* PHOTO COURTESY SYLVIA A. EARLE

During sea trials in Santa Barbara in 1994, I admire one of a pair of Deep Rovers, *each able to take two people to 1,000 meters depth. Built for the French cable television company Canal Plus, they are designed for ocean exploration, research, and filmmaking.* PHOTO COURTESY DALE ANDERSEN

Presently, the two-person Deep Rover *subs are part of Ray Dalio's Ocean X research and exploration expeditions. In 2016, I joined a BBC film team in the northern Gulf of Mexico and witnessed deep-sea biologist Samantha Joye in one of the* Deep Rovers *observing life within brine pools 600 meters beneath the surface.* PHOTO COURTESY BUCK TAYLOR

The first DeepWorker *subs were used during the five-year Sustainable Seas Expeditions (SSE) sponsored by National Geographic, NOAA, the Goldman Foundation, and more than fifty partners, including the Harte Research Institute (HRI) and DOER Marine. Student summits were held at each of the locations, mostly in areas designated as National Marine Sanctuaries. The Mexican government collaborated with the HRI and SSE for an expedition to Yucatán and Belize in 2003.* PHOTO COURTESY KIP EVANS

DeepWorker *subs often operated as a pair of "buddy diving submarines," here exploring Pulley Ridge, a deep coral reef 100 miles off the coast of Florida.* PHOTO COURTESY TIM TAYLOR

A new generation of three-person, 1,000-meter Deep Hope *subs is underway at DOER Marine, the company my daughter, Elizabeth Taylor, and her husband, Ian Griffith, have developed to build advanced underwater systems.* PHOTO COURTESY DOER MARINE

Most of the ocean has yet to be explored, but exploitation by industrial fishing and deep-sea mining are nonetheless commanding widespread investment. This puts rich communities of life at risk and destabilizes the basic life support functions governed by the ocean. Shown here is a field of manganese nodules and some of the local residents in the Clarion-Clipperton zone, 5,000 meters deep. PHOTO COURTESY UNIVERSITY OF HAWAII

Remotely operated vehicles (ROVs), while no substitute for "being there," can take cameras and instruments deeper and stay longer than divers. Shown here is the Lu'ukai (meaning "sea diver"), a 6,500-meter system designed and built by DOER and used by University of Hawaii scientists to explore critical deep ocean areas before mining occurs. PHOTO COURTESY TAYLOR GRIFFITH

Humpback whales are one of more than seventy species of dolphins and whales, highly social creatures fine-tuned during 65 million years as expert divers. PHOTO COURTESY SYLVIA A. EARLE

Sperm whales can descend at least a mile while holding their breath for more than an hour. The six young males above at the last whaling station of Albany, Australia, were killed in 1978 before commercial whaling was banned. PHOTO COURTESY AL GIDDINGS

Some say the only good shark is a dead shark, but stripping these ancient predators from the sea exacts a heavy price on ocean ecosystems. Like many land predators, most sharks are long-lived, slow growing, produce very few young, and so can rapidly be depleted. Here, conservationist and explorer Sharon Kwok Pong holds fins taken to support the luxury market for shark fin soup.   PHOTO COURTESY SHARON KWOK PONG

The fish market in Tokyo distributes more than a million tons of wild caught creatures, mostly predators, from the sea every year. The tuna above represent a tremendous investment from ocean ecosystems. It takes about fifty thousand pounds of microscopic plankton working their way through a food web to make a pound of ten-year-old tuna. In comparison, twenty pounds of plants yield a pound of yearling calf; while two or three pounds yield a pound of chicken.   PHOTO COURTESY SYLVIA A. EARLE

Near areas of high population, many reefs have succumbed to environmental changes brought about by the recent influx of air and waterborne fertilizers, pesticides, herbicides, and siltation from agriculture, land development, and the large-scale extraction of fish, lobsters, conch, and other vital components of a reef system. Even in areas far removed from human impacts, the rise in temperature globally is driving widespread decline. PHOTO COURTESY SYLVIA A. EARLE

Since 1980, increasing numbers of "bleaching" events are happening globally. This whitening of corals is caused by expulsion of microscopic algae that normally live within the coral's tissues. Death often follows such loss. Prolonged increase in water temperature has been correlated to many coral bleaching incidents worldwide, but pollution and other stresses are also involved. PHOTO COURTESY KIP EVANS

Debris in the sea is not only unsightly, it is deadly. Entanglement with lost and discarded fishing gear yields a tortured end for millions of fish, birds, mammals, and turtles each year. This windrow of trash was drifting near Cocos Island, a protected area of Costa Rica about 300 miles offshore. PHOTO COURTESY KIP EVANS

Trash, largely fishing gear and discarded plastic, dumped into the sea is insidiously appealing to many creatures, who become victims of entanglement or death by swallowing indigestible plastic pellets, bags, and fragments. PHOTO COURTESY SYLVIA A. EARLE

In 1989, 250,000 barrels of crude oil flowed from the tanker Exxon Valdez into the pristine waters of Prince William Sound, Alaska, swamping rocky shorelines and engulfing millions of creatures with lethal oil. PHOTO COURTESY SYLVIA A. EARLE

Oil seeped deep into the ground between the rounded beach stones, and much of it is still there. PHOTO COURTESY SYLVIA A. EARLE

In 1991, more than forty times the amount of oil spilled in Prince William Sound in 1989 flowed into the Persian Gulf, a deliberate act of terrorism by Iraq against Kuwait. As chief scientist at NOAA, I repeatedly dived along the oil-soaked beaches of Saudi Arabia. <small>PHOTO COURTESY SYLVIA A. EARLE</small>

Instant fossils imbed an oiled beach at Ras Abu Ali. In some intertidal areas that had hosted as many as half a million small, burrowing creatures per square meter, nothing survived. Marshes and beaches of Saudi Arabia were hardest hit by what some have called the "mother of all oil spills." More than a year later, researchers found few living things in oil-smothered beaches and discovered liquid oil more than two feet under the sandy surface of intertidal flats that once had been rich feeding grounds for millions of migrating birds. <small>PHOTO COURTESY SYLVIA A. EARLE</small>

In April 2010, the Deepwater Horizon oil spill in the northern Gulf of Mexico resulted in the loss of eleven men who were on the rig, a tragedy magnified by the death toll of whales, turtles, birds, fish, and countless other sea creatures. PHOTO COURTESY US COAST GUARD

Oil spills do not have to be large to be lethal. This seabird in San Francisco Bay floats in one of the 16,000 or so small spills that occur in the United States each year. Spilled oil is messy and conspicuous, but it is not as grave a danger to ocean health, overall, as the continuous flow of toxic materials into the sea from the sky, rivers, groundwater, and deliberate dumping.

In 1975, teenagers John F. Kennedy Jr. and Timothy Shriver witnessed the fate of debris on a grand scale—ships lost at sea—during a research project in Truk Lagoon. With Al Giddings, we observed and documented the effect of decades of natural processes acting on thousands of tons of pollutants dumped into a tropical lagoon. PHOTO COURTESY SYLVIA A. EARLE

This ship, disguised as a coral reef, was among the more than sixty Japanese warships sunk in Truk Lagoon in 1944 by Allied forces. During dives in 1989, I confirmed that forty-five years after sinking, oil still leaked from the sunken fleet and live munitions posed a continuing threat. PHOTO COURTESY SYLVIA A. EARLE

Momentum is growing to protect the land and ocean through a global network of parks and reserves vital not only for the protection of wildlife and wild places but also to the existence of humankind. When President George W. Bush designated the Papahānaumokuākea National Monument in 2009, it became the largest protected area on Earth. At the time, only a fraction of one percent of the ocean was under significant protection. PHOTO COURTESY GEORGE W. BUSH PRESIDENTIAL LIBRARY AND MUSEUM

When President Barack Obama increased the size of the Papahānaumokuākea monument, he did so with the clear understanding that climate and weather are directly tied to the state of the ocean. Other countries are stepping up commitments with the goal of having at least ten percent of the ocean safeguarded by 2020 and thirty percent by 2030. PHOTO COURTESY WHITE HOUSE, BY PETE SOUZA

Wisdom, the sixty-five-year-old Laysan albatross who returns with her mate to a nest on Midway Island, is cause for hope. PHOTO COURTESY SUSAN MIDDLETON

This map shows the Hope Spots established in the last ten years. Many more communities are working to establish Hope Spots of their own across the globe. With knowing comes caring for the natural systems that support life on our blue marble.

As a child, I had puzzled about the value of bioluminescence while watching fireflies magically flashing brilliant yellow signals against dark evening skies. It was so easy to catch them! I often held them briefly cupped in my hand, eerie glints of light seeping through the cracks between my fingers as they blinked on, then off, delicate antennae and slender legs softly probing possible avenues of escape. Predators with a taste for fireflies would have an easy time of it, I thought, and I wondered how any of the enchanting, glowing creatures survived long enough to get together and do whatever it is fireflies do to produce the next generation. Much later, I learned that the advantages in terms of finding and signaling mates significantly outweigh the predation liabilities. The same is true for many spark-in-the-dark creatures in the sea.

More than a century ago, Charles Wyville Thomson, the distinguished biologist mentioned earlier, who spent many months exploring the depths of the North Sea with nets and trawls and other instruments, speculated about the mixed blessings bioluminescence might bestow. Describing one evening's catch of starfish, snared in deep water off the coast of Scotland, he noted, in *Depths of the Sea*:

> *The tangles were sprinkled over with stars of the most brilliant uranium green. The light was not constant, nor continuous over all the star, but sometimes it struck out a line of fire all around the disk, flashing . . . up to the centre . . . or the whole five rays would light up at the ends and spread the fire inwards. . . . It is difficult to doubt that in a sea swarming with predaceous crustaceans . . . phosphorescence must be a fatal gift.*

Fatal for some, perhaps, but advantageous enough that about 80 percent of the animals in the dimly illuminated midocean depths, at 650 to 3,900 feet (200 to 1,200 meters), have the ability to produce light, sometimes to locate food or use as a lure, sometimes to signal a prospective mate, and perhaps sometimes to frighten, distract, or decoy a determined pursuer. Nearer the surface and in deeper water, the percentage of light bearers appears to be less, but clearly, bioluminescence is a widespread, fundamental phenomenon of profound importance to life in most of the biosphere.

With *Jim*, I stood in a place where it is more normal for the local residents to have some kind of bioluminescence than to be without it. Of those that do not, many are equipped with eyes more exquisitely light-sensitive than those of cats and owls, able to register clear images in illumination that to me appears nonexistent. To them, *Jim* and the *Star II* submersible must have been a blindingly dazzling sight, aliens with "photophores" more powerfully brilliant than those of the largest squid, fish, or crustacean ever to venture into their domain. One such creature, a sinuously lovely cat shark with glowing green eyes, seemed disoriented by the sub's lights, bumping into a sponge as it careened away into the dark sea beyond.

With the sub's light turned off, I willed my eyes to adjust quickly to the barely visible seascape, wishing that I could enjoy even for a while the enhanced light-gathering skills of a hatchetfish, and longing to catch a glimpse of one of the deep-sea squids that emits clouds of glowing ink when frightened. Various deep-sea creatures employ a similar "squirt and run" strategy, using a burst of luminescence as a decoy while they make good their escape. Tiny crustaceans known to afficionados as ostracods use a dash of luminescence for other purposes. These tiny creatures manufacture in a special gland one of the critical chemicals needed for light production, luciferin, and in another, an enzyme, luciferase, that is required to activate the light-producing process. When the ostracod is excited (usually a male seeking to impress a nonluminous female, or outshine several rivals), both glands empty their contents into the water, where they react to form a miniature puff of astonishingly brilliant blue light.

In vain I looked for evidence of ostracods and squid, but minute sparks of light from disturbed microorganisms flashed as I moved *Jim*'s arm, and then a startling burst of blue fire erupted from a surprising source. I had brushed against a spiraling bamboo coral taller than *Jim*, initiating a light show that would have impressed even the most discriminating female ostracod.

I had landed with the sub in the midst of at least a hundred coral colonies, miraculously managing not to crush any, although several were within easy reach. Named for their resemblance to jointed stalks of bamboo, the stately formation appeared pale and unobtrusive until touched. The most gentle nudge of my "claw" provoked ring after ring of blue light to pulse from the point of contact down the full length of the coral, small bright circles cascading in neatly spaced sequences, fading to darkness after nearly a minute. Curious, I touched one of the spires at the base and near the tip and watched, anxious to see what would happen when descending rings collided with those racing upward. Would they cancel each other out, or what? Serenely, the miniature fiery blue doughnuts merged, then each passed through the other and continued onward, apparently unperturbed by what appeared to me to be a setup for a truly scintillating encounter.

I have to wonder about the usefulness of such powerful light-generating capability for a colony of animals that is firmly rooted in place. Is touch required to turn on the lights? If so, is it useful as a means of signaling clumsy fish to be more careful about where they swim? Might those venturing a bite of the coral's clusters of soft, delectable white tentacles be deterred by the equivalent of a deep-sea burglar alarm? Are there times, perhaps at the moment of spawning, when the lights come on for reasons obscure to human observers but of vital significance to the hundreds of interconnected polyps that make up a single coral colony? Or, I ask myself, is it possible that there is no special, serious function requiring a practical explanation of the sort that will satisfy rational human inquiry? Maybe bamboo coral just is that way. Period.

I had but two and a half hours to explore the enticing mysteries of the bamboo coral forest, not long enough to thoughtfully frame, let alone find, answers to an avalanche of questions stimulated by encounters with the twilight dwellers, creatures such as the seven large, spiny rays that fluttered by in formation. Where did they come from? Where were they going? Is it a species known—or not yet noticed by humankind? How old are they? How many more exist? Were they attracted to the sub's lights? Did they come over to check out a disturbance in the electromagnetic field? Or what about the single, slender gray eel that looped itself around the base of a low-growing golden-plumed coral. Is this a typically solitary creature, or was this particular one a stray? What does it eat? Is there a special association with that kind of coral, or was it just a pleasant place to perch for a while? Was the eel forever a resident of the bamboo forest, or are there times when it travels to other places? I do not know, nor is it likely that I, or anyone, can soon find answers to hosts of questions that would be somewhat easier to resolve if it were not such a big deal to arrange for thoughtful time one fourth of a mile underwater.

Reluctantly I made a final entry in my logbook and signaled that I was ready for "liftoff" and reentry to the topside world of creatures equipped with lungs, and senses tuned to airborne sounds and smells—and gloriously brilliant sunlight! Immediately, there were more questions, mostly directed my way:

"What did you see?"

"Were you afraid?"

"Any problems with *Jim*?"

"Did you see any *sharks!*"

"What's it like down there?"

I was delighted that people wanted to listen, and tried every way I knew how to explain not only how beautiful it is in the deep sea, but also, how urgent it is for us to understand as much as possible about the nature of the complex ocean systems that occupy so much of Earth. Like astronauts, charged on behalf of humankind to bear witness to sights not yet within the reach of most, it seems to me that those who travel into the deep sea also have significant news to share. Unfortunately, the results of most ventures into the depths are communicated to a small, select scientific circle. On this occasion, however, there appeared to be a wide and eager audience.

The dive was featured prominently in a two-hour television special produced by Al Giddings for the American Broadcasting Company, and in a National Geographic book that I authored, with photographs by Al and others, *Exploring the Deep Frontier*. Al and I also teamed up for an article for *National Geographic* magazine, but months passed before such accounts were released. Meanwhile, within days, newspaper and magazine articles appeared in Spanish, French, German, Dutch, Japanese, Chinese, Russian, Arabic, and several variations on the theme of English about the female scientist diving in the *Jim*, and much was made about this being the deepest solo dive ever made by a woman or man without a tether to the surface. There was little mention of the thoughtful insights I wanted to share, but I hoped that my dive had aroused at least some interest that might cause people to take the oceans more seriously.

When I arrived home, my three children eagerly led me to the kitchen and there, plastered across the refrigerator door, was a lively centerfold account from the supermarket tabloid *The Star*, headlined BRAVE MOM'S HISTORIC DIVE TO BOTTOM OF THE WORLD. For a moment, I thought I heard music transporting me into that grayness between fantasy and reality—the other "twilight zone." It was great, I thought, to demystify ocean exploration and encourage people to go see the sea for themselves. It was maybe *not* so great for me, as a scientist, to be featured in a publication notorious for wild accounts of giant grasshoppers, greenfolk from Mars, and lurid sex scandals. Caught between laughter and despair, I was pushed by my three trusty reality checks—Elizabeth, Richie, and Gale—into the right mode. Grinning wickedly, they urged, "You can handle it, Mom. Just—be brave!"

Bravery, lots of it, is just what I needed in the next venture—starting from scratch a company whose purpose would be to develop new equipment for ocean exploration. The "perils of the deep" I have found to be mostly nonexistent; the monsters mostly mental. The world of business, I was soon to learn, is less forgiving, and predators truly await the unwary. Still, the idea of creating something new, a productive source of income—and a place where new equipment could be developed for ocean exploration—well, that was a lure worthy of the risks, even though it meant figuratively and literally getting in over my head, stepping into the unknowns of accounting, tax, and business law, negotiating contracts, raising capital, filing reports, making payroll . . . making a difference while making a profit.

Chapter 8

OUR BUSINESS IS GOING UNDER

What kind of man would live where there is no daring? I don't believe in taking foolish chances, but nothing can be accomplished without taking any chance at all.

Charles A. Lindbergh

Showers of blue-green light cascaded from the clear sphere around me and red numbers glowed from the instrument panel above my knees, the only illumination marking my solo descent to 3,000 feet along the Coronado Escarpment offshore from San Diego, California. Ten minutes ago I had turned off the outside lights of my small craft, the tiny bubble-like submersible *Deep Rover*, to witness more clearly the luminescence of the thousands of small creatures responding to contact with the sub.

Like falling into a galaxy! I thought, as light from an array of single-celled creatures, small jellyfish, salps, comb jellies, crustaceans, annelids, and an occasional small squid sheathed my inner-space suit with faint silver-blue sparks. I paused in my descent through the night sea to revel in the sheer pleasure of witnessing the living light show. In the thousand feet just traveled, I had come in contact with more diverse broad categories of life than can be found anywhere on land, from the minute planktonic young of several categories of bottom dwellers to full-grown midwater fish and free-swimming octopuses.

Twelve miles to the east, nearly a million people slept. Above, the support crew monitored my progress and spoke to me using through-water acoustic communications from the bridge of *Egabrag III*. The name, "garbage" spelled backward, was a sardonic reference to the ship's unmistakably low-slung Mississippi mud boat origin; but within the homely shell, instrumentation bristled that would make a Tom Clancy hero sigh with envy. A roomful of sophisticated electronics equipment assured accuracy of positioning and other measurements, mostly for dark assignments from military sponsors. Like a dragonfly seeking out a lone lily pad, a helicopter from time to time homed in on the ship and touched down on the top deck.

It was improbable that this whole ship should be supporting the *Deep Rover* operation, that I was alone, gliding through a night sea far below, and even that *Deep Rover* had come to be at all. Improbable, but clearly not impossible, although from time to time I had to remind myself, *This is not a dream—I am really here!* It *was* a dream, in a way—one of those nice ones that had come true, thanks to substantial quantities of sheer pigheadedness (mostly mine), some engineering genius (from my friend Graham Hawkes, who had been the technical adviser for our *Jim* project), and a few lucky breaks.

Five years earlier, in May 1980, the plan for developing the little sub was born during a discussion with Graham in a Washington, D.C., restaurant. We were having one of many earnest conversations about how to overcome the problems of gaining access to the deep sea, well below where free-swimming divers could venture. The *Jim* project had opened my eyes to more than the nature of life in the deep sea; suddenly, I was propelled into the community of ocean engineers and commercial divers, and through them, the robust realm of ocean industry. Most influential in expanding my underwater horizons was Graham, whose quiet competence repeatedly surfaced during the *Jim* expedition. My first serious conversation with him did not bode well for an enduring friendship, however. My attention was focused on *Jim*'s arms and hands and I remember saying, quite clearly, "That *stupid* pincer! All I can do is reach out, and things just fly through its *jaws!* It's like having a pair of pliers on the end of a *stick!*" He looked at me sympathetically, and agreed that the manipulators really weren't very good and should certainly be modified. He quizzed me, earnestly asking what character-istics I thought a really effective arm should have, and carefully explained the difficulties of keeping joints flexible, with 14.7 pounds of pressure per square inch on the inside, and 600 pounds per square inch on the outside.

Shortly thereafter, I discovered to my great chagrin that *he* had designed *Jim*'s "stupid" arms.

Some months later, Graham called from England and said he had something he wanted to show me. Could he stop by? He arrived several days later, a myste-rious brown envelope in hand.

"If you were to devise a test for a mechanical arm—a manipulator—to demonstrate that it really could do what you told me in Hawaii an ideal arm should be able to do, what would that be?" Graham asked. His expression was unreadable, but I suspected that he might or might not show me what was in the envelope, depending on my response.

I said that one demonstration of dexterity would be to be able to write my name, in letters of normal size, holding a pencil or pen. This was obviously a good answer; Graham withdrew from the envelope an elegant pencil sketch of an Antarctic krill, *Euphausia superba*. The name was written in neatly formed block letters and at the bottom appeared the signature of the artist, Graham S. Hawkes, written in script in letters of normal size.

I was *really* impressed. For the first time I could envision using a submersible in a manner comparable to free-diving, that is, to travel deep in the sea inside a comfortable, one-atmosphere shell, and control arms that could perform with precision close to what human hands can deliver. In fact, the concept might be used in any environment, from space to burning buildings. Graham called his new device the "human-equivalent arm."

At our May 1980 meeting in Washington, Graham and I reviewed the advan-tages, and limitations, of *Jim*. Clearly, I was delighted with the concept of a "per-sonal submersible"—a one-atmosphere system that I could control myself for descents far below where I could go as a free-swimming diver. Typically, research submersibles are built for two or three people, including at least one highly trained pilot who drives the sub while one or more scientist-observers watches and, when necessary, asks the pilot to stop or maneuver or pluck some small object from the sea floor with a mechanical arm. Often, as scientist-observer in a small submarine, I asked too late for a pilot to turn the sub, and we passed right over, lost, or crunched the very thing I wanted to see.

In *Jim*, my eye-mind-hand coordination was comparable to what I enjoy while working in a laboratory—or walking through the woods. I did not have to ask someone else to please go left or right or pause to look at something that caught my attention; I could just do it. As smitten as I was with the do-it-myself concept, I was nonetheless frustrated by the cumbersome nature of *Jim*, the restrictions on depth, the usual mode of tethered deployment. Graham had arranged for me to try two of his other designs, both highly successful one-person systems widely used in the offshore oil and gas industry. *Wasp*, the bright yellow and black diving suit used as backup for the *Jim* dives in Hawaii, "flew" at the end of a tether, and had earned a reputation for its superior maneuverability. With electrically powered thrusters, it could move vertically up and down as well as forward through the water. Eighteen were then in use worldwide.

Even more popular was *Mantis*, another insectlike submersible, also yellow and black with thrusters and two large hydraulically powered manipulators at the front end. Both had clear hemispherical ports for the pilot, and both operated with power supplied through a long, heavy cable. I admired these systems for their innovative and effective working capabilities, and could see why they had been so successful for maneuvering around rigs and for other site-specific assignments. But I longed for something that did not then exist—and, I thought, might not even be possible.

Graham listened patiently while I described my dream machine: a one-person system that I could wear like a comfortable suit of clothes, with which I could travel freely (no tether!) from the ocean's surface to the greatest depths 35,800 feet down, and that had arms and hands that could be used to touch, feel, and effectively retrieve objects encountered. I also thought it would be nice to be able to stay for more than a few hours—or come back quickly, if I wanted to.

"Look," he said, reaching for a napkin. It was one of the rare occasions when he did not have a drawing pad close at hand. "Arms and legs are fine for running or climbing trees, but they're really not designed for traveling underwater." On the napkin he quickly sketched a pair of small, clear spheres, with a person sitting in each one. "I've been thinking about something that you would probably like. To go ultradeep, a sphere is perfect. Pressure is distributed evenly from all sides, and one size fits all—with room enough for lunch, cameras, even a pillow and a book or two."

In effect, he explained, the little subs would look like submersibles, but behave like individual diving suits, one for each person, equipped with "instinctive" controls and articulated, sensory manipulators operated outside the sub as extensions of the pilot's own arms and hands.

"See, no tethers," Graham pointed out. "Batteries can supply enough power for many hours of operation, and—how about a week of life support?"

A week! I thought. Submersible dives typically do not last more than a few hours, but it was nice to have a generous safety margin. I nodded approvingly.

"How deep?" I asked.

"To have a clear sphere, the only material that has been properly tested now is acrylic. Depth rating depends on how thick it is, how large the sphere, how often it gets exposed to pressure, and so on."

"How deep?" I repeated.

"Well, to start, not deep enough—a few thousand feet. But with this design, when better materials are available—clear ceramics, various glasslike compounds—we'll be ready."

The concept for one-person "bubble subs" seemed so reasonable and practical for scientific research, we thought it should be possible to find sponsors to build at least a pair, one as the primary system and a second for "buddy" diving work as well as for safety. Graham and I presented the vision of a program of exploration and research using the subs to numerous would-be backers in 1981. Many endorsed the concept and wanted to be included in the action—once the program was under way; but finding support for the "bricks and mortar" phase proved elusive. Jerry Stachiw, the world-recognized guru of glass and acrylic pressure hulls, urged us not to give up and introduced us to the U.S.

Navy's long-retired *Nemo* and *Makakai*, among the first subs built using acrylic pressure hulls.

We were greatly encouraged by Ed Link, friend of many years, and designer of the two state-of-the-art manned two-person submersibles, *Johnson-Sea-Link I* and *II*—also sometimes referred to as "bubble subs" for their clear acrylic spheres, coupled with a special steel chamber to transport two divers. The *Johnson-Sea-Link* subs, then in operation for a decade, had set a new high standard for subs used for scientific work. My several experiences in these subs had convinced me of the advantages of maximum visibility. In deep-diving subs observations typically are made through tiny portholes that are located in deviously awkward places, requiring imaginative contortions to attain a viewing position. Seated inside an acrylic sphere, there is an illusion of sitting warm and dry in a magic aquarium, where the fish are on the outside. Not surprisingly, Link had already considered the possibility of a system similar to Graham's concept: a simple sphere equipped with manipulators, lights, and cameras, but for two people.

Just before Link departed for a trip to England with his wife, Marion, he wrote a letter asking if we could meet when they returned. He was optimistic about the possibilities for the new subs, and thought "something could be worked out." But it was not to be. Soon after the Links returned from Europe in August 1981, Ed Link suffered a heart attack, was hospitalized, and did not recover.

Discouraged with the prospects of attracting philanthropic backing after months of serious effort, Graham and I decided to resort to the do-it-ourselves approach, and in the summer of 1981 pooled our resources and cofounded Deep Ocean Technology, Inc., and later, Deep Ocean Engineering, Inc., to bring the small subs and other underwater designs to reality. To do so meant a move for Graham from his home and family in England; for me, a major shift in priorities was required, from science to technology, and from the role of active field researcher to that of diligent student of accounting, law, corporate structure, marketing, tax requirements, insurance and labor policies, as well as janitor and owner of the property where the business began—the garage adjoining my home in Oakland, California. I moved my laboratory, library, and lifetime collection of marine plants from the California Academy of Sciences to a place at home, and continued to work on research projects on a catch-as-catch-can basis. I am not a businesswoman by training, but Graham's logic made sense to me: "Better to go with something *in* our hands than with our hands *out*," he reasoned.

Graham became the founding president and chief engineer, I was co-founder, vice president, secretary, treasurer, and whatever else was needed to get things done. I soon leaned on some of my friends, who knew more about running a business than either the president or "other officers," to join us as members of a small board of directors. When Surgeon Vice Admiral Sir John Rawlins, a distinguished "statesman of the sea" and former medical director-general ("surgeon general") of the Royal Navy and senior consultant in aviation and diving physiology, agreed to join Deep Ocean Technology as chairman of the board, we felt we could not fail. The businessman and conservationist Alan Weeden and a brilliant, nimble-minded New York attorney, Robert Beshar, who became my mentor, friend, and vital corporate strategist, rounded out the expert team needed to get under way.

For me, the company was a digression from my lifelong passion for research on marine plants and animals—one that would, I thought, be balanced by creating new tools to improve access to the sea. Consequently, I debated accepting an invitation to go to Washington to head the recently established National Underwater Research Program. If I said yes, I thought I might be able to help generate a greater national commitment within the U.S. government to ocean science and technology. But in the end, I was drawn irresistibly to the prospect of starting something new and, while making a living, perhaps also making significant contributions to ocean exploration, research, and the understanding that underlies effective protection of natural resources. At least, that was the goal. Getting there was something else.

In 1981, we could find no interest in funding the development of a new manned submersible. However, unmanned systems, underwater robots, were beginning to win wide support in industry to substitute for divers and to perform a wide range of special tasks, from basic inspection using television and still cameras to heavy-duty work with special tools. We developed a business plan around one of Graham's ideas for a large, remotely operated vehicle (ROV) called *Rig Rover* dedicated to oil-rig inspection, maintenance, and repair. Graham and I celebrated the signing of Deep Ocean Technology's initial contract—$1.8 million to design, build, and operate the new ROV—with starry-eyed enthusiasm, then set about to turn dreams into machines. I had no concerns about the technical success of the ROV design, but competing successfully with well-established offshore operating companies was a formidable challenge. Track record was vital, and we had none. Connections within the industry, like networks everywhere, were also crucial, and we were newcomers. One of our first attempts to interest a prospective customer, Chevron, made clear the need to think hard about our marketing strategy.

We prepared carefully for the meeting with Chevron's man, mindful that a contract to operate *Rig Rover* on an offshore oil rig was necessary to demonstrate the device's practicality and, most significant, to provide revenue beyond initial research funding. We were especially excited about the performance of a totally new, powerful sensory manipulator system that Graham had designed, inspired by the "human-equivalent arm" but based on a different approach. To illustrate the system's "surgical precision," Graham used the prototype arm to grasp a pen and draw another detailed version of the shrimplike creature, *Euphausia superba*, again signing his name with a flourish. The merits of an arm capable of being operated with such a high degree of accuracy and sensitivity seemed self-evident, but then, we had not reckoned with the mind-set of the crusty Texan who had been assigned by Chevron to meet with us. He listened politely and seemed especially taken with Graham's manipulator drawing, but we knew we had erred badly in emphasizing this aspect of the ROV system when he said, "Looks like a great machine. When we need to draw shrimp underwater, we'll give you a call."

Soon thereafter, we lured onto our team Steve Etchemendy, a man comfortable with Cajun, calculus, the technology of oil rigs, and the sociology of those who operate them. With him as our vice president for operations we finally succeeded in winning that first vital engagement with a major customer—Shell Oil. For ten months, our prototype (renamed *Bandit* on Steve's advice that a tough-sounding name would be more acceptable to Gulf of Mexico roughnecks) proved its worth and more than earned its keep for Shell and for our young company. Other customers followed (ten *Bandit* systems were ultimately built) and the prospects for a sound, stable source of income appeared promising—until the worldwide collapse of oil prices caused the sharp reduction in expenditures for new rig-support equipment.

By 1983, despite abundant evidence that we had a successful system that we could sell and license to others,* tough times for the offshore oil industry meant tough times for our venture as well. We kept going by economizing however we could and employing many strategies familiar to small business ventures. We were able to sell some of the stock, until then closely held by the founders and directors. I mortgaged my home to gain some working capital, and built an addition on the house to provide more office and working space. For a while, plans to build *Deep Rover* had to be shelved, but it did not stop Graham from making endless sketches and calculations that were vital to the final design.

I quickly discovered that there is nothing like fighting for survival to sharpen one's wits in dealing with businessmen who, as it turned out, usually tried to eat our lunch. So I was enormously grateful when Bob Beshar invested hundreds of hours discussing strategies and educating me concerning the fine art of negotiation.

My subsequent appearance at meetings throughout the 1980s in otherwise all-male sessions involving executives of oil companies, diving operations, and equipment manufacturers often proved advantageous because no one took me—a woman, a scientist, a Ph.D., and a mother!—very seriously. Underestimating my ability to follow the winks and nods, as well as the numbers being tossed about, these men sometimes proceeded with negotiations as if I were not even in the room. It was easy to delay signing contracts until there was time for "consultation with other directors," especially Beshar, and return to the table with revisions that clearly indicated that we were not pushovers. I felt that I had begun to earn my spurs as a businesswoman when our chairman, Sir John, began referring to me as "our secret weapon" and "the titanium dwarf." (Titanium is the metal of choice for most deep-diving submersibles because it is strong, light, and able to withstand great pressure. I felt I had graduated to a new plane when

---

* Tenneco agents told us that the *Bandit* once saved them more than half a million dollars with two hours of dexterous work that cost them less than $3,000.

my name was changed to "the ceramic dwarf," when Graham began considering the properties of new ceramic materials for the pressure hull of submersibles to go the ocean's greatest depths.)

In due course, we leased critical work space from a friend who had an on-site machine shop, and instead of renting posh offices, Deep Ocean Technology economized by gradually taking over every square foot of space at my home. My children did their homework and ate meals amid piles of blueprints and engineering discussions. My mother, who often visited following the death of my beloved father in 1981, became a popular member of the "Deep Ocean" team with her swift sense of humor and freshly baked cookies. Even the naming of a new black Labrador retriever puppy, Blue Prints, reflected the spirit of the time. As someone observed, "A lot of people have an office in their home; you have your home in your office."

Our fortunes began to change in 1985, when Graham conceived of a new vehicle—small, portable, and easy to operate—that could accomplish much of the work of the large ROVs then in vogue, but for a fraction of the cost. Collaborating with what proved to be a durable and highly creative team of engineers, David Jeffrey, Dirk Rosen, and Philip Ballou, a system was completed from first sketch to working prototype in exactly three weeks—just in time to be exhibited at a trade show in San Diego. Developed under the auspices of a second company, Deep Ocean Engineering, and dubbed *Phantom* because the operator's "spirit" was projected via the vehicle to underwater destinations, the little ROV enjoyed instant success. One of the first went to work far from offshore oil fields, however. It was acquired by a friend, Kym Murphy, originator and manager of Disneyworld's Living Seas Pavilion, to greet visitors to the pavilion's huge aquarium with blinking lights and real-time images of the fish and fish-watchers.

In 1986 the prototype *Phantom* made history as the first submersible to dive under the ice in the Antarctic. We had sent it in the care of Phil Ballou on a scientific goodwill mission to explore the nature of the dry valley lakes near McMurdo Sound. It was the first of many subsequent research expeditions where a *Phantom* proved to be a valuable research tool for tasks such as locating and exploring the nature of ancient shipwrecks offshore from the coast of Florida, or bringing back the first video images of the depths of Crater Lake, Oregon, a deep, volcanic caldera lake said to have the clearest natural water in the world. *Phantoms* were also adopted as professional film platforms by the British Broadcasting Company (who insisted that their little sub be camouflaged with silver and blue paint to vaguely resemble a mackerel) as well as other television networks and film producers.

As our track record grew, so did the list of customers; the versatile little vehicles were snapped up for uses in more than thirty countries. Companies selling their services to the offshore oil and gas industry bought them, and others found them useful for search, survey, and salvage. Police departments put them to work recovering evidence of accidents and crime from deep water disposal sites, including now and then a body and stashes of drugs. Nine navies acquired *Phantoms* for mine countermeasures, hull inspection, and numerous other ingenious applications. Perhaps the most extraordinary use to date was devised by scientists at the NASA Ames Research Center engaged in the study of Mars. In the spring of 1993, a NASA scientist in California sat at the "controls" of a *Phantom* remotely operated vehicle that was cruising around under the ice in Antarctica. Via laser, microwave and satellite connections, signals were beamed from the *Phantom* to the operator-scientist, who was wearing a virtual-reality-style helmet supplied with small video screens. He could not only see in real time what cameras on *Phantom* were recording, but also position the camera by turning his head, with instant response on the vehicle, thousands of miles away. "Reality-reality"—not synthesized images—were delivered to the pilot, and encouraging progress made toward a useful approach to both underwater and space exploration.

John Edwards, a natural mechanic who has personally supervised the construction of all but one *Phantom*, delights in their varied exploits. He can account for at least as many uses in the past nine years as there are systems, now more than three hundred, and it seems unlikely that the list will end anytime soon. *Phantom*, in more than fifteen iterations, has accomplished precisely what we hoped it would: improved human access underwater, thereby making a difference in the way our species views the world, and our place within it. It has also provided the basis for a profitable venture, one that has enabled us to support ourselves while keeping alive the dream of developing manned—and "womanned"—access throughout full ocean depth.

The solution to the problem of how to fund the back-of-a-napkin concept for a submarine—the one that caused Graham and me to consider starting a company in the first place—came from an old friend who shares that dream, Phil Nuytten. Phil did not have to be convinced of the merits of the *Deep Rover* design. For years, he and Graham had brainstormed the "next steps" in submersible development. Phil wanted to build a lighter, more flexible version of *Jim*, a "swimming *Jim*,"* but he was also intrigued with the *Deep Rover* design and thought he could make a convincing case for funding a commercially oriented system to work in depths to 3,000 feet.

In June 1983 we signed a contract with Phil's Canadian company, Can-Dive Services, to build one *Deep Rover*. Funds were provided by a grant to Can-Dive from the government of Newfoundland and were further secured by a contract from the Canadian oil company Petro-Canada, who would have use of the sub once it was built. We thought a second "buddy sub" might eventually follow.

The *Deep Rover* launching ceremony at the edge of Halifax Harbor in June 1984 was an occasion for celebration that Deep Ocean Engineering, Can-Dive, and the local and regional government officials and citizens took quite seriously. As the ceremony progressed, one person, Graham, was conspicuously absent—and so was the "honored guest," Deep *Rover*. Just as murmuring questions began—"Well, where *is* this machine?"—*Deep Rover* burst into view from its submerged hiding place a few feet offshore, elegantly piloted by Graham, who, with James Bond 007 aplomb, was attired in an immaculate tuxedo. It was my pleasure, then, to crack a bottle of champagne across *Deep Rover*'s "bow," with Graham still inside, giving the "reverse launch" a splashy grand finale.

During the following year, *Deep Rover* was used first for various industrial applications and a highly successful series of research dives near Monterey, California. But by mid-1985 no task had taken it to its maximum rated depth, 3,300 feet (1,000 meters), and there were no plans to do so.

Running a start-up business and a full household, plus maintaining an active scientific career left me with little time for eating and sleeping, let alone

---

* He subsequently succeeded and launched the "Newt Suit," an ingenious futuristic "hard suit" used for diving to 1,000 feet.

dreaming up expeditions for small submersibles, but the vision of actually doing what we had talked about, using *Deep Rover* for exploration and research, spurred several months of after-hours fund-raising calls and proposals. To demonstrate *Deep Rover's* maximum-depth capability and use the system ourselves, it was necessary to raise support needed to hire it back from Can-Dive and pay for a ship and other operational costs. We proposed using the sub to explore the Coronado Canyon and the deep slope offshore from San Clemente Island and to compare the direct, on-site approach with the use of standard net sampling techniques. Also, we wanted to conduct the first open-water trials for the sub to the maximum allowable depth. Such dives would be significantly deeper than anyone had ventured solo before, though far from our ultimate goal of seven miles (11,000 meters).

Fortunately, the project caught the imagination of Dr. William Evans, then director of Hubbs Sea World Marine Research Center in San Diego. The center served as our coordinating base and Evans made available a boat to be used for the duration of the mission. Several small grants ranging in size from $1,500 to $15,000 were compiled from various sources, including one check from a sponsor in exchange for a ride in the sub. Rolex U.S.A. helped in the spirit of exploration—and enterprise—and we, in gratitude, agreed to strap four of their chronometers on *Deep Rover's* mechanical arm for deep-sea testing. *Discover* magazine signed on as a sponsor and sent a reporter to document the expedition's findings. A small band of volunteers agreed to serve as working divers, vital for launch and recovery operations. Everything was in order. Then, the owner of the support ship we had engaged backed out in favor of another, higher-paying job. That's when I learned about *Egabrag III*.

The good news was that *Egabrag III* appeared to be more than adequate for the venture, and the price was less than for the original ship. Also, she was in San Diego and available. The bad news was that the reason *Egabrag III* was in port was because the captain planned to get married during the time we hoped to sail. Some artful juggling on both sides, plus mutual curiosity about what *Deep Rover* could do, led to a honeymoon and champagne at sea—with the captain, his bride, and the *Deep Rover* crew celebrating together.

I was not the first to take *Deep Rover* to 3,300 feet. As the sub's designer, Graham felt obliged to do the first deep checkout dive himself, to demonstrate that the sub could and would meet its specifications. He had personally tested the prototypes of his previous designs, now including a fleet of forty or so small submersibles in use worldwide, and had amassed thousands of successful underwater hours, but always with a tether connecting the systems to a support craft above, and never in depths greater than 2,300 feet (710 meters). Graham's first dive in *Deep Rover* was his first dive in an untethered system at sea, and would be the deepest anyone had ever descended, solo—and returned. More than anyone, Graham knew the many subtle things that individually or in combination could cause trouble, things unrelated to basic design or construction, but more insidious for all that: a flake of rust jamming an air valve, a few drops of saltwater landing, unnoticed, on a critical connector, an undetected structural weakness . . .

Nonetheless, he appeared confident and cheerful as the two halves of *Deep Rover*'s clamshell pressure hull closed, and he and his sub-suit were lifted into a dark sea over a steep dropoff bordering San Clemente Island. As soon as the last flicker of light from *Deep Rover* disappeared from sight, I ran to the ship's bridge to help monitor communications between Graham and the surface crew.

I had packed along my treasured copy of William Beebe's *Half Mile Down*, and we pored over what *he* had found half a century earlier in the very depth Graham was now penetrating:

> The only other place comparable to these marvelous nether regions, must surely be naked space itself, out far beyond atmosphere, between the stars . . . where the blackness of space, the shining planets, comets, suns, and stars must really be closely akin to the world of life as it appears to the eyes of an awed human being, in the open sea half a mile down.

More occupied with the sub's performance than the grand view before him, Graham responded patiently to the questions I fired into the through-water communications microphone. I tried to show restraint, mindful that he was trying to concentrate on the serious job at hand, especially when he passed into aquatic space deeper than anyone had traveled solo before. But for the record, as well as to satisfy my bursting curiosity, I couldn't resist asking from time to time "What do you see? Have you encountered any squids? Any sign of a deep scattering layer?"

As his replies came back, I made notes, logging time and depth. Kathy Sullivan, oceanographer, skywalking astronaut, and longtime friend, monitored his progress on an electronic tracking device. As Graham neared the maximum depth, people crowded onto the bridge and clustered around the communication desk. We listened intently and cheered when Graham confirmed that he had reached 3,000 feet. Lee McEaren, a news reporter for Channel 4 in San Francisco, was the first to extend congratulations. "Graham," he said, "now that you have reached your goal, and you've gone deeper, alone, than anyone ever has gone before, tell us, what does this mean for you?"

Kathy groaned and shook her head. Phil Sharkey, head of the diving volunteers, struck his head with his fist in mock dismay. Lee motioned for us to hush and listen. There was a pause and Graham responded in a quiet, clear voice:

"Well, Lee, it means . . . I got my sums right."

Right. I had no doubts, but it was nice to hear the one who had made the calculations say so, especially since it was my turn next.

Spare batteries were waiting to be changed out after Graham and *Deep Rover* returned to the surface. A few hours later and a short distance farther down the coast near San Clemente Island, I too went to *Deep Rover*'s maximum rated depth, paying minimum heed necessary for safety to dials and gauges, and giving maximum attention to the action outside. Our intention was to make a series of deep dives to demonstrate forcefully that in *Deep Rover*, going to 3,300 feet should be as safe and easy as going to 330 feet.

The first set of batteries were charged up during my excursion and were ready to be swapped out when I returned, to enable the next person, Phil Nuytten, to enjoy a full dive. Phil flew in by helicopter just in time to be briefed and make his descent. Phil landed *Deep Rover* along a steep slope and followed the terrain downward. As he did, he reported encounters with ominous-looking bales of wire, lengths of cable, and what appeared to be intact munitions. I had observed some junk—cans and wire, mostly—but Phil apparently had happened on a major dumpsite. Concern about entanglement or unintentionally triggering something explosive caused him to return well before the projected six-hour dive time was up.

Rough weather brewing and a desire not to find another dive site like Phil's prompted a move inshore for several days of shallow dives. Finally, the seas calmed and we could return to a deep-water location to implement plans for my second dive, and the beginning of our research, which would compare conventional oceanographic methods, and those made possible by *Deep Rover*. Scientists aboard a Hubbs research vessel would drag a net in the midwater area at about 1,500 feet to sample life—a conventional oceanographic method—while I observed from my most unconventional perch below. Later, we would compare results. Dr. Bill Evans stayed aboard *Egabrag III* to communicate both with the Hubbs boat operating at the surface, and with me in the sub below.

We moved into position, about 12 miles offshore San Diego, above a place along the Coronado Escarpment indicated on an up-to-date navigation chart as a steady, steep slope into deep water. I planned to hover for a while at 1,500 feet, drop down to 2,000 feet, then follow the bottom to the sub's maximum depth. As I neared 1,000 feet, Bill's voice crackled over the small speaker inside the sub. "Sylvia! We have visitors! The sea is alive with dolphins! Spinners! Hundreds of them. Can you see them?"

I leaned into the darkness, hoping to glimpse a slim, black-and-white form streaking by or a dolphin's smiling face peering at me from the sparkling blackness. But, no. Small lanternfish, flashing silver; a translucent reddish jelly, pulsing gently; a scattering of iridescent blue copepods . . . but no dolphins. I switched off the lights then, thinking that I might at least see luminous turbulence that would mark their presence. Again, I could not see dolphins—but suddenly I could hear them! The audible rhythm of a dancing city with fluid boundaries bathed the sea, its moving symphonic edge marked with soft chirps, wheeps, and staccato probing.

"We have some other visitors here, too," Bill broke in. "Looks like a Soviet ship, coming in close." I wasn't concerned about a Soviet presence, although at the time, November 1985, their proximity to a major U.S. Naval facility in San Diego was not welcome. The ship had no identifying markers, but with binoculars, observers on *Egabrag III* could make out the faint outline of a hammer and sickle, painted over with white, on the ship's stack. The ship had come in fast, refused to acknowledge radio contact, and hovered nearby, evidently tuning in to the conversations between *Egabrag III* and me. They came close enough to smile and wave at *Egabrag III*'s crew—and to receive uneasy smiles in return.

As I continued my descent I wondered what Bill might be catching in the nets. Reaching overhead, I nudged a lever to adjust the sub's buoyancy and in a moment *Deep Rover* hovered in the water, a transparent planktonic creature in its own right, like many of the small, naturally buoyant animals I had come to see. So far, I had encountered thousands of tiny beings, mostly too small to identify clearly, but now, suspended weightless midwater, I could focus on individuals: a miniature, speckled octopus clinging precariously to *Deep Rover*'s sphere, a wary red squid darting into view, then a shimmering fish, and dozens of minute jewel-like jellies, edged with glittering threads. So far, though, I had found no well-defined layer of life. With a gentle touch of downward thrust, I continued down slowly, occasionally turning on the outside lights to identify creatures glimpsed first as blue flashes. More action was concentrated close to the bottom, where several small fish darted through the light beam, and dozens of red shrimp skated majestically by, barely twitching their long, graceful antennae.

According to the navigation charts of the area, the landing place ahead should continue to slope gradually downward to depths about three times as great as *Deep Rover* could safely travel. *So much for the charts*, I thought. In the small area illuminated by the sub's lights, I could see a level area bordered by a sharp drop, *not* a gradual slope. The acoustic methods used to map the area had averaged the readings, "connecting the acoustic dots" to produce an image of smooth transition from shallow to deep.

Gently I eased my arms forward, an instinctive gesture that activated the sub's thrusters. I wanted to get a better look at a strange object on the sea floor ahead. It seemed to flash when the light beam crossed it. I thought of certain deep-sea fish with large, luminous eyes. Had I encountered one of these? The mysterious creature looked reddish—not unexpected for deep-sea animals. Closer, now. Many fish in the deep sea are slow moving; others are stunned by the sub's lights and hold their position in an almost trancelike state. This whatever-it-was did not move, but I did not want to cause undue alarm or startle it into a fast escape.

I leaned forward, pushing the manipulator control as I did so. One of the sub's arms responded, reaching toward the still motionless, red, sparkling object, nestled on a soft gray-brown blanket of silt. At last I could identify the mysterious creature. A soda can, RC Cola.

I should not have been surprised. The sea has been used as a convenient place to dump trash throughout our history. Debris from human activity litters the coastal areas of the world, but most of what goes into the sea remains unseen by those who put it there. Rather, I probably should have been surprised that in the course of my brief visit to the Coronado Escarpment I saw only a dozen or so cans, several bottles, a metal box, a spool of heavy wire, a rusting crate, and what appeared to be remnants of a large stove. Although to be expected, perhaps, close to a well-populated area, the junk on the sea floor was a jarring reminder that there is no true wilderness anymore, no place unaffected by the not-so-gentle touch of humankind.

Chapter 9

OCEAN EVEREST

We shall not cease from exploration
And the end of all our exploring
Will be to arrive where we started
And know the place for the first time.

T. S. *Eliot,* Four Quartets

M ountain climbers justify their passion for ascending to the highest place above sea level, the top of Mount Everest, with the explanation "because it's there." For some, the same rationale is adequate to explain why they want to descend into the deepest ocean, the dark, cold realm between 20,000 and 35,800 feet, into depths far beyond the reach of most submersibles. Don Walsh, pilot of the bathyscaphe *Trieste* during its once-and-nevermore descent to the deepest part of the ocean in 1960 suggests that those enamored with deep-diving "adventures for the record books" have "missed the point." But what, then, is the point? What motives can justify development of the ways and means to gain convenient, repeated, effective working access throughout the sea, top to the bottom, worldwide? Why isn't current capability good enough?

Those questions—and lots of inspired answers—dominated conversation over a dinner that Graham and I hosted for a dozen friends at home in Oakland in the spring of 1988. Mutual admiration and a shared vision of ocean exploration had blossomed into a desire to do more than work together as professional colleagues, and two years earlier, with romantic flourish, Graham and I had married.* Gradually, the business moved to a proper office facility in nearby San Leandro, but the Skyline Boulevard residence remained the principal design office—and headquarters for brainstorming sessions.

That night, Bruce Robison, veteran deep-sea biologist, homed in on some of the critical issues of deep-ocean exploration. "A lot of people think that since *Alvin*† can reach the average depth of the ocean, most of the deep-sea floor is within our reach. Actually, nearly half of the ocean's total *range* of depth is below twenty thousand feet. And, while some open-sea animals occur throughout the ocean, knowledge of those living below twelve thousand feet is profoundly poor. And that's not all," he grumbled. "Safety regulations prohibit use of *Alvin* for mid-ocean observations over depths greater than twelve thousand feet, so the majority of the water column—every cupful occupied by life of some sort—is off-limits and remains unexplored."

I chimed in with a story about the reaction of a distinguished Naval oceanographer to my repeated earnest questions about why the U.S. Navy has not continued its once vigorous commitment to vehicle development for deep—really deep—ocean research.

"Well," came the reply. "You're only talking about two or three percent of the ocean that is out of the reach of present submersibles. Several can go to nearly twenty thousand feet—and if all of them worked full-time day and night for the next several centuries, there would still be a lot of ocean that no one had yet seen. When so much remains to be done in the ninety-seven percent currently within range, why bother with three percent? After all, if it costs fifty million to build a submersible to go to six thousand meters, it will probably take twice that

---

* In 1989 we separated domestically, but the friendship and shared vision endure.

† The most active submersible ever built, in continuous operation by Woods Hole Oceanographic Institution since 1964.

amount to go twice as deep. It just isn't worth it for only three percent of the ocean." His seductive logic probably had won the hearts of certain accountants and administrators who thought they knew how to deal with questions such as mine.

I laughed and said that I knew several line engineers who were convinced that half his proposed cost estimate for a 20,000-foot system would be sufficient to build manned submersibles for full ocean access—with funds to spare—if new materials and creative engineering were applied. Still smiling, I murmured, "Let's see, that's only about four million square miles, a little bitty part of the planet about the size of the United States or Australia or China, that no one can get to because it's so far from the top of the ocean—three and a half to seven miles. Right?"

He nodded, looking a little more thoughtful, a little less amused. We were standing outside a government office building in Washington, D.C., a pale full moon clearly evident against a blue summer sky. I glanced at that distant sphere, 240,000 miles from the top of the ocean—in the other direction—shook my head, and said no more.

Around the dining-room table, my friends chewed on the "it's-only-three-percent" rationale. We tended to agree that the 97-percenters had a point . . .

- If the ocean everywhere were the same. (It isn't, of course, especially with increasing depth.)
- If the area involved were insignificant. (It isn't, unless an aquatic chunk of real estate roughly the size of the U.S. can be regarded as trivial.)
- If the unknown, unreachable area in question were not inextricably tied to the planet's other biological, physical, and chemical processes. (Of course it is. Some of the most interesting aspects of plate tectonics relate to the nature of the bottom-of-the-trenches subduction zones and some of the most interesting questions concerning the origin and adaptability of life relate to the nature of organisms—in the unique high-pressure environment of the deep sea.)
- If it could be absolutely guaranteed that the three percent would remain undisturbed for the foreseeable future. (It will not. Chemical

pollutants are distributed pole to pole, over land and sea, and into the depths below, and proposals repeatedly are made to dump high-level toxic wastes into the trenches as the "most remote disposal sites on earth.")

- If no other nation had plans for developing underwater vehicles to gain access to full ocean depth. (Several do—Japan, France, and Russia—and others will follow.)

Returning to the cost question as a reason given by some for *not* going ultradeep, Graham emphasized new materials—composites, ceramics, and special kinds of glass—all with the potential for making possible the production of deep submersibles for substantially less money than the cost of existing steel or titanium vehicles. "The real key is to build small or modular systems," he pointed out—a familiar Hawkes theme.

"So what kind of sub would *you* build to go to the deepest sea?" Lynne Carter teased. Lynne, a marine policy expert and a red-haired dynamo, has pushed hard for the U.S. to build deep ocean vehicles. She knew as well as anyone who had been in the vicinity of Graham or me for the past four years that we had embarked on what some regarded as a wild and crazy dream, to build two sleek one-person submersibles that we had dubbed *Deep Flight*. The first two, "his and hers," would be able to go about as deep as *Deep Rover*, and would prove the concept of underwater "flight," free movement in all directions. The next step would be to construct two or more one-person units that would provide effective, including cost-effective, working access to full ocean depth. Unlike other research submersibles, which behave like underwater dirigibles, descending heavy, then dropping weights or displacing water ballast to become buoyant for the return trip, *Deep Flight* would always be buoyant and require power to move—up, down, or horizontally, like a small, fixed-wing aircraft. *Deep Flight* would go about ten times as fast as most research subs, and, depending on the batteries available, the range could be twenty miles or more. Its maneuverability would even let me do barrel rolls with the whales!

Development of *Deep Flight* had been painfully slow. The first sketch that resembled the now half-finished subs appeared in 1984, soon after *Deep Rover* was launched, but we could not find a financial backer for the *Deep Flight* concept, so finally, in 1987, both of us opted to sell some of our hard-won Deep Ocean Engineering shares with the idea that we would take a leave of absence from the company for six months and work full-time on the new subs. In fact, we simply doubled up, working full-time for Deep Ocean Engineering without salary while squeezing in additional hours for *Deep Flight* as "recreation." Some vital materials were donated, and a growing number of enthusiastic volunteers made real progress possible.

Not long before the dinner-table discussion in 1988, we had had a major boost in backing for *Deep Flight* when IMAX films agreed to help support the construction costs in exchange for an opportunity to use the systems for underwater filming.* It seemed to be a clear win-win agreement, and the new resources had resulted in significant progress, but much remained to be done before the units would be ready for their first bath in saltwater.

Perhaps the most frustrating issue that we dissected during that memorable "Ocean Everest" evening concerned the question I had posed to my naval oceanographer friend: Why had this country abandoned its valiant engineering and research efforts to operate effectively in the deepest parts of the sea, which had culminated in the famous *Trieste* dive in 1960?

"Just suppose," I suggested, "that after Sir Edmund Hillary and Tenzing Norgay returned from their ascent to the top of Mount Everest, it was generally agreed that no one would ever have to go back, that one brief visit would be sufficient to satisfy the curiosity of humankind for all time." In fact, more than a hundred expeditions have reached the summit of Everest since the first ascent in 1953. Meanwhile, no one has returned to the subsea counterpart—to the deepest sea, to "Ocean Everest."

Dr. Andy Rechnitzer, a marine scientist and vital member of the *Trieste* team, has suggested that the retreat from ultradeep excursions was deliberate. Having demonstrated that the U.S. Navy *could* go to the ocean's maximum depth if necessary, the focus of effort was then directed to what many regarded as more useful, practical efforts above 20,000 feet. In fact, although *Trieste* was later used

---

* Later, similar agreements were made with National Geographic Television and TV New Zealand.

for various undersea operations, she was never again used to explore deep ocean trenches.

A few intrepid seagoing scientists have tried taking samples from the bottom of these deep-sea canyons while perched on the deck of a ship miles above. One of these is physical oceanographer Willard Bascomb, who explains in *Crest of the Wave* some of the impressions he garnered while working near Tonga some years ago:

> The Tonga Trench is 15 to 30 miles wide and 7 miles deep, with a steep walled gorge . . . in its bottom. As the seismic soundings produced evidence of this great chasm beneath us, we sought words to express it properly: "a mile deeper than Mount Everest is high," "as deep as seven Grand Canyons but with much steeper sides," "higher than a stack of 30 Empire State Buildings." None of these seemed adequate to explain this heroic crack in the crust. . . .

To sample this "heroic crack," a six-foot-long tube, weighted with lead, a gravity corer, was lowered over the side; when pulled up it contained a small chunk of basalt. If similar techniques were applied—blind, randomly taken chips of rock—to explore the Grand Canyon or Wall Street, what would be known about those places?

Fortuitously snared samples and indirectly acquired environmental data have been immensely helpful in providing the little information that we now possess about the nature of the deepest areas in the ocean. Those involved in research based on such techniques are, in truth, deep-sea detectives solving grand puzzles from tiny fragments of information.

Enthusiasm is growing for the concept of "telepresence"—exploring and working in the oceans while sitting in an armchair in Indiana or England, pushing buttons and watching television monitors while a mechanical device responds via satellite links. Still, when I think of watching a video of Paris as compared to going there myself, or of looking at a photograph of a meal with friends rather than sitting down at the table, or of viewing through the camera "eyes" of a remotely operated vehicle whales gliding through a clear blue ocean—I am reminded that there is no completely satisfactory substitute for *being* there. To really decipher the nature of this unique part of the planet, direct access, with human eyes and brains as well as instruments is essential.

For "being there" operations in depths to 20,000 feet, there is a small global fleet of manned submersibles: the U.S. Navy's *Sea Cliff*, a French vehicle, *Nautile*, two Soviet submersibles, *Mir I* and *Mir II* (launched in 1987), and Japan's *Shinkai 6500*, the deepest-diving manned system now in operation, capable of descending to 21,450 feet (6,500 meters).

An unmanned towed array rated for 20,000 feet (6,000 meters) is frequently used by scientists from Woods Hole Oceanogrophic Institution, and there are at least two deep-tow systems capable of being operated to 23,000 feet (7,000 meters). A few remotely controlled tethered vehicles can descend and work along the tantalizing upper edge of the deep trenches, and one, the recently developed Japanese system, *Kaiko*, is designed to go all the way, to 35,550 feet (11,000 meters). In March 1994, this marvel of human ingenuity proved its capability by successfully plunging into the deepest crack in the ocean, the Marianas Trench, setting a new record for deep-diving unmanned vehicles, but not quite reaching the maximum depth achieved by *Trieste*. A few feet short of touching bottom, power was lost and the vehicle had to be brought back to the surface for repair.

To eliminate the problems of managing several miles of powered cable, battery-operated unmanned vehicles that can go to 20,000 feet, guided by computer "brains," have been developed. These systems have the advantage of being able to range over long distances unencumbered by cables—or by the need to return a pilot to the surface after a few hours—but they are not as able to respond to surprise as those that have a human operator directly in the loop, either on board or connected via a tether allowing remote control.

Japan is a nation that has always been heavily dependent on ocean resources, and has sharply stepped up the pace recently with respect to engineering and research. In less than a decade, Japan has firmly taken the global lead in deep-diving capability. I have long been aware of Japan's commitment to extracting fish, whales, and other living creatures from the sea. A hefty 44 percent of that nation's food supply comes from the ocean, compared to the global average of about 16 percent. I have also been aware for years that Japan's population has exploded, climbing to approximately 120 million, or half that of the United States, 244 million, with a land area four percent that of the U.S. More people translates to the consumption of more food and other resources, and Japan looks largely seaward for the source for both. Moreover, as a resident of earthquake-rocked California, I have become increasingly sympathetic with Japan's concerns about the consequences of tectonic activity, surrounded as it is by a series of deep ocean trenches. As the Pacific, Philippine, and North American tectonic plates underlying the Pacific Ocean slowly move and are subducted into these trenches, Japan literally feels the crunch of resulting earthquakes.

In 1990 I had an opportunity to go to Japan and visit the Japan Marine Science and Technology Center (JAMSTEC), the facility where much of that nation's *in situ* undersea engineering and research activity is focused. I had been invited to present a lecture to the staff of about 160 people and share information about using *Deep Rover* and *Phantom* from dozens of ships, rather than the method favored at JAMSTEC, where each submersible is operated from its own dedicated "mother ship." A lively discussion followed my presentation, and my listeners showed much interest in the cost advantage of being able to engage a boat, small or large, only when needed, and of using systems compact enough to be transported by plane, truck, or whatever, thus giving swift access to any body of water in the world. The cost of operating a manned submersible such as *Alvin* or a large remotely operated vehicle from a dedicated support ship is more than $25,000 per day.

I was told that the costs of highly sophisticated ship-sub arrangements in Japan exceeded $100,000 per day, and therefore was not surprised that I was quizzed very closely about the description I gave of recent dives with *Deep Rover* in Crater Lake, Oregon. There, the little sub was flown, suspended from a helicopter, to the mountainous location, more than 100 miles inland, then lowered into the water. A simple floating platform and portable winch were used for repeated launch and recovery during weeks of diving to the lake's most remote, craggy depths, as much as 1,932 feet below the surface. The same platforms were used to launch and recover a *Phantom* small enough to be carried by two people, but robust enough to fly more than 1,000 feet underwater with cameras, a mechanical arm, and a tracking mechanism to pinpoint and map observations.

After the talk, a young marine scientist from NOAA's National Underwater Research Program, Greg Stone, met me in JAMSTEC's main lobby, a brightly lit room filled with pushbutton displays showcasing animated miniature versions of submersibles and other underwater equipment. On special assignment as a resident scientist at JAMSTEC, Greg had become fluent in Japanese and was often called upon as an "ambassador" and tour guide when English-speaking visitors such as I appeared.

"Come with me and I'll show you the real thing," Greg said, leading me out the door and across an open lawn to huge workshops, laboratories, and two large research ships. I was especially intrigued with the new deep-diving submersible *Shinkai 6500*, a gleaming blue-and-white creation that, from a distance, resembles a small, chubby whale. I did not dream, then, that a year later, I would be inside the sub, descending to more than 13,000 feet into the dark canyon known as the Nankai Trough, offshore from Japan's southeastern coast.

A few months after my visit, I was invited by JAMSTEC officials to submit a proposal to dive during a special summer cruise in 1991 aboard the 330-foot research vessel, *Yokosuka*, using *Shinkai 6500*. The fact that I had meanwhile been appointed NOAA's Chief Scientist no doubt enhanced my credentials and helped overcome whatever resistance there might have been to the fact that I would be the first woman scientist and one of the few "foreigners" to dive in Japan's extraordinary new machine.

What I really wanted to do was explore without the constraint of a predetermined agenda, but it is hard to justify just "looking around" at a cost of $100,000 a day, so I diligently puzzled over what I might propose that could yield "worthy" results—*and* allow latitude for the exploratory freedom I cherished. I thought this might be a good opportunity to investigate the behavior of bioluminescent creatures, and try some new techniques to identify the sources of various sparkles, flashes, and glowing patches of blue light witnessed from deep-diving submersibles. Dr. Michael Heeb, a former NOAA marine scientist then working for the Department of Defense, suggested using light-enhancing night-vision goggles to electronically magnify their "living light," and thus spy on them using their own illumination. Such goggles had been used effectively by U.S. troops during the Persian Gulf War. Perhaps we'd be able to see and identify the fish, jelly, or shrimp responsible for the mysterious light.

Wearing goggles, I thought, I might also be able to better understand what it is like to be one of the many deep-sea creatures with eyes exquisitely tuned for amplifying light—like cats or owls, only more so. I have seen hauntingly beautiful eyes of certain deep-sea fish at the Tokyo fish market, great translucent orbs backed by silver, even in death responding to light with an otherworldly glow. Such eyes are miracles of design, making maximum use of the small amount of light available to enable animals to maneuver effectively in what appears to be utter darkness.

To try out the concept above water, in the spring of 1991 Mike arranged for us to go to Fort Belvoir, Maryland, to visit the night-vision lab and participate in an experimental white-knuckle "map of the earth" helicopter flight: The pilot would fly the craft close to the ground with no lights other than ambient starlight, amplified by the goggles.

"We have to pick a time when there is no moon," Mike told me. "The pilots and passengers, you and I, will all be wearing goggles that amplify light about thirty thousand times. Starlight provides enough illumination to make midnight look like high noon. A full moon is almost blinding."

There are times when I find myself intently doing something that until that moment has seemed to be quite reasonable; then, a curious tingling detachment settles in and I pause and wonder, *What on earth am I doing here?* Standing with Al Giddings in 100 feet of water hundreds of miles offshore from the Australian coast being circled by more than fifty curious gray reef sharks, for example. Or diving along the edge of a steep subsea "wall" in the Bahamas with television hero and friend Hugh Downs at night, with no lights, in the midst of a violent thunder-and-lightning storm. Or standing before a Senate confirmation committee charged with evaluating my qualifications to be NOAA's Chief Scientist, and earnestly trying to articulate answers to killer questions about global warming by the committee's chairman and recognized authority on the subject, Senator Al Gore. Or whizzing along at 50 miles an hour, ten feet off the ground, among dense stands of trees in the hills of Virginia, in pitch darkness . . .

In some ways it would have been better for my psyche had I *not* been able to see what the pilots could see: the needles of pine trees a few feet from our plane; a deer, white tail flashing as it leaped out of the way; and distant lights, barely visible without electronic enhancement, suddenly amplified into brilliant points with shining halos. But once under way, I shrugged, relaxed, and thoroughly enjoyed the sensation of being a creature of the night—a great owl, swooping effortlessly through darkness magically made visible.

Before leaving Fort Belvoir for Japan, I signed papers stating that I would accept responsibility for the care and return of three night-vision units, two binocular systems for general observations, and one special monocular set to be used with a Cohu low-light-level camera and portable recorder, with which I hoped to document deep-sea creatures in action.

It was past midnight when Mike dropped me off in Alexandria, Virginia, where I was staying with a friend and fellow oceanographer, Linda Glover. I strolled in wearing the goggles and handed her the second pair, and together we descended to her windowless basement study, for the first time on equal visual terms with her cat, Fulton, a notoriously nimble creature who typically gallops up and down the stairs at top speed. Now, in a convincing demonstration of the importance of night vision, to Fulton's amazement and our own, we raced her in the dark and won.

When my proposal to make a dive in *Shinkai 6500* was officially approved, I was pleased that it would include my friend Dr. Bruce Robison, whose specialty is the behavior of deep, midwater life. He advised putting red filters over two of the sub's outside lights to observe creatures who are insensitive to red illumination—unlike humans, the night-vision goggles, and my special camera. In theory, the filters would allow us to observe and document creatures adapted to the dark without dazzling them in a blaze of brilliant lights.

In Japan, aboard the support ship, *Yokosuka*, we met the expedition leader and head of scientific operations at JAMSTEC, Dr. Hiroshi Hotta, Greg Stone, and the sub's two pilots for discussions about diving protocols. Normally, *Shinkai 6500*'s lights are kept on continuously for observation and filming, but for my dive concessions were made. All the lights would be turned off during the descent so I could watch for bioluminescent life throughout the water column. On the bottom, we would scout for a while using the regular lights, then park where I could use my electronic night-vision eyes.

Having experienced some of the smallest research subs, I now had a chance to clamber down the hatch of one of the largest. *Shinkai 6500* is a sturdy but sleek 30-foot-long, 10½-foot-high, 26-ton submersible for three, bristling with cameras, sensors, and recorders and navigation, sensing, and collecting equipment. The one-person *Deep Rover*, in comparison, is 7 feet long, 7 feet high, weighs 3 tons, and typically carries very basic instrumentation. In the vastness of the ocean, however, even *Shinkai 6500* seemed tiny . . . a small, bright haven for creatures not designed by nature to venture into the planet's nether regions.

At precisely 10:00 A.M., the sunlit surface disappeared in a wash of blue, and the slow fall into darkness began. There was no sensation of movement, but with each passing minute, the surface became one hundred feet more distant.

"CD?" one of the pilots asked, pointing to a neat rack of choices.

"Sure," I said. "Why not?"

Moments later, my mind was fixed on the sight of strange jellies colliding with the sub in luminous explosions of blue, so it took a while for an insistent, surreal message from my ears to penetrate. "Hotel California!" Suddenly I recognized the music, and the pilots, waiting expectantly, beamed.

As we neared the bottom, an hour and a half later, it was necessary for the pilots to switch on "landing lights," and for a while, I concentrated on the view of the deep-sea floor at precisely 13,065 feet, two and a half miles beneath the surface.

A casual glance at the moonscape-terrain outside my small porthole seemed to confirm what I had heard from various geologists, physical oceanographers, and several entrepreneurs, that the place is generally barren, devoid of any "useful" forms of life—an attitude that has helped promote deep-sea dumping. "Why not?" the argument goes. "There's nothing but mud down there." Right. Nothing but mud and strange fish and sponges and sea cucumbers and brittle stars and amphipods and polychaete worms and other mostly small life forms, including billions of largely unappreciated microbes whose contribution to the healthy functioning of the planet remains at present unexplored. I for one would not wish to risk the consequences of massive disruptions to this enormous living system—a real possibility given the recent addition of toxic substances to the sea now found in deep-sea sediments.

Recent studies have made it clear that the deep sea is a dynamic place, home to complex ecosystems and biological diversity that rival even incredibly rich terrestrial systems, especially when broad genetic categories—phyla—unique to the ocean are balanced against land areas that are "species rich" but "phylum poor."* While the importance of life in the deep sea to everyday activities of residents of Chicago, Singapore, and other terrestrial places is at present not well defined, it is certain that there are connections. Living creatures alter their surroundings, whether with bulldozers or minute chemical interactions. Today is different from yesterday and all the tomorrows of the future, in part because of the presence of every living thing, large and small, interacting in subtle, and sometimes not-so-subtle ways with the surrounding ingredients that make life possible. Whether the sum of the changes brought about each day yield a planet that continues to be hospitable for humankind or instead favors an array of species better adapted for the modified circumstances, only time will tell. Until the role of life in the deep sea in maintaining the present (and, for us, comfortable) circumstances is better understood, it makes great good sense not to risk upsetting or destroying it.

Cruising the bottom of the Nankai Trough offshore from Japan, I admired clusters of small clamlike mollusks and white crabs concentrated around warm mineral-laden water seeping through small cracks in the sea floor, and watched while the two *Shinkai 6500* pilots maneuvered the sub to collect mud and water inhabited by deep-sea microorganisms. The samples were destined for JAMSTEC's ongoing Deep-Sea Environmental Exploration Program Science and Technology for Advanced Research (DEEP STAR), a well-funded commitment to explore the potential of biotechnology applications of new strains of bacteria and other microbes obtained. Japanese scientists are looking for ways to take advantage of

* Dr. Fred Grassle, director of the Institute of Coastal and Marine Studies at Rutgers University, has estimated species richness in what appeared to be a homogenous deep-sea region ("just mud") designated as the Eastern United States Continental Slope and Rise Study offshore from New Jersey and Delaware. In samples taken from 233 sites, 798 species in 14 phyla were obtained (not counting plankton and rock-hugging species). Intensively studied shallow areas nearby had yielded fewer species than the lightly sampled deep-sea areas.

the unique metabolic processes of deep-sea creatures for industrial and pharmaceutical applications.

During the dive, we followed a predetermined course, making observations
and measurements that fit into an ongoing survey of the area by JAMSTEC scientists, then chose a place to park and shut off all the lights except for two
that were covered with red filters. The large claw-like manipulator held in its
metallic grasp a small bouquet of mackerel offered to us by *Yokosuka*'s chef so
that we in turn might offer it to passing deep-sea creatures. Within minutes,
small white crustacea—amphipods—began circling the bait, then settled in to
leisurely enjoy the unexpected feast. Twenty minutes passed with no sign of anything larger than my thumbnail, when suddenly, a silver-gray meter-long eel
slithered into view and swam around slowly, appearing to deliberate on whether
or not to try a bite. Then another eel, not at all inhibited, barged right in and
began eating. Moments later, both eels were displaced by a much stockier fish,
a beautiful sleek creature designated by soulless scientists as a "rat tail." The aft
end of the graceful creatures so named *does* taper for about half the body length
finally into a slender, sinuous "tail," but I thought more of ribbons and silk than
furry mammals as I watched the newcomer artfully engulf a large mouthful of
silver-blue bait, an event that I was happily able to record on tape using the low-
light camera armed with its single light-enhancing "goggle."

A week later, back in San Francisco, I described my experiences aboard *Yokosuka* and in the Nankai Trough to Graham as we walked around the gleaming
white cylinders that would serve as pressure hulls for the new *Deep Flight* subs.
Our mood was philosophical. It isn't exactly appropriate to compare enthusiastic, well-supported, and nationally endorsed ocean-engineering programs in
Japan to the modest commitment evident in the United States, but some points
are relevant. According to a study made by Greg Stone, in that year, 1991, the
total Japanese budget for ocean research and development (R & D) was approximately $375 million, while the U.S. ocean R & D funds, not counting military
expenditures, amounted to $580 million. The U.S. budget included contribu-
tions from NOAA ($210 million), the National Science Foundation ($205 million), NASA ($106 million), and the Department of the Interior ($37 million),
with $22 million coming from several other sources. In the U.S., the ocean

R & D funding supports individual and institutional scientific research efforts of the sort that have helped make the U.S. a world leader in ocean science, but in recent years little has been forthcoming to further the development of new technology for civilian deep-ocean access. When I arrived at NOAA as Chief Scientist in 1990, I discovered that ambitious new programs invoking billions of dollars were under way to modernize and expand technology to observe the ocean from the sky via satellites, and from the surface, with a long-overdue fleet-modernization proposal, but virtually nothing was being considered for observing and conducting research *under* the sea.

Much of this nation's hope for underwater engineering and research rests on a small program within NOAA, the National Underwater Research Program (NURP). For many years NOAA has recommended zero funding, but so far, every year, through congressional intervention, a modest budget of $16 to $18 million has been funded. In 1993, a jarring example of the nation's priorities was given when Congress awarded NURP the unprecedented budget of $19 million not long after it gave approval for a $26 million toilet for the space shuttle.

In Japan, undersea engineering, especially *deep*-ocean engineering, is precisely the focus of strong and growing support. Their new remotely operated vehicle *Kaiko*, coupled with *Shinkai 6500* and the magnificent ship *Yokosuka*, which supports both, gives that country unique access to a major feature of the planet and the opportunity to take the lead in establishing policies concerning their fate. The cost of acquiring their unique underwater systems is estimated to be about $60 million for the *Shinkai 6500*, somewhat more for *Kaiko*, and in excess of $70 million for *Yokosuka*, now operating, with submersibles, at a cost of about $150,000 per day.

Graham was quick to notice that for an investment of less than $250 million, Japan has acquired unique access to an area roughly the size of the United States. With a small grimace, or maybe a touch of pride, Graham caressed one of *Deep Flight*'s shining acrylic domes. "Not a penny of government money has been used on these," he said, then added, "That's one of the wonderful things about this country. Here, a couple of people, with private money and a lot of help from their friends, can make something like this happen—and maybe, soon, reach the deepest oceans of the world side by side with the superpowers." He continued, "Japan is a small, island country . . . and so is England. By mastering the surface of the sea, my English ancestors were soon able to dominate and control much of the world. Japan is now master of capability for access to what lies *beneath* the surface. That doesn't necessarily lead to control, but it does give them a powerful edge." As Harlan Cleveland, astute political scientist and visionary, puts it, "The oceans have long been an unregulated highway for those with the technological prowess to travel it."

William Broad, the science writer for *The New York Times*, assessed the advantage to countries with technological leadership in the sea in a 1994 article, "Plan to Carve Up Ocean Floor Riches Nears Fruition." Fie observes, "Half the Earth is up for grabs in the culmination of a long-running dispute between rich nations and poor ones. At issue is how the high seas will be developed in the decades and centuries to come, most especially their depths."

Until recently, the high seas have been regarded as the common property of mankind, a limitless territory where "anything goes." The attitude gradually crystallized into the notion "freedom of the seas," a legal concept that evolved during the age of imperialism for practical, political reasons, primarily concerning movement on the surface of the sea; it was later applied to fishing and other resource use.

In the decade of the seventies the United Nations Law of the Sea Treaty was drafted to set guidelines for the behavior of nations with respect to the oceans. Japan signed the treaty sooner than most countries, perhaps because the development of a stable, legal framework may well prove to be favorable to that country's clearly ambitious plans for expanding its substantial ocean interests. Despite the high costs involved with recovering minerals from the deep sea, according to William Broad, "Countries like Japan, which has few mineral deposits of its own, are seen by experts as possibly wanting to mine the oceans as a way to help achieve greater resource independence, even if such labors initially had poor commercial returns." The United States and most other industrialized powers opposed the treaty early on, primarily because of provisions concerning the deep-sea bed, which the treaty proclaimed to be "the common heritage of mankind." The treaty's framers foresaw that this vast area would be developed under the authority of the United Nations in a manner that would require industrialized countries to share their technology—and claims—with other nations.

U.S. industry has viewed such provisions as unfair, a "sure way to kill interest in investing in ocean mining technology," according to an engineer from Lockheed's Ocean Minerals Division. Another ocean-mining expert regards the terms of the treaty as disastrous for U.S. interests. He said it's simply a matter of "Give me half of what you've got. Tell me everything you know."

In the 1980s, U.S. corporations were encouraged by the Reagan administration to ignore the proposed treaty and mine freely in accordance with U.S. law. Four American consortia were formed to stake out claims and mine an area of the eastern Pacific between Hawaii and California. The focus was on manganese nodules, lumps of rock rich in nickel, copper, iron, cobalt, and, of course, manganese. First discovered during the famous *Challenger* expedition in the 1870s, the potato-sized nodules are enormously abundant in some areas of the deep sea below 16,500 feet depth, carpeting the sea floor for thousands of square miles. The process by which they are formed is not fully understood, but it is thought that bacteria are involved in a gradual deposition of minerals, usually resulting in concentric rings of metal ore formed around an organic core—a bit of bone, a shark's tooth, or a squid beak. To develop a lump the size of a walnut is estimated to take more than a million years. While there are billions of them, and more may be developed in the next few million years, for all practical purposes, they constitute a "nonrenewable" resource.

A group of seventy-seven countries favoring the Law of the Sea provisions for U.N. governance of mining the deep-sea bed formed a coalition known as the "Group of 77." They vigorously opposed the U.S. policies and took the position that the claims by the U.S. consortia were illegal. Still, those claims and more by other nations, including some preregistered with the United Nations, represent precedents that will be considered as mining the deep sea develops.

In 1979, marine scientist Bob Wicklund, then serving on the staff of Senator Lowell Weicker, called me to describe provisions in proposed U.S. deep-sea mining legislation. When he finished, I remarked that it appeared that the bill as drafted for the Senate was a fine formula for carving up the goose that laid the golden eggs.

"Has anybody thought about the potential importance of the deep-sea environment for anything *other* than the nodules?" I asked. "The single-minded focus on minerals without regard for whatever else might be there brings to mind aliens with a taste for cement not seeing anything else of value in New York or Chicago."

"What do you suggest?" he asked. "The bill will certainly pass, but maybe something can be added about protecting the living systems. I'm sure you know that most people either think there's nothing there but mud and rocks, or if they are aware of deep-sea life, don't think it matters—or don't care."

Soon thereafter, Wicklund and I met in Washington and in a one-afternoon session drafted wording that would require establishing "stable reference areas," regions that would be designated as protected sites to be kept free of mining activity, and monitored as experimental "controls" against which changes brought about in other areas might be assessed. The idea was inspired by a resolution concerning protection for deep-sea areas that I helped draft during a meeting of the International Union for Conservation of Nature (IUCN), held in Ashkabad, USSR, in 1979. Most resolutions at international meetings are forgotten soon after they are issued, but the spirit of this one took on form and substance in the U.S. deep-sea mining bill that ultimately became law.

Deep-sea mining is an inevitable development. On November 16, 1993, South America's Guyana became the sixtieth country to sign the Law of the Sea Treaty, triggering the minimum number needed for the United Nations Convention on the Law of the Sea to go into effect and become binding on all signatories.

Other aspects of the use of the open ocean besides mining will be influenced when the treaty goes into effect. Some policies have already been adopted, such as the provision in the 1982 United Nations Convention on the Law of the Sea recognizing that coastal states have jurisdiction over the waters within a zone of 200 nautical miles (230 miles), which is their exclusive economic zone (EEZ). At the same time, the policy of free access and use of the remaining "high seas" was reaffirmed. Although in theory, there are limits—based on respecting others' interests—on the freedom to use the high seas for navigation, overflight, scientific research, laying submarine cables, construction of artificial islands, and, perhaps most significant, taking fish and other living resources, in practice, the frontier mentality (i.e., anything goes) is very much in evidence, especially with respect to fishing.

In the decade since the 1982 convention, advances in fishing technology, from the deployment of thousands of miles of lightweight, inexpensive drift nets to the use of sophisticated sonar and even satellite observation techniques to locate populations of fish and squid, have led to swift and devastating reductions in what once seemed to be "limitless" populations. As creatures become depleted in one area, fishing effort has been shifted to other species and to previously untouched places, including the deep sea and parts of the Arctic and Antarctic that were until recently true ocean wilderness. In 1992 the United Nations imposed an unprecedented constraint on the "freedom to fish" concept by banning the use of large drift nets on the high seas, largely because of the outrage generated over indiscriminate killing of incidentally caught victims—birds, turtles, whales, dolphins, seals, and numerous nontarget kinds of fish.

"Freedom of the seas" historically has also included the freedom to pollute, a global policy that follows from the notions that the ocean has a limitless capacity to assimilate even the most toxic wastes and, through dilution, render them "safe," or that toxins can be placed far enough away from human activity to be harmless. Such concepts have been shown to be dangerously misleading. Pollutants work their way through the system in due course and some, notably heavy metals such as mercury, pose significant direct risks for human consumers, not to mention other organisms in the food chain.

Many are questioning the ability of the Law of the Sea Treaty in its present form to address successfully the complex and rapidly changing issues of our time. Especially worrisome are policies that will bring about swift changes in parts of the sea that have not yet been seen, let alone studied with care or evaluated in terms of their highest, best use for the future of mankind. It may well be that deep-sea ecosystems, including vast fields of manganese nodules, serve mankind's interest best by remaining exactly where they are, continuing to provide a part of the basic framework of living planetary systems upon which our species is utterly dependent.

One thing is sure: "freedom of the seas" is no longer an acceptable doctrine. Freedom works and everyone wins if everyone takes care, but clearly, all lose if even one misbehaves. The concerns were recently summed up by atmospheric scientist Dr. Robert White, President of the National Academy of Engineering and the first administrator of NOAA. Recognizing the dangers to those systems—ocean, atmosphere, space—that everyone uses but nobody owns, he observed,

What is at stake in all of this is the fate of the global commons. We are all dependent on maintaining the habitability of the planet. . . . This is the quintessential challenge for mankind in the next century.

Part II

~~~~~~~~~~~~~~~~~~~~~~~~~~~~~~

PARADISE LOST

*Nobody knows, for sure, why the dinosaurs became extinct, but it
is known, for sure, that the dinosaurs are not to blame.*

*John Knauss, Administrator of NOAA,
from a speech given at the Coastal Zone '90
conference in Long Beach, California*

Chapter 10

NO FREE LUNCH

The farmer has his rent to pay.
Haul, you joskins, haul.
And seed to buy, I've heard him say.
Haul, you joskins, haul.

But we who plough the North Sea deep
Though never sowing, always reap,
The harvest which to all is free . . .
Haul, you joskins, haul.

        Old Yarmouth sea chanty

A deft slice with a freshly honed knife exposed dark red flesh and pale bone near the tail of a struggling silver and blue fish, a young amberjack that had been transported alive to the world's largest fish market, Tsukiji, in Tokyo and was one of more than a million sea creatures offered for sale there each day. With surgical precision, the shopkeeper rammed a long, needlelike skewer down the fish's spinal column, causing instant, quivering paralysis, but slow death. Freshness brings premium prices, and this mode of killing ensured that the fish's eyes would remain bright and clear, the flesh less bruised than if it were allowed to flop about.

I watched as another fish was netted from a large tank shared with a dozen others of its kind, huddled in one corner, each destined to be kept alive until the last possible moment before being sold to a wholesaler, retailer, or individual shopper. Each year, nearly a million *tons* of seafood are sold at this market alone. With amazing skill and detachment, the shopkeeper continued using swift, rhythmic motions, barely glancing at the fish as he skewered them, his eyes fixed on an equally active man exchanging rapid-fire dialogue with him from an adjacent stall.

Sidestepping quickly to miss being squashed by a shark-laden cart pulled by a muscular young man in a hurry, I then artfully dodged a motor-powered vehicle so packed with a tower of crates that the driver—and his view—were virtually obscured.

Thus I discovered that it is impossible to stroll leisurely among the hundreds of stalls within the huge, covered structures that make up the market. The pace at Tsukiji is laced with urgency, everyone moving at a blurring fast-forward speed. Pausing in one place for long means getting crushed in the steady, swift flow of people and carts, all intent on finding and distributing tons of live or freshly killed and occasionally frozen fish, crabs, lobsters, shrimp, krill, prawns, mantis shrimp, octopus, squid, sea hares, clams, cockles, conchs, oysters, whelks, winkles, abalone, limpets, barnacles, sea squirts, urchins, starfish, jellyfish, roe (at least seventeen kinds), seaweed, and even dolphins, sometimes whales, and much much more.

Collected from all over the world, brilliant red and iridescent silver fish from the deep sea lay in sparkling ice-lined crates side by side with artistically arranged rows of tropical green-blue parrotfish, slender yellow trumpetfish, spiny box fish, mottled brown filefish, pale orange sea robins, small speckled flounder, spotted eels, grunts, grouper, wrasse, and rays. Stiff, inside-out octopuses filled bins with bouquets of pale, arched arms, while fresh, silk-skinned squid, eyes staring, seemed still contemplating distant seas. If I squinted slightly, the rows of boxes filled with ocean life merged into precise geometric patches of gold, silver, violet, red, and blue, stacks of collapsed rainbows, an orderly avalanche of fish as colorful as cut flowers.

The driver of the taxi that brought me to the market was amused but nervous. When he discovered at 3:30 A.M. that I wanted to go to the legendary "Tokyo fish market," he gallantly insisted on accompanying me as a guide at no cost, but now had difficulty tracking me in the orderly but fast-moving frenzy of people unloading, shifting, selling, and buying that day's offering of thousands of tons of seafood. More than once his alert warnings saved me from being the flattened victim of dazed concentration as I tried to get my mind around the concept of an ocean large and productive enough to support the taking of such enormous bounty, day after day, decade after decade. Each year, approximately one million tons of animals transit through Tokyo's market to supply Japan's taste for hundreds of variations on the theme of raw, smoked, steamed, fried, dried, grilled, baked, braised, boiled, and broiled creatures captured and taken from the sea.

In fact, I knew from hundreds of ominous signs—the collapse of most major fisheries and sharp declines in many others—that the ocean *cannot* sustain the massive removal of wildlife needed to keep Japan and other nations supplied with present levels of food taken from the sea. Within a generation, the vision of limitless fish in an infinitely productive ocean has been shattered. Greatly expanding demand to feed growing populations of people coupled with the means to find, capture, and transport animals from all parts of the sea to distant lands have drastically altered the nature of ocean ecosystems in a few decades. Even familiar staples of the early part of this century—cod, herring, haddock, pollock, halibut, several kinds of salmon, and tuna—are in sharp decline after years of heavy fishing pressure, echoing the fate of the once abundant populations of buffalo, passenger pigeons, and dozens of other tasty but vulnerable terrestrial species of birds and mammals. In Tokyo I was witnessing a small sample of the global consequences of primitive hunting and gathering techniques accelerated, magnified, and multiplied many times over by modern technology—the large-scale commercial tracking, catching, and marketing of *wild-caught game*.

Just as on the land, various factors are responsible for the collapse of wild ocean populations, from pollution and habitat destruction to high-tech capture techniques and government subsidies that perpetuate otherwise unprofitable ventures. But there is no doubt about the main cause of the problem: *too many fishermen, not enough fish.* In the Tokyo fish market, as I moved among the trays of fish concentrated there from the oceans of the world, I was mindful that what I saw represented but a fraction of the amount that was that day being offered for sale in thousands of markets from San Francisco to Samoa—and that that one day represented but a small part of the total of some 90 million tons for the year. I wondered about the cost—not the asking price in the markets, but the cost to the ocean of such wholesale removal of the elements of a system hundreds of millennia in the making, and of the effects of techniques that can be likened to using bulldozers to catch squirrels, and nets in the sky to blindly snare ducks—and whatever else might be flying by.

"Living off the land"—and the sea—as hunter-gatherers is a way of life deeply rooted in human nature. It is seductive and challenging—to have the freedom to roam, using one's wits to take whatever comes along, to have the freedom to cleverly seek out and take, without asking permission from anyone, the goods required for clothing, shelter, food, and pleasure. When our numbers were no greater than those of other hunter-gatherers such as bears and wolves, the land and sea ecosystems could tolerate us and incorporate the impact of human actions without great disruption to the healthy functioning of the system as a whole. But in a geological second, three factors have forever altered the traditional "freedoms" coincident with the hunter-gatherer way of life: swift population increase (from 1 billion in 1800 to 10 billion early in the year 2000), the rapid advance of technology, and, most influential and potentially dangerous of all, a perverse self-centered mind-set that shapes the character of many human actions.

On the land, in a few places, a few people are still able to engage in some of the old ways full-time, but it is increasingly difficult for rugged individualists or for communities of native people to be "self-contained," maintaining their traditions as hunter-gatherers,* even when coupled with subsistence farming,

---

* In North America, native people once dependent on taking a small number of now-protected species of whales and other highly regulated fish and wildlife are encouraged to try to maintain elements of their cultural heritage with a privileged "subsistence take." Ironically, most have accepted a modernized version of traditional hunts, putting aside

while all around human societies are speeding in other directions. It has proved to be more realistic for many more people to indulge the urge to be independent and "free" part-time, as recreational sportsmen, hunting and fishing in ways that reflect deep roots, although most such sports enthusiasts pursue their passion armed with modern twists that tip the scales much in favor of the hunters: fiberglass boats, radios, motor-powered vehicles, Styrofoam coolers, exotic plastic lures and lines, precision firearms, and high-tech synthetic clothing that makes occasional excursions into the "wild" comfortable, even fashionable. Only through careful management of areas designated for sportfishing and hunting is it possible to sustain even such limited use; areas must be suitably large and sometimes "enhanced" with feeding and stocking programs, and the number of users limited through licensing or limited access. No one expects such activities to provide more than token sustenance, however. Only through thousands of years of learning how to cultivate certain plants and domesticate selected kinds of wild game has it been possible to support the large numbers of people now populating the planet.

Many who would recognize the absurdity of a plan to sustain large and growing numbers of people by hunting and gathering from the land buffalo, deer, wild birds, rabbits, squirrels, roots, and berries seem to disengage their power of reason when it comes to the sea, apparently believing, somehow, that ocean systems are fundamentally different from those on the land—that they *can* year after year yield huge, commercially viable takes of wild-caught organisms and rebound indefinitely. Some refer to commercial fishing as "harvesting the sea," but this misleadingly implies that fishermen, like farmers, have planted and tended a crop; rather, fishermen are aquatic hunters, catching what they can from natural ecosystems—ever reaping, never sowing.

Modern catch techniques, while effective in bringing in fish and other targeted sea creatures, are rarely discriminating and often are crudely catastrophic. Nets dragged along the sea floor, "bottom trawls," scoop up everything in the way, the subsea equivalent of collecting the entire farm when the goal is to

---

hand-held spears and skin boats for explosives, rifles, and faster, motor-powered boats, snowmobiles, and trucks. At this point in history, it is virtually impossible to avoid mixing cultures, or be far from the influence of modern technology.

bring in a bushel of apples. Thousands of miles of drift nets, vast webs of nearly invisible monofilament each up to forty miles long and several hundred feet deep, became notorious for nonselectively "strip mining" the sea of life in the 1980s and early '90s. Drift nets are now outlawed for use on the high seas, but smaller and just as lethal gill nets are used closer to shore, where they entangle birds, turtles, dolphins, whales, sharks, and numerous nontarget species in the process of snaring those intended. The pain inflicted is incalculable.

Little is being done to change laws that permit agonizing death by suffocation, strangulation, crushing, drowning, panic, shock, slicing, spearing, or other modes of modern fishing. No one doubts that dolphins, whales, seals, and birds feel the burn of rough webbing on exquisitely sensitive skin, the slashing bite of knives and gaffs, the searing shock of separation from close-knit societies . . . and no one *should* doubt that fish do as well.

In the eastern tropical Pacific, where approximately one quarter of the world's tuna catch is taken, thousands of dolphins are inadvertently captured and killed in purse seines (nets that can be pulled closed like a drawstring purse) in the process of taking tons of yellowfin tuna. Although U.S. tuna fishermen have been able to reduce the annual number from more than 200,000 dolphins killed a year in the 1970s to a number now restricted by law to be no more than 20,000 per year, the casualties elsewhere remain high. To some, taking *any* dolphins is unacceptable, and public pressure on the U.S. tuna industry stimulated an unprecedented effort by the major players to guarantee consumers "dolphin-free" tuna starting in 1992.*

There is no comparable constituency to bring about similar changes concerning the "bycatch" associated with the taking of shrimp, however. For every 10 pounds of Gulf of Mexico shrimp scraped from the sea floor, 80 to 90 pounds of "trash fish"—rays, eels, flounder, butterfish, redfish, batfish, and more,

---

* There is no easy solution to this problem. While "dolphin-safe fishing" has been heralded as a step toward reducing dolphin mortality, growing use of an alternative method, "log fishing" (where nets are set around rafts of logs and debris), has increased the incidental killing of other species. According to estimates from the Inter-American Tropical Tuna Commission (IATTC), average bycatch from 1,000 net sets for tuna around dolphin herds includes 500 dolphins, 52 billfish, 10 sea turtles, and no sharks. In contrast, bycatch from the same number of nets using the log-fishing method includes only 2 dolphins but a much larger number of other species: 654 billfish, 102 sea turtles, and 13,958 sharks.

including juveniles of many species—are mangled and discarded, in addition to tons of plants and animals not even considered worth reporting as "bycatch," i.e., starfish, sand dollars, urchins, crabs, turtle-grass, seaweed, sponges, coral, sea hares, sea squirts, polychaete worms, horse conchs, and whatever else constitutes the sea-floor communities that are in the path of the nets.

Objections have been raised by fishermen dependent on redfish, snapper, grouper, and other commercially valuable species destroyed as juveniles in shrimp nets, and from conservationists and government agencies concerned about the toll on turtles. The turtle specialist Archie Carr identified shrimp fishing as the single most damaging threat to the future of sea turtles in Atlantic waters in recent years. The mandatory use of "turtle excluder devices" on shrimp nets by U.S. fishermen has helped reduce the number of casualties, but the goal of "turtle-free" shrimp remains elusive, and no modification can alter the inherently destructive nature of bottom trawls. Their use continues, despite the high cost to the environment, largely because such costs are not taken into account, because there is a continuing strong market for shrimp, and because the shrimping industry has powerful political support in Congress.

Not all shrimp that now appear in supermarkets and shrimp cocktails are caught in trawls, however. In Honduras, Ecuador, Nicaragua, and other Central and South American countries, shrimp farms that resemble huge ballparks covered with several feet of greenish water are being carved out of coastal mangrove forests. Young shrimp are netted from natural nurseries, the flooded roots and tangled network of undisturbed mangrove vegetation, where, like trawlers offshore, the nets indiscriminately snare and destroy a "bycatch in miniature"— the young of numerous other species trapped with the larval shrimp. When food is abundant, it takes about a year for the wild-caught larvae to reach marketable size. Shrimp farming, generally, is profitable—but not all of the costs are accounted for. The wild-caught young are regarded as "free," and no loss is noted to reflect the reduced value of a shoreline depleted of its protective, productive mangroves. Some shrimp fishermen blame their declining offshore catches on the shrimp farmers, who destroy the natural mangrove "nurseries." And, they resent the competition.

These are among the things I find myself contemplating when someone places before me a shrimp cocktail: Was this farmed or trawled? How many of what kind of other creatures were killed so that I might have this small bite of pink prawn?

Some successful sea farming—aquaculture and mariculture—is taking place, but most of the fish and other "seafood" species now marketed are not good candidates for cultivation, given the volume of food, and time, it takes to raise them. The sea is, after all, mostly a fish-eat-fish world; the bins and baskets crowding the Tokyo fish market are crammed mostly with carnivores, the ocean equivalent of lions, tigers, hawks, and wolves—top-of-the-line predators, not herbivores. Thus, indulging a taste for tuna or swordfish is comparable to dining on mountain lion steak or eagle pie.

The progression is straightforward. It takes many plant eaters (mice, rabbits) to sustain one predator (wolf, cougar), especially one that lives for many years. A ten-year-old wolf or tuna or a fifty-year-old whale—or human being—has eaten many times its weight every year in the course of an active life. For any young, rapidly growing creature, about 90 percent of the food consumed is converted to metabolic energy that keeps creatures healthy and moving about, and roughly 10 percent is retained for growth. These proportions vary considerably depending on the nature of the food, the efficiency of the creature's processing mechanisms, and the level of activity involved in living.

It takes about twenty pounds of plants to make a pound of beef—if the animal is young and taken to market before it has reached the "leveling off" phase of its growth. To make a pound of ten-year-old Atlantic bluefin tuna, a magnificent high-energy open-ocean predator that may consume as much as 25 percent of its body weight in food a day, it takes thousands of pounds of fish that in turn have eaten tens of thousands of pounds of smaller fish that in turn have eaten hundreds of thousands of pounds of small crustaceans who in turn have derived their energy from millions of minuscule plants that have translated the sun's energy through photosynthesis into simple sugar—the ultimate food for most of life on Earth. To make an entire fish, not just a single pound, multiply these figures by 500 to 1,500—the number of pounds an adult Atlantic bluefin may weigh. The numbers tell the story: Twenty pounds of plants to make a pound of marketable yearling calf; perhaps 100,000 pounds of phytoplankton working through complex food webs to make a pound of ten-year-old marketable tuna.

Marine biologist Lionel Walford observed in his book *Living Resources of the Sea* that in 1948, fishermen in the North Sea were able in 100 hours to capture on average 58.6 tons of herring, a small fish that feeds directly on microscopic animals and plants strained from the surrounding seawater. He adds:

> To collect plankton equal to that amount of herring, it would be necessary to strain over 57.5 million tons of water! Indeed, the herring must do much more than that. They work very hard at it and it takes three or four years of feedings before they come to useful size.

Certain passive, mostly herbivorous creatures—oysters, mussels, and clams—are efficient gatherers of energy produced by the ocean's microscopic plants and have more promise as an effective food source for humankind than fish carnivores. In 1969, marine scientists John Ryther and G. C. Matthiessen reviewed aquaculture activities worldwide and analyzed the yields of various filter-feeding "shellfish." They found that high numbers were being produced in the nutrient-rich waters of Spain, and in Japan, cultivation of oysters yielded 46,000 pounds per acre—not including the weight of the shells. This compares very favorably to the 5 to 250 pounds of beef obtained from an acre of pastureland, and is clearly a more direct route from plant to protein than high-speed carnivores such as tuna.

Standing in the Tokyo fish market, surrounded by the distillation of billions of pounds of planktonic food energy concentrated into protein that would mostly be consumed before the end of the day, I was forced to consider a tough reality: If we humans are to make a successful, sustainable place for ourselves in the world—whether our population holds steady or doubles—it is mandatory that we establish effective food sources that minimize impact to crucial natural systems and maximize efficient use of food energy. The chances of making ecological ends meet on a diet of tuna, swordfish, grouper, snapper, sharks, and other large predators are about as great as the likelihood that we could feed our growing millions by eating panthers. Rather, it makes sense to "eat low on the food chain," consuming plants or animals that make efficient use of plants.

The concept of eating low on the food chain has long been accepted and practiced with respect to the land, where great selectivity has resulted in the cultivation of a handful of plant and animal species. For instance, of the 250,000 known species of flowering plants of which 75,000 are believed to be edible, about 2,000 have been domesticated and approximately 150 are commercially cultivated—yet at the present time only 20 of these supply 90 percent of the world's food needs. Of these 20, just three grass species—wheat, rice, and corn (maize)—supply over half of human caloric intake. Similarly, most bird species are edible, nearly all are palatable, but very few have been domesticated, and the burden of translating plants to protein for human consumption has fallen to a few, with chickens, ducks, and turkeys far and away the most common. The same is true of mammals. Beef, lamb, and pork dominate meat consumption worldwide.

Part of the attraction of seafood is the diversity of choices now available, and one reason for the diversity is the seafood's wild origins. However, no consumer, whether purchasing bay scallops at a gourmet restaurant in Paris or cans of "whitefish" catfood in Chicago, pays the full cost of wild-caught seafood. If, on top of the catch and transport costs, fishermen, like farmers, had to pay for and pass along the costs of the space, food, and other circumstances required to produce what they bring to market, the price of wild-caught seafood would be unaffordable for most. If, in addition, the damage inflicted to the sea through the "bycatch" killed or the ecological impact of removing entire groups of species were put on the balance sheet, it would be impossible to justify commercial fishing as a profitable enterprise.

For years I have been fascinated with the rationale behind fisheries management—utterly mystified by popular policies that leave little leeway for consideration of any value for fish except as a commodity, and little hope that even as highly valued resources, large populations of edible species can long survive the profound unknowns and wishful thinking underlying these management policies.

During the late 1970s and early '80s, while based at the California Academy of Sciences as a research biologist, I had a chance to delve into some of these issues when I agreed to serve on the council of the International Union for Conservation of Nature (IUCN) and the board of trustees of the World Wildlife Fund (WWF) International.* Those appointments came just in time for me to become part of an ambitious effort by scientists, administrators, business leaders, economists, governmental officials, and conservationists to face up to the problems essentially caused by "too many people, not enough planet." In collaboration with the United Nations Environment Programme (UNEP), the IUCN and WWF sponsored development of the "World Conservation Strategy," a bold global initiative that takes into account the planet's entire stock of natural assets and seeks to understand what it will take for humankind to prosper—to use, without using up the resource base upon which we are all utterly dependent. Naturally, my interest focused on ocean issues, and the curious disparity between the broad appreciation shown for terrestrial fauna and flora—and the relentless, single-minded attention paid to ocean creatures by most of my scientific colleagues as primarily *something to eat!*

The truth is, I am fond of seafood, and admit to a personal sense of conflict when offered a chance to munch on a fish that I know tastes terrific—but that I also know is on my plate at great cost to the ocean. It is also not easy to view with detachment species of fish, lobsters, conchs, and clams that I have come to know on their own terms, in the sea. Even so, as one who grew up in a family where blue crabs are considered the ultimate in fine dining, with cherrystone clams, Apalachicola oysters, sweet bay scallops, and a dozen kinds of fish vying for second place, I appreciate the enthusiasm of those who want to find solutions to the challenge of maintaining healthy populations of wild sea creatures for the purpose of getting them onto a plate. Each savored memory of steamed crab and succulent shrimp provides a delectable reason to lament the

---

* Formed in 1961, the IUCN consists of a network of more than 500 governments, government agencies, and nongovernmental organization members. It works closely with and derives significant funding from WWF International and its governmental and nongovernmental members. WWF, renamed the "World Wide Fund for Nature," is a private international conservation organization based in Switzerland, with more than 30 national affiliates. The largest is the World Wildlife Fund-U.S. in Washington, D.C.

decline of the productive ocean ecosystems that once yielded a dazzling array of catch-them-if-you-can options to professional fishermen, as well as those who just wanted to go out on a weekend and try to snag something for dinner.

Recently, I was reminded of such dilemmas at a favorite restaurant, Sam's Seafood in Dunedin, Florida, where my large extended family mulled over a menu that included blackened redfish, cultivated catfish, mullet, and other catch-of-the-day creations served with okra, hush puppies, black beans and rice, fried green tomatoes, cheese grits, slaw, Greek salad, and other choices, with homemade key lime pie (the real thing) or Phyllis Hart's peach cobbler and fisherman-style black coffee as a topper. These choices, and the fact that Sam's fishing and culinary skills are legendary along Florida's Gulf coast, repeatedly draw the Earles and other regulars. Sam, who began fishing so long ago that he knew Clearwater, Florida, when it actually had clear water, has recently become uneasy about the future. Some fish are in short supply, regulations intended to protect fisheries are becoming increasingly complex, and large commercial enterprises seem to be favored over individuals and small outfits. Through Sam's eyes, I can see more clearly that the future for fishermen, as well as the fish, is as uncertain as a sea breeze.

"Remember when we found fifty-two clams in one place in Tampa Bay without moving our feet," my mother asked, eyeing the clam chowder. "That was in 1960! As soon as word got around that clams were there, the rush was on. Now they're gone." Sam's 1995 clams are from distant waters.

Others chimed in. "I remember getting enough scallops from the bay for dinner just by wading out on the grass flats at low tide armed with a bucket and a fast-grab technique. Those grass beds are gone—and so are the scallops." . . . "How about the red snapper we used to catch—right off the dock!" . . . "Or those little coquinas that were so common over on the beach. Whatever has happened to them?"

Eyes turned to me. I, the marine biologist, was supposed to have answers. I was tempted to say what seems obvious: *Too many fishermen, not enough fish; too many people, not enough planet.* But straight answers are not quite so simple and I said so. With my eye on my four-year-old nephew, Reade, then artfully arranging a French fry on his sister, Dee Jay's, neck, I added, "It is not just how

many people there are that causes problems; it is also how we behave, or misbe-
have, that determines our success, or lack of it."

Human behavior relative to the ocean's living resources is the main focus of one
of NOAA's five divisions, the National Marine Fisheries Service (NMFS), the
agency responsible for maintaining the good health of the U.S. fishing industry and the ocean systems upon which they are based. As NOAA's Chief Scientist and "sturgeon general" from 1990 to 1992, I had a front-row seat for
the debate on numerous unresolved fisheries issues and a chance seriously to
study the scientific basis for policies that had led *not* to the widely acclaimed and
highly desirable goal of "sustained use" of natural ocean ecosystems, but rather,
to precipitous "crash and burn" declines. I felt a special sense of urgency, and
frustration, in face of these thorny problems relating to use of resources. I also
felt anxiety in the realm of ethics: We must take care of systems that take care of
us, whether we are required to do so or not.

Among the intractable, hot topics requiring action that frequently came up
at management meetings were the following: How, other than with heavy fines,
can we encourage Gulf of Mexico shrimp fishermen to comply with regulations
requiring them to use "turtle excluder devices" to avoid the killing of sea turtles,
now on the edge of extinction? How can we come to terms with the conflicts
between sportfishermen and commercial fishermen competing for diminishing
resources (both camps have compelling arguments)? How can we resolve the
needs of northeastern U.S. fishermen in light of clear news that the groundfish stocks that support them are gravely depleted? How can we balance the
demands of West Coast fishermen against recommendations for new restrictions
on taking Pacific salmon? (The fish are caught in a hopeless squeeze between
hydroelectric development that greatly damages their spawning success, and
offshore pollution and fishing pressure that inexorably reduces the number of
adults.) Every day new problems arose, and rarely did any go away.

At one morning meeting in 1991, we were given the latest assessment of
western Atlantic bluefin tuna, which showed that the adult breeding population
had declined *to ten percent* of what it had been twenty years ago, when regulation

of the taking of these wide-ranging creatures began. The population as a whole had been reduced by more than half, and most of what remained were immature. Stunned, I blurted out, "Are we trying to exterminate them? If so, congratulations! We're making great progress."

I was assured that, no, the United States, in collaboration with twenty-one other nations who are members of the International Commission for the Conservation of Atlantic Tunas, ICCAT, were attempting to regulate catches to ensure healthy populations of fish—and thus a healthy fishing industry. Unfortunately, the quotas were set too high and there was evidence that more tuna were being taken, both by commercial and sportfishermen, than were being officially reported, i.e., "pirate tunafishing" was taking place. Few question the need for limits of some sort on the number to be taken if there are to be such fish in the future, but it has been extremely difficult to reach accord among commercial fishermen, sportfishermen, conservationists, and others who are involved from several nations. Despite diminishing numbers, several factors make overfishing—of tuna and other species—tempting, the most common being the rueful plea: "Hey, if *I* don't get them . . . someone else will."

Combating this myopic philosophy will require far-reaching, fore-sighted planning. Because the wide-ranging migratory habits of tuna and their relatives take them across the boundaries of many nations, conservation measures and management of fishing operations are extremely complicated but necessary if populations are to survive increasing fishing pressure. Tuna, swordfish, marlin, sailfish, bonito, and mackerel are considered by many to be among the most aesthetically pleasing, elegantly formed creatures on the planet. Canned tuna marketed for use in sandwiches, casseroles, and salads is usually from the group of fish known as the "true" tuna, which includes albacore, yellowfin, skipjack, or bigeye tuna. At least fourteen other less well-known tunas are in this group, as well as another fifty or so species ranging from the relatively petite six-pound silver-blue frigate tuna to the enormous 2,000-pound *Xiphias*, or black marlin, prized by sportfishermen for their legendary speed, size, and agility, and by many others simply because of their taste when served with lemon slices and butter. I have listened to engineers sigh with envy while admiring the streamlining and hydrodynamic refinements that allow some members of this group to speed through the sea keeping pace with nuclear submarines. It is no wonder they wonder! Atlantic bluefins, for example, are able to travel in excess of 50 miles per hour and navigate over thousands of miles of open sea. They are efficient ocean nomads who retrace their pathways year after year by means still mysterious to human observers.

Significant numbers of tuna and their relatives are taken from tropical and temperate waters of the world by about seventy island and coastal countries, but more than half of the total catch is taken by only two—the U.S., which does much of the fishing, and Japan, which provides most of the market. The taste for fresh, raw bluefin tuna, coupled with diminishing numbers of fish, has sent the price per fish into the realm of precious jewelry. A single 900-pound bluefin might bring a fisherman $40,000 dockside. At the market, such a fish may sell for more than twice that amount, and by the time it is turned into delicate, fresh slivers artistically presented to diners in Japanese restaurants, its value will have more than doubled again. In Japan, two slim slices of prime bluefin sushi may cost $75. It is little wonder that the sale of tuna, usually by skilled auctioneers, is a serious, high-voltage event that commands the rapt attention of everyone involved. It was a phenomenon I did not want to miss during my 1991 visit to the Tokyo fish market, especially after the recent news of the pathetic numbers of large bluefins remaining in the northwestern Atlantic.

At five A.M. I tried to ask my self-appointed fish-market guide, the taxi driver, to "take me to the tuna," but he seemed not to understand. In desperation, I stretched my arms wide and said, "Sushi! Sashimi! Toro?"—then made a motion with my hand like a fish swimming. At that moment a flatbed cart zoomed past bearing what appeared to be a 500- or 600-pound bluefin tuna! I pointed and began running in the direction from which the cart had come, followed closely by my would-be guide.

A hundred-yard dash put us in the vicinity of the loud, singsong voice of an auctioneer lilting high above the noise of crowds and carts. Like an acoustic beacon, the distinctive cries led us to a special section of the market where serious business transactions involving millions of yen were in full swing.

Carefully, and with great respect, I stepped along a row of frozen, misty carcasses, dusted with frost, each one tagged and numbered. Several men moved methodically from fish to fish, aiming the beam of a flashlight into the gills, then peering intently at a sliced section near the tail, looking for subtle clues to freshness and quality. The real prizes were in a separate row: dozens of sleek, metallic silver-blue fish that seemed to have leaped straight from the sea onto the platform. In fact, they had been airlifted from distant oceans, and treated with great care to maintain their lifelike condition. These, too, were tagged and each was carefully scrutinized by prospective buyers.

Fresh or frozen, each of these bluefins, I mused, is worth a small fortune—with stakes held by the fishermen, the auctioneers, the buyers, the restaurant owners. But even at a price of $100,000, $1,000,000, or, in fact, any price, duplicating such a fish is presently beyond human know-how. Who, I wondered, notices the cost to the ocean, now missing creatures that have taken perhaps twenty years to reach their Tokyo destination?*

By mid-1991, the years of fishing pressure on western Atlantic bluefin tuna had created a future so bleak that Sweden proposed having the species declared endangered, a suggestion later withdrawn because of objections raised by Japan, the United States, and other countries with strong economic interest in taking tuna. In November of that year at a meeting of ICCAT in Madrid, it was decided that Japan, Canada, and the United States, who profit most from the northwestern Atlantic bluefin catches, would each reduce their take by 10 percent—277 tons—in 1992 and 1993. Greater reductions were set by ICCAT in 1994 after heated debates concerning whether or not the western Atlantic populations overlapped significantly with those in parts of the Atlantic where regulation is less stringent. The commercial fishing journal *National Fisherman* quoted an often-heard fisherman's lament: "All the fish we're conserving and letting go are swimming over and being caught in the Mediterranean and eastern Atlantic."

---

* Some attempts are under way at the New England Aquarium in Boston to discover how to cultivate these ocean giants, to better understand how to protect them in the wild and, perhaps in time, to provide an alternative to wild-caught tuna. In Japan, wild-caught young yellowfin are being raised in special ocean enclosures to market size. These and other efforts to grow top-of-the-food-pyramid predators as food for humans or replacement stock for depleted natural areas are encouraging, but costly, given the volume of food, and time, required to raise these creatures. Meanwhile, fishing continues.

Whether there is one large, intermingling population or two or more separate sets of bluefin tuna gliding through the dark waters of the North Atlantic is only one of the many unresolved issues concerning the habits of these great ocean rovers. What is happening to bluefin tuna highlights the broader issue: If the oceans of the world are to continue to provide wild-caught food, we must acquire a better understanding of natural populations and apply this knowledge before they deteriorate past the point of recovery.

Maybe some formula can be devised that will take into account the complexities of natural ecosystems, and make possible the calculated removal of a small number of individuals from certain populations without dooming them to decline. Or, it may be that careful study will show that there is no "safe" level that both is commercially interesting and allows wild populations to resist decay from a long list of "natural" impacts—storms, disease, decreased food supply, increased predation, increased competition, natural and human-induced changes in the environment, and so on.

So far, the prospects are not promising. It is startling to some to discover that to date, not a single example exists of commercially successful "sustained use" of a wild population, terrestrial or marine. There are, however, hundreds of examples of wild species whose numbers have been reduced sharply, or, like dodos, great auks, Carolina parakeets, and passenger pigeons, have been hunted to extinction. This poor track record of achieving sustainable takes has not slowed the quest for "free lunch" from the wild sea, however. Not only does overfishing continue; some nations are gearing up for increased effort.

To find fish today, acoustic sensors are used to locate and even discriminate among schools below; aircraft, and sometimes even spacecraft, spot fish from above; power boats, factory ships, and catch gear do the rest, usually without finesse or precision. Ships as large as football fields deploy nets capable of engulfing tons of fish in one sweep. Iceland boasts of trawls large enough to engulf twelve 747 aircraft. Longlines—floating arrays of hundreds, sometimes thousands of baited hooks—extract large fish from many miles of open ocean, leaving critical voids in ocean ecosystems. Japanese fishing fleets, so large and conspicuous as to be visible to space shuttle astronauts, illuminate the night sea to attract for capture millions of tons of squid. In the process of eliminating populations of some of the planet's least-known creatures and taking away the "lunch" of species who have no other choices, they disrupt the rhythm of the night with massive fields of blinding light.

The problems associated with overfishing began more than half a century ago, when the annual take of living creatures from the sea was already huge—about 20 million tons worldwide, with fishing largely conducted as local enterprise. But following the end of the second world war, a growing world population, an increasing appetite for "food from the sea," and new technology quickly altered the odds against the fish, and forever changed the nature of the living systems that dominate the planet. By the 1960s, the world catch had grown to more than 60 million tons a year, then continued more slowly for the next two decades. In 1989, the catch was 86 million tons, and hope was expressed by some that the often-named goal of "100 million tons a year" would soon be achieved, perhaps by increasing the numbers of krill removed from Antarctic waters, or by using new technology to find as yet "underexploited" stocks in deep water or some-where no one had yet thought to try, or through employing some new and clever technique to extract more from existing areas, or perhaps by finding uses for the enormous tonnage of presently discarded "trashfish" or "bycatch" species. But it was not to be.

In 1990 and 1991, after a fivefold increase within forty years, the annual take began to decline, despite use of various new technologies and increased fishing effort. Experts in the United Nations Food and Agriculture Organization (FAO), as well as many fisheries scientists—and savvy fishermen—at last began to artic-ulate what some had long sensed. The sustainable limit of wild-caught fish had been exceeded many decades previously. As reported soberly by *The Economist* in March 1994, "Fishermen are living off capital, consuming the resource that should yield their catch."

Cycles of greater and lesser abundance of fish and other wild species occur naturally, but there is no doubt that huge catches have nudged individual pop-ulations and entire ecosystems toward precipitous, sometimes irreversible, declines. Atlantic halibut were so heavily fished at the turn of the century that commercial fishing stopped in the early 1900s. The species still has not recov-ered. So many sardines were taken off the coast of Monterey in the 1930s (as many as 550,000 tons in one season—about half of the total number thought to be present) that the population collapsed, taking the livelihood of numerous sea creatures—and fishermen—with them. After fishing for sardines stopped, their numbers slowly began to increase, but still remain far from the legendary abundance that made Cannery Row famous.

Dramatic depletions have occurred in every ocean, but the best statistics are available for the North Atlantic, where some nations have kept catch records for a century or more. Systematic fisheries research began in 1902 with the formation of the International Council for Exploration of the Sea (ICES), when concerns were already being voiced about the causes for fluctuating stocks. Even then, it was recognized that care needs to be exercised if sustained use of a resource is to be achieved. Bear in mind that this was at a time when pollution was not a major complicating factor, when coastal habitats were in relatively good, sometimes pristine, condition, when fishing techniques were less destructive, and the number of fishermen fewer. Enormous fleets of government-subsidized fishing vessels did not exist, and a world population one fifth the present size did not demand wildly unrealistic returns from natural ocean ecosystems.

According to the National Marine Fisheries Service, as of 1993 nearly 80 percent of the commercially valuable fish whose status is known were overfished or "fished to their full potential." A report to Congress by forty scientists stated that virtually all of the important finfish stocks off the northeastern, southeastern, and Gulf of Mexico coasts of the U.S. are either overfished or on the way to that condition. The 1991 *National Fisherman Yearbook* noted: "At present, 45 of the 190 fish stocks that provide 90% of the world's fish production are overexploited. For instance, in North America, half of the species are considered to be overfished, and off West Africa, 70% of commercially caught species are classified as overfished."

A 1989 FAO report concludes that commercial fleets have driven some fish to "commercial extinction," that is, to a level so low that it is not economically attractive to try to extract the few that remain. In fact, deliberately taking a species to the point of commercial extinction is sometimes a matter of policy, the rationale being that when one kind is reduced to the point where it is not economically attractive to continue fishing for it, the focus then would shift to other species. In theory, the depleted stocks either would recover to be fished again—or would be replaced by other fish that could in turn be exploited. This "self-adjustment" theory, consistent with current accounting practices, assumes that fish and other sea creatures have value only when dead, and that live fish in the sea represent a "wasted" resource.

People rationalize this scheme by saying that the sea is so vast, life so abundant, and populations subject to so many natural variables, that there is little that mere human beings can do to disrupt ocean production. Unfortunately, it has been obvious for a long time that the "take until it hurts" (i.e., until populations collapse) policy of fisheries management does not work, if the goal is to have sustained use of a resource. It "works" only if the goal is short-term profit for those doing the taking without regard for long-term losses to those who inherit what remains. Although the inherent fallacies have stirred considerable criticism, the concept of "self-adjustment" haunts fisheries policies. Profits may be artificially magnified when government subsidies are involved, and if the price is right, a fishery may continue to be profitable to fishermen even though the target species is reduced to a few individuals. An image comes to mind of the *very last* California abalone or Atlantic bluefin tuna or New Zealand orange roughy being sold for an astronomical—and profitable—figure to satisfy a wealthy gourmet, and forever terminating the fishery.

Since the majority of fishing takes place within 200 nautical miles (230 miles) of shore, and thus falls within the jurisdiction of a bordering country, there should, in theory, be ways for governments to protect their respective fishing interests through smart management of the living resources in coastal waters. Taking too many fish is not smart, but many countries, including the United States and Japan, subsidize fishermen through low- or no-interest loans and outright payments that have encouraged many to go into—or stay in—commercial fishing who otherwise would not. *The Economist* reports:

> Japan, Norway, and the former Soviet Union, among others, have poured money into the fishing industry. The European Union increased fishing support from $80 million in 1983 to $580 million in 1990. A fifth of that went to build new boats or to improve old ones. . . . The EU's solution to Spain's demand to fish Norway's waters?

Buy Spain the right to catch another 8,000 tonnes of fish elsewhere. Instead of protecting scarce resources, governments subsidize their destruction.

There is an even more fundamental reason why too many fish are taken from the sea, even when it is clear that the resource may be permanently damaged in the process. That reason concerns the "tragedy of the commons," the phenomenon elegantly articulated in 1968 in an essay by Garret Hardin, who used as an example cattle grazing on public lands where no limits are set concerning use. Since no one owns the property held in common, there is incentive for each user to gain as much benefit from the resource as possible—before another member of the group takes it instead. When too many people add too many cattle, the capacity of the land to support the herd is exceeded, and tragedy—an exhausted system incapable of supporting any—follows.

Until recently, "the sea and the fishes," like the air, sunlight, and rain, have been treated like unowned resources—"free goods," accessible to all for the taking. Freedom worked when fishermen's capacity to catch fish did not outstrip the ability of living systems to recover. But treating fish as free goods has led to the assumption among both fishermen and nations, "If we don't take it, somebody else will." The tragedy of the commons has followed.

As recognition grew that too much fishing pressure can in fact have long-term, permanently depressing impacts on wild populations, earnest efforts were devised to manage fisheries to ensure "maximum *sustainable* yields." The concept of maximum sustainable yield, often referred to as "MSY," is seductive. It is defined as the greatest amount that can be taken from a self-generating wild population of animals year after year while still maintaining a constant average size of the population. Or, as summed up by the biologist P. A. Larkin in his keynote address to the American Fisheries Society in 1976, it is an optimistic dogma that assumes that "any species each year produces a harvestable surplus, and if you take that much and no more, you can go on getting it forever and ever. . . ." In discussions of MSY, there is often an implied, and sometimes openly stated, *obligation* to take the maximum, so as not to "waste" a resource.

In theory, a related group of animals (referred to by fisheries managers as a "stock") normally increases in size to a point where the rate of growth slows through natural limits of food, space, or other factors, eventually reaching a "final size" that reflects the so-called "carrying capacity" of the habitat for that particular population. Supposedly, at this stage, a "steady state" is reached where births are balanced by deaths. To fisheries managers, this hypothetical steady state is referred to as the "initial" or pre-exploitation stock. When fishing begins, this initial stock is reduced, but if the take is not too great, the population will continue to reproduce, allowing for repeated takes.

Among the assumptions that underlie the theory of maximum sustainable yield are that the population of individual species under consideration are of known size and do not mix with other populations of the same species (i.e., are "self-contained"), that the carrying capacity of the habitat remains stable, that the population when first exploited is, in fact, the peak "final size," that natural fluctuations in the population are not large, and that an "initial stock" has the potential to recover from reduction by fishing. Most important and most often overlooked is the assumption that the populations being exploited somehow live in isolation, not linked to complex systems that are modified when changes are brought about in *any* species, not just the "target" species.

In fact, never has reliable information on which the above assumptions are based been available prior to exploitation of a population. Nor has adequate knowledge of the life history of any species ever been worked out beforehand to determine whether or not timing, fishing methodology, or other special finesse might be applied to ensure the continued health of the fishery. In fact, the concept of maximum sustainable yield snares good minds, creates unrealistic expectations, and encourages the setting of unattainable goals.

Larkin acknowledged that the concept of MSY had served a useful purpose in helping to organize the way managers deal with wild populations, and that despite its limitations, fish populations would probably have suffered greater declines had there been *no* attempt to manage fishing. Now, however, he proposed the following: "Like the hero of a western movie, MSY rode in off the range, caught the villains at their work, and established order of a sort. But it's now time for MSY to ride off into the sunset." He concluded with the following epitaph for the concept:

M.S.Y.

1930s–1970s

Here lies the concept, MSY.
It advocated yields too high,
And didn't spell out how to slice the pie.
We bury it with the best of wishes,
Especially on behalf of fishes.
We don't know yet what will take its place,
But we hope it's as good for the human race.

Alas, the concept was laid to rest in name only. The notion that humans can and should become the principal predator in wild systems, skimming off some imagined "surplus" each year or season, is a notion still firmly entrenched in fisheries management policies, and thoroughly woven into the thinking of those wrestling with the goal of sustained use of other wild living ecosystems.

In 1991, the National Marine Fisheries Service issued *The First Annual Report on the Status of U.S. Living Marine Resources, Our Living Oceans*, an account that included a vision of "reducing overfishing . . . while maintaining currently productive fisheries." Built into the plan are elements that sound hauntingly like the underpinnings of "maximum sustainable yield" concepts. For example, "surplus production" is defined as "the weight . . . of fish that can be removed by fishing without causing a change in population size." A new term, "Long Term Potential Yield," is defined as "the maximum long-term average yield (catch) that can be achieved through conscientious stewardship, by controlling the fishing mortality rate to maintain the population at a size that would produce a high average yield. . . ."

This dressed-up version of maximum sustainable yield sounds virtuous, but it is fundamentally flawed by the assumption that humankind can blindly swoop down on an ecosystem that has been building for many thousands of years, indiscriminately remove major chunks of it, and expect that it will recover to its former fine health, despite continued high-level predation (fishing). It doesn't take many excursions underwater to see what fish actually *do*, to become extremely depressed by the brash claims made about fish behavior, population size, and ecosystem "health"—claims based on dragging nets through the water and counting what is caught! Chances of recovery of heavily fished areas are greatly diminished because in the process of taking, ecosystems are shattered and the population dynamics and social structure of targeted, and untargeted, creatures are shaken in ways that have never before been a part of ocean dynamics—not in the entire three and a half billion years of the history of life on Earth.

It is a curious twist of human nature that makes us apparently unable to raise the possibility that there may be no such thing as a "surplus" just waiting to be taken by one of the newest arrivals on the planet. It seems not to have occurred to those who dutifully calculate population size, reproduction rates, natural mortality—and "maximum yield"—that the right number for us to remove to ensure the sustained health of a wild population might be something very close to zero.

The shortcomings of the "maximum sustainable yield" concept came under sharp review in a landmark 1978 summary, *New Principles for Conservation of Wild Living Resources*, by biologists Sidney Holt and Lee Talbot. They identified alternative approaches, suggesting:

The privilege of utilizing a resource carries with it the obligation to adhere to the following . . . :

1. *The ecosystem should be maintained in a desirable state such that*
   a. *Consumptive and nonconsumptive values could be maximized on a continuing basis.*
   b. *Present and future options are ensured.*
   c. *Risk of irreversible change or long-term adverse effects as a result of use is minimized.*

2. *Management decisions should include a safety factor to allow for the facts that knowledge is limited and institutions are imperfect.*
3. *Measures to conserve a wild living resource should be formulated and applied so as to avoid wasteful use of other resources.*
4. *Survey or monitoring, analysis, and assessment should precede planned use and accompany actual use of wild living resources. The results should be made available promptly for critical public review.*

Many enthusiastically endorsed these recommendations in principle. In practice, though, quotas for taking from wild populations continue to be based on what many freely admit are "scientific *wags*"* or educated estimates at best.

To an ecologist, the most enigmatic of the flaws inherent in MSY and related fisheries management policies is the way investigators typically regard each exploited species independently, and treat each commercially exploited population as though it lived separate from the complex communities of other fish, invertebrates, plants and an ever-changing physical and chemical environment. Such an approach is convenient for statisticians, but is far from realistic.

The "single-stock" model cannot deal effectively with mathematically untidy factors such as competition among species for food, disruption of the "target" species' close interactions with other organisms, variables in times of maturation, variable social structures that influence reproductive rates, changes in the habitat because of pollution, climate shifts, shoreline modification, and natural or human-induced stress. Policy-makers' focus on short-term economic and social issues coupled with their overly optimistic, wishful thinking has led to overfishing and habitat destruction on a grand scale, worldwide. The vision of sustained use of wild populations in the sea has failed catastrophically, as is evidenced by the sharp drop or utter collapse of first one species, then another, that once seemed "endlessly abundant" and were earnestly "managed": More than one hundred species—including herring, cod, haddock, salmon, menhaden redfish, pollock, sharks, and several kinds of tuna—are now in serious trouble.

---

* Wild-ass guesses.

Among the first to wrestle with the problems of "managing multispecies fisheries" were biologists Robert May and John Beddington. In 1979, they attempted to analyze interactions among various kinds of organisms in Antarctic waters—whales, krill, squid, fish, birds—and, as they said, "using simple models [to] discuss the way multispecies food webs respond to the harvesting of species at different trophic levels [of the food chain]." Unfortunately, nature rarely performs on cue. Often, models reinforce commonsense conclusions that in the absence of understanding how complex ecosystems work, great caution should be exercised in disrupting them. At worst, trends reflected by models based on incomplete data may be dangerously misleading and raise false hopes, for example, about "surplus" numbers of one species (such as krill) when a known predator (such as the blue whale) is depleted.

In the United States, Congress made an attempt in 1976 to control overfishing in the nation's coastal waters by passage of the Magnuson Fishery, Conservation, and Management Act (MFCMA). This legislation protects U.S. fishermen by prohibiting foreign fleets from fishing within the Exclusive Economic Zone (EEZ), which reaches from shore 230 miles seaward, without special permits. The act also provided for low-interest loans to fishermen, and as a consequence, the U.S. fishing fleet grew significantly. Although the act contained provisions for setting quotas and other restrictions, the net result was fleet expansion, increased fishing pressure, and larger catches.

The Magnuson Act created eight regional councils, each charged with setting policy for taking fish and "shellfish" within their respective areas. Since the councils are composed primarily of representatives from the commercial fishing industry (a case of the barracuda guarding the fish-coop), it proved difficult for them to impose on themselves reasonable limits on takes.

Robert McManus, who served as general counsel for the National Oceanic and Atmospheric Administration (NOAA), had firsthand experience with problems that developed as a consequence of the Magnuson Act's provisions, including some that had less to do with fish populations than with jurisdictional "turf." In 1988 he wrote:

The Magnuson Fishery Conservation and Management Act of 1976 (MFCMA) enshrines a peculiar—some might say bizarre—apportionment of regulatory and enforcement authority between the states and federal government. . . . The states retain virtual plenary authority within their boundaries—in most cases, out to the three-mile limit. Furthermore, under Section 306 (a) of the MFCMA, the states may continue to exercise jurisdiction over state-registered vessels, even those engaged in fishing beyond state boundaries. Meanwhile, the regional fisheries management councils established . . . have virtually plenary jurisdiction with respect to fisheries conducted beyond state waters and within 200 miles of the coast. The peculiar jurisdictional seam thus created by the MFCMA has been a source of abiding conflict.

Thus, while individual states retained authority within their boundaries (in most cases, out to the three-mile limit), they could also keep control over their vessels even when they were out of state boundaries. This allowance clashed with regional councils' authority over all waters within 200 miles of the coast.

At first the Magnuson Act seemed good for fishermen. In the early 1980s, Alaska fishermen made $1.5 billion a year for catches of pollock alone. In New England, without competition from the Soviet Union, Germany, and Spain, record catches quite literally netted rich profits for the expanded taxpayer-subsidized fleet. In less than twenty years, however, the short-term boon for fishermen led to a long-term boondoggle for the resource: a squandering of natural assets deliberately encouraged by national policies that sought to protect the interests of fishermen without adequately taking into account the interests of the fish upon whom those fishermen were utterly dependent—or the long-term interests of the nation as a whole. The effect of the Magnuson Act was, essentially, to replace overfishing in U.S. waters by foreign fishermen with overfishing by domestic fishermen. Frank Mirarchi, a North Atlantic fisherman, noted in *Fish for the Future*, a guide to fisheries management:*

---

* By Suzanne Fowle: Center for Marine Conservation, 1993. This guide also contains a useful review of U.S. regulations and their implications.

The great diversity . . . which gave such resiliency to the groundfishery is gone.
Haddock, which I caught by the thousands, are gone. The mid-winter run of spawn-
ing whale cod is gone. The spring run of dabs the size of a hatch cover is gone. The
fall pollock are gone. I now rely on three species: cod, yellowtails and dabs. Even
with these . . . we must now measure each and every one, as most are legal by only
a fraction. These are not characteristics of a healthy fishery. I fear they are omens
of disaster.

Populations of fish in U.S. waters have collapsed, despite avowed attempts to achieve "sustainable" takes. Fish, fishermen, fish eaters, and fish admirers—and the ecosystems to which fish contribute health and living luster—all now suffer the consequences. Proposed revisions to the Magnuson Act are intended to help correct some of the problems, but it is not likely that the legendary Grand Banks and other heavily overfished areas can ever rebound from the unfortunate onslaught of the past half century, especially of the past two decades.

To get on track toward recovery of species, and ecosystems, revisions are needed in the complex maze of regulations that now govern exploitation of living marine resources. In U.S. waters, for example, size limits are often set for certain fish, shellfish, crabs, and lobsters without meaningful understanding of life history or other biological considerations. Annual quotas for some species—the "total allowable catch"—are based on total weight, not number of fish caught, a policy that encourages fishermen to cull their catch, dumping small individuals that are less valuable or under the minimum size in favor of pricier large fish. Limiting the total catch volume, but not the number of catchers, also leads to highly competitive, often dangerous "derbies" where individuals race to obtain as large a portion of the quota as possible. There are also "bag limits" and "trip limits" that regulate sport and commercial fishing, but it is again difficult to safeguard against discarding small fish in favor of later-caught, larger, more desirable individuals.

One approach that is receiving some thoughtful consideration is the use of individual fishing quotas, a management strategy that involves dividing up a total allowable catch among designated individuals or vessels. A refinement of this approach is the "individual transferrable quota," whereby quotas may be bought and sold. Once assigned, no one else can enter the fishery except through pur-chase of some of the original shares. Variations on this scheme are being used, but it is clear that no one solution will work for all species, and that none can be applied independently without influencing the overall system. While managers puzzle over how best to work through the complexities of regulations and fish-eries theory, the catching continues without the benefit of a comprehensive plan that will maintain the resource, let alone restore what has been lost.

A century ago, it may have been possible for wild populations to recover naturally from the level of fishing then drawing on the ocean's living capital. The amount of seafood consumed by a world population of about one billion was modest compared to today's insatiable demands, and technology for catch-ing fish gave the fish a better than even chance of getting away much of the time—usually enough to allow exploited populations to grow back. Modern commercial fishing technology has tipped the scales strongly in favor of the fishermen, however—at least for a while. Clearly, it is in the best long-term interests of both fishermen and fish not to take so many of any species that the populations cannot recover, just as it makes great good sense from the stand-point of both fishermen and fish to make sure that the habitats required for the populations to prosper are kept in good shape. Unfortunately, these "com-monsense" guidelines have rarely been followed. Like small, exuberant children entranced with myriad bright lights and colored wires on a giant computer, we tug and twist and pull without the faintest idea of the consequences to the machinery—or to ourselves.

Chapter 11

CRITTER CRISIS

Biological diversity—"biodiversity" in the new parlance—is the key to the maintenance of the world as we know it. Life in a local site struck down by a passing storm springs back quickly because enough diversity still exists. Opportunistic species evolved for just such an occasion rush in to fill the spaces. They entrain the succession that circles back to something resembling the original state of the environment. . . . This is the assembly of life that took a billion years to evolve. It . . . created the world that created us. It holds the world steady.

Edward O. Wilson, The Diversity of Life

Something was terribly wrong with Bob Wicklund's favorite reef, a pristine part of the Exuma Islands in the Bahamas. We anchored on a sandy patch offshore, then slithered our way through stands of coral, sponges, lavender sea fans, and feathery soft corals, hoping to see the usual lively crowd of damselfish flickering over Rainbow Reef's crown jewel: an immense dome of brain coral as high as an elephant and more than twice a well-fed pachyderm's girth. Instead, I saw what appeared to be a giant snowball looming from the sea floor. Normally a great colorful lump, formed during centuries of action by thousands of interlocking flowerlike animals, the living city now had the aspect of a ghost town.

"Coral bleaching," Bob said, as we surfaced to talk, treading water. "The dome has had some pale spots before, but this is the worst I've seen."

"This is awful!" I sputtered, a wave of saltwater lubricating my indignation. "If any reef in the world should be healthy, it's this one." (The reef is located at Lee Stocking Island, home of the Caribbean Marine Research Center.) "What's going on?"

Bob shook his head and shrugged. "The problem seems to be related to a rise in water temperature," he said, pointing to a cylinder tucked into a crevice at the coral's base. "We've been using battery-powered thermometers like that one to track temperature continuously for several years here and at other locations around the Caribbean. The change isn't much—a degree or two—but high temperatures sustained in the late summer appear to trigger bleaching. There is some concern that corals may be casualties of a global warming trend. Whether global warming is true or not, the water is warmer and corals are dying."

We submerged and moved in for a close look at the individual crowns of coral tentacles, clear as glass, and focused on the stark white calcium carbonate structure below, normally tinged a rich golden brown by the presence of symbiotic algae. The algae, zooxanthellae, not only give color to the animal's tissue and to the otherwise white, rocklike coral skeleton, but also produce carbon compounds that are believed to nourish the coral. In return they get a protected place to live and are supplied with vital nitrogen and phosphorous. When disturbed or stressed by water too warm or too cold, by salinity too high or too low, by pollutants, disease, high turbidity, or even increased ultraviolet radiation, the coral may expel the algae. Often, a combination of factors seems to set off the process we now were witnessing. Sometimes, when circumstances stabilize, bleached corals recover; other times, they simply die. When they succumb, seaweeds grow over the dead areas, boring animals tunnel into the limestone, and in time, the coral skeleton crumbles and the underlying structure for one of the planet's most diverse ecosystems comes unraveled.

Reef corals are notoriously fussy, living within a narrow band of just-right circumstances that have existed in tropical seas worldwide for eons: an average water temperature between 65 and 86 degrees Fahrenheit (18 to 30 degrees centigrade), well-illuminated, clear water, and salinity that is maintained close to 35 parts per thousand. While some bleaching occurs as a result of natural stress from storms or normal variations in weather and climate, since the early 1980s, these "rainforests of the sea" from Belize to Australia's Great Barrier Reef have shown unprecedented signs of decay. Even before the recent insidious wholesale deterioration, reefs were being physically assaulted by people who took coral for the tourist trade or hauled away truckloads of coral and underlying limestone rock for road-building material, destroyed reefs to make way for shoreline development, and dynamited them to kill then take the abundant and diverse fish that live among the coral's sheltering branches.

Without thought for the consequences, intricate strands connecting coral to plants, to fish, to thousands of invertebrate animals, to birds, to marine mammals, and even to ourselves, were snapped. No one can ever know how many kinds of creatures have been lost forever through reef destruction, and no one knows how to put them back together again, once destroyed. Some people are deeply concerned about the consequences of damage to the network of life that ultimately supports humankind, but others are more casual.

"So we lose a few species of fish and bugs? So what? What the hell is all the fuss over biodiversity. I just don't get it!"

The source of the question, a friend, avid sportfisherman, and staunch advocate of protecting marine habitats, startled me almost as much as the question itself. I *knew* he recognized and appreciated the importance of the varied forms of life that make up the underpinnings of favorite fishing haunts. Ecosystems work when all the small elements are in place, and they don't do as well when pieces are pulled out. Ultimately, the survival and well-being of people, as well as fish and other wild creatures, are dependent on maintaining the health of complete natural systems. Worldwide, loss of species and destruction of terrestrial ecosystems has advanced with dizzying speed; now ocean species, populations, and ecosystems are also being destroyed. The nature of the sea itself is being modified radically; in the process, our chances for a bright future are becoming increasingly precarious.

I knew that my friend really *did* "get it," even if he, and probably many others, didn't know what to call "it." If asked, "Are you concerned about the swift decline in the number and kinds of creatures with whom we share the planet, and are you worried about the consequences to the species we all care about the most—ourselves?" I suspect he would say "Yes." (Actually, this particular friend would say, "Well, hell, yes!")

Instead, I took a shortcut to get to the point, as I had years before when asked, "Who needs the ocean?"

"Think of Mars!" I suggested. "Zero biodiversity there! Even if there were water, rocks, sunlight, and wind, we couldn't live there without space suits and food from home. Our survival is utterly dependent on the existence of life on Earth—of biodiversity."

My friend's questions deserved a more thoughtful response. I have wrestled with the issues raised by biodiversity—and its loss—seeking insight from experts who share my worries that ignorance, not malice, often drives fateful decisions to blow up productive reefs to obtain a few dozen fish, and allows fertilizers, pesticides, and herbicides from agricultural areas, golf courses, and lawns to poison adjacent marine waters. I am convinced that if people know more about the nature of biodiversity, what it is and why it matters, more respect might be shown for taking care of it.

For inspiration, I turn to Dr. Peter Raven, chairman of a national committee overseeing a survey of the biodiversity of the United States. He defines biodiversity as:

> the sum total of all the plants, animals, fungi and microorganisms in the world, or in a particular area; all of their individual variation; and all of the interactions between them. It is the set of living organisms that make up the fabric of the planet Earth and allow it to function as it does, by capturing energy from the sun and using it to drive all of life's processes; by forming communities or organisms that have, through the several billion years of life's history on Earth, altered the nature of the atmosphere, the soil and water of our planet; and by making possible the sustainability of our planet through their life activities now.

Species are the basic units of biodiversity, a fundamental, natural level of organization that goes far beyond giving things names. The biological definition of a species is: *a population of organisms whose members are able to interbreed freely under natural circumstances*. Genetically manipulated organisms escaping to the wild have created some unnatural blends of uncertain position in the greater scheme of things, such as hatchery-reared salmon and wild salmon crosses, domesticated dogs and wolves, and a number of cultivated and wild grass hybrids. The bottom line is that some creatures are "clear cut" as species, and others are highly variable, evidence that the concept has soft edges. But despite the fuzziness, species nonetheless represent natural, fundamental building blocks of diversity.

I am often asked how the naming system works, and suggest a brief summary here to set the stage for other issues.

One or more distinctive, related species are grouped into genera and designated by a two-part Latin name, written in italic type. The genus-species name system neatly shows that *Felis domestica* (the house cat), *Felis margarita* (the small, desert-dwelling African sand cat), and several North American cats including *Felis concolor* (the North American mountain lion) are variations on a common theme.

In the family of cats, the Felidae, there are numerous genera, including cats different enough from *Felis* to merit their own monikers, such as the genus *Lynx*, which includes such species as *Lynx canadensis*, the short-tailed, tufted-eared cat that lives in the northern U.S., and its southern counterpart, *Lynx rufus*, the bobcat, with its subtle distinctions. Lions, leopards, tigers, and jaguars are in another genus, *Panthera*, as *Panthera leo*, *Panthera pardus*, *Panthera tigris*, and *Panthera onca*.

None of these is to be confused with another bewhiskered creature, *Arius felis*, the hardhead catfish, a distinctive denizen of the western North Atlantic and Gulf of Mexico, or with any of the more than two thousand species of freshwater catfishes clustered in thirty distinctive fish families grouped under the order Siluriformes.

The first systematic attempt to enumerate all of the planet's plants and animals was made by the Swedish biologist Carolus Linnaeus in the 1700s and by 1758 he had described and named 4,236 kinds of organisms. Other scientists got into the spirit of things, using the genus-species method of naming introduced by Linnaeus; within a century, more than 130,000 animals and plants were on record, and by 1900, more than half a million were acknowledged. According to the Harvard University biologist and biodiversity specialist E. O. Wilson, 1,413,000 kinds of organisms* had been named by 1992. No one knows for sure how many species there are, or how many have come and gone in ages past, but all of them seem to fit into one of five "kingdoms": Monera, Protista, Fungi, Plantae, and Animalia. Each kingdom is divided into a hierarchy of categories: phylum, genus, class, order, family, and culminating with individual species.

As an example, two members of the animal kingdom, housecats and hardhead catfish, are members of the same kingdom (Animalia) and phylum (chordata), but of different classes. They are classified, generally, as follows:

Phylum Chordata: includes all animals that have a rodlike "notochord" for support of the body, a single tubular nerve cord running down the back, paired gill slits (evident in the embryos of lizards, chickens, and mammals), and a tail (present in embryonic stages of all and the adults of most). Within this phylum are about a dozen classes of diverse animals including sea squirts, certain burrowing worms, sharks, rays, amphibians, reptiles, birds, mammals, and of course, cats, catfish, and you and me. For my household cats, the ranking then goes:

> **Class Mammalia**: Chordates with hair, teeth, and various glands in the skin, characteristics shared by moles, bats, cats, horses, whales, and humans, among others.
> **Order Carnivora**: Clawed, fanged creatures including dogs, cats, bears, and seals.

---

* Those with an orderly turn of mind might like to explore more about the thoughtful efforts that have gone into discovering the variations on the theme of life on the planet, and trying to make sense of how one thing relates to another, and in the end, where we human beings fit in. A fine place to start is with Edward O. Wilson's *The Diversity of Life*, an important and penetrating review of the past, present, and future of life on Earth, and David Attenborough's television series and book by that very name, *Life on Earth*, presented by the British Broadcasting Company. There are, as well, numerous scholarly volumes about plants and animals that show in detail how genetic or biological diversity is measured not just in terms of species, but also of the larger categories that embrace clusters of related species thought to share a common ancestry.

Family Felidae: All cats.

  Genus *Felis*: A related group of a dozen or so species of cats.

    Species *domestica*: All of the various cultivated derivatives of cats once common in the deserts of Africa and the Middle East, now distributed in barns, fields, and on laps of humans.

For the hardhead catfish, the formalities are as follows:

Class Osteichthyes: All bony fishes (does not include sharks, rays, and other cartilaginous fishes).

  Order Siluriformes: Catfish great and small, freshwater and marine.

  Family Ariidae: Certain marine catfish notable for their naked-looking scaleless hides and the way the males brood clusters of eggs and young, several dozen at a time, in their mouth.

    Genus *Arius*: A related group of marine catfish species.

      Species *felis*: That particular species that dwells in inshore Atlantic waters, including a group of about 10,000 that lurk next to a certain seafood restaurant in Key Largo, Florida, for nightly handouts of leftover garlic bread.

On the basis of the number of species currently inventoried, there is more diversity on the land than in the sea, but it is well to remember that more than 95 percent of the sea is still unexplored. It is also important to recognize that even if the precise accounting of species were known, consideration of numbers alone can be a misleading measure of diversity. For example, the group with the largest number of species is the class Insecta, but all share basic genetic identity and body structure with 750,000 or so other insect species. A genetic legacy of comparable magnitude is represented by just four remaining species of horseshoe crabs that make up the class Merostomata. If only one of these species becomes extinct, the loss in some ways might be compared to the extinction of 187,500 kinds of beetles, bugs, and butterflies. Furthermore, at higher taxonomic levels (class, order, phylum, etc.), marine ecosystems have a significantly higher degree of genetic diversity than terrestrial ones. Almost every major division of plant and animal kind has at least some representation in the sea, and many are principally or wholly marine. In contrast, only about half of these large categories occur on land. This observation supports the conclusion that the greatest diversity is unquestionably in the sea, if the presence of individuals representing various broad categories of life is given somewhat greater weight than the splintery ends of diversity known as species.*

We can see that the land's ecosystems vary from deserts to grasslands to rainforests; the ocean, too, has extremely rich and diverse habitats, ranging from the easily recognizable coral reefs (called by some the "rainforests of the sea," to hammer home the point about diversity) and lush stands of kelp (the stately "sequoias of the sea") to rich communities of life surrounding deep-sea hot vents to vast plains of soft sediment chock-a-block full of living creatures. There is a widespread misconception that much of the deep-sea floor is devoid of life, but given the high number of newly discovered forms in areas thus far explored, and the enormous territory remaining to be investigated, the actual number of deep-sea species could range in the tens of millions.

The group chaired by Peter Raven to explore and inventory biodiversity in the United States must try to account for all the elements of a habitat: trees, birds, insects, microbes; coral, fish, mollusks, algae. It is easy to see why each ecosystem has its own distinctive signature. Raven concludes: "Ecosystems and communities . . . of living organisms are also comprised within the concept of biodiversity, which is, therefore, a shorthand way of referring to the world's living endowment."

From prehistory to the edge of the year 2000, great chunks have been removed from Earth's treasury of species and ecosystem diversity as a result of habitat

---

* Dr. Elliot Norse, Chief Scientist of the Center for Marine Conservation and editor of the recent comprehensive review "Global Marine Biological Diversity" observes, "The sea hosts almost the entire extant variety of basic animal body plans, whereas the land and freshwater animals comprise myriad variations on the theme of insectdom and just a smattering of other body plans."

destruction, overkill, and the introduction of exotic species, joined in the last half century by the lethal addition of toxic chemicals and other pollutants. Large creatures tend to be particularly vulnerable to human predation. In North America, more than 70 percent of the large mammal genera that existed at the time human hunter-gatherers crossed the Bering Strait from Siberia are extinct. E. O. Wilson remarks that as human populations spread, "Mankind soon disposed of the large, the slow, and the tasty."

A similar pattern is emerging in the sea, where marine mammals, large fish, and anything edible, no matter how critical to other species or systems, are most vulnerable to deliberate human predation. The principal difference is the swiftness with which extinction may be brought about now on a global scale, a phenomenon partially counterbalanced by the speed with which new knowledge can be communicated and drastic actions possibly averted.

The pattern of annihilation of large animals has been reenacted recently in the sea with the largest mammals ever to live, the nine species of great whales and their cousins, more than fifty kinds of "small" whales, dolphins, and porpoises. As a direct consequence of deliberate killing, with conscious effort or even through negligence, these species could follow the mastodons, mammoths, and other large, slow, or tasty creatures who survived millions of years prior to the fateful conjunction that put our kind in touch with theirs.

It is worth elaborating a bit about the plight of the whales and dolphins at the end of the twentieth century, both as vital components of ocean ecosystems and as powerful symbols that people can more readily relate to than, say, krill, cod, or some equally important but nondescript burrowing worm. Human empathy for these intelligent, warm-blooded, seagoing creatures was succinctly summed up by six-year-old Andrea Soros, who wrote in *There's a Sound in the Sea*:

> *Whales are mammals. I am a mammal.*
> *We hunt whales. We kill our nature.*

In 1979, I attended an international meeting of whale specialists at Guerro Negro near Scammon's Lagoon, Baja California, Mexico, to help develop strategies that would restore and protect these wide-ranging marine mammals. For a while we tried to "think like a whale," to imagine what it would take to turn around the devastating plunge in numbers of the large cetaceans (the group of animals including dolphins and whales), and many of the small species. There was talk at the time in the international organization that regulates whaling, the International Whaling Commission, of a total ban on commercial whaling, worldwide, and it seemed to our small group that this would be an obvious and necessary first step—but that protection for the whales alone was not enough. Like people, we reasoned, whales need space, food, uncontaminated water, a place where their young can be safely born and cared for—in short, a healthy environment. This concept had been learned on the land long ago for creatures such as tigers and deer; to save such animals meant saving forests. To save whales, it is necessary to protect the ocean. An ocean supporting healthy populations of whales must be an ocean in reasonably good condition; conversely, an ocean devoid of whales may well be an ocean in trouble.

For the California gray whale, Scammon's Lagoon represents a critical component of their environment and thus for an enduring recovery and survival plan. It is one of the few remaining places along the coast of North America where whales migrating from as far away as the Bering Sea come together for months in the winter to mate and give birth. One hundred twenty-two years ago, the now peaceful embayment had been a sea of slaughter as hundreds of whalers, led by Captain Charles Scammon, attacked whales of all sizes, often deliberately wounding a calf to ensure close approach to the distressed and distracted mother and thus more easily set a harpoon in her.

In his classic account *Marine Mammals of the Northwestern Coast of North America Together with an Account of the American Whale Fishery*, published in 1874, Scammon matter-of-factly reveals some of the best and worst attributes of human nature. While describing gray whales as "unusually sagacious," "playful," a species who manifest "the greatest affection" for their young and "a power of resistance and tenacity of life that distinguish them from all other Cetaceans," he is methodical in his description of strategies to take advantage of their playfulness and curiosity, to outwit them and kill the greatest numbers. It is also clear that Scammon could foresee that the whales that he found and ruthlessly exploited for their precious cargo of oil had a precarious future, caught as they were between predation in the north from seagoing Indians who depended on them for food, oil, trading materials, and even clothing, and along the coast of California southward; it was an unremitting gauntlet of slaughter from whalers. As Scammon noted:

> *None . . . are so constantly and variously pursued . . . The large bays and lagoons, where these animals once congregated, brought forth and nurtured their young, are already nearly deserted. The mammoth bones of the California gray lie bleaching on the shores of those silvery waters, and are scattered along the broken coasts, from Siberia to the Gulf of California; and ere long it may be questioned whether this mammal will not be numbered among the extinct species of the Pacific.*

The gray whale was not alone in suffering from overexploitation. Although the method of killing was crude and dangerous for whalers as well as whales, by the mid-1800s, right, humpback, and sperm whales were also showing signs of serious depletion. Soon, modest twists of technology caused things to get much worse for all of the large whales. Development in Norway of an exploding grenade attached to a harpoon forever altered the odds against whales, as did the introduction of steam-powered catcher boats and factory ships in the late 1800s. Nothing in the 65 million–year history of whales had prepared these creatures for this mode of pursuit and killing, and the relentlessness with which it was applied. Within four decades, most of the large whale species were decimated.

The swift worldwide decline of whales, including fast-swimming species previously not often captured, caused the League of Nations in 1925 to call for regulations to protect the badly shattered populations. Clearly, the technology for killing whales had reached a point where the end of whales—and therefore, of *whaling*—was in sight, a situation not unlike the present state of many fish and their respective *fisheries*.

A bureau to keep track of the number and kind of whales killed was established in Norway in 1930, and the following year, the Convention for the Regulation of Whaling was signed by twenty-two nations. Modest protective measures were proposed, including no killing of immature whales, calves, or females with calves. However, some of the largest whaling nations—Japan, Germany, Chile, and Argentina—did not sign the agreement, thereby signaling their disdain for efforts to conserve the resource for the benefit of all, and behaving, instead, like outlaws. Just as in present dilemmas concerning overexploited populations of fish, the whales "belonged" to no one—and everyone. Any whaler or whaling country that showed restraint had no guarantee that another would do the same. Rather, it was generally believed, as many do now with fish, "If I don't get them, someone else will." In 1931, with increased numbers of whalers using more sophisticated methods, more than 43,000 whales were reported killed (no one knows how many more were not accounted for), and production of oil reached a level ten times what it had been ten years before.

Reduction in the number—and size—of caught whales was becoming increasingly obvious, and in 1932, in an attempt to "do something about it," whaling nations devised a truly bizarre method of self-regulation, one that warmed the hearts of number crunchers but completely ignored the nature of whales. A value was assigned to the oil content of each species relative to the largest, most valuable kind, the blue whale; that is, each species was assessed in terms of "blue whale units," or bwu's. Only the amount of oil was considered; thus, two finback whales, 2½ humpbacks, or 6 sei whales were equivalent to one bwu. The slim, dolphinlike minke whales were too small to be relevant; more than 20 would be required to yield the amount of oil contained in a big blue. In effect, it was comparable to valuing Saint Bernard's for the weight of their bones; 2½ Labrador retrievers, or 8 cocker spaniels, or, say, 110 Chihuahuas might equal one "sbu." Blue whale units were used as a form of measurement until 1972, when individual species quotas were substituted, but by then only a fraction of the original number of large whales remained.

Meanwhile, though, an attempt was made to sort out who would get how many of what through a refined international agreement in 1937. So few gray whales remained by then that total protection was recommended. Depletion of Antarctic whales led to further meetings. Whaling pressure continued to increase—more boats, more shore stations, more factory ships—but catches were steadily declining. During World War II, the killing of whales slackened while humans concentrated more on killing one another, but new and more sophisticated methods were brought to bear when whaling resumed.

The noted whale researcher Scott McVay observed, "The whale has no more chance than a bull in the ring as it is scouted by helicopter, scanned by sonar, and run down by mechanized ships designed to travel three knots faster than a finback's top speed."

In 1949, fourteen nations established a new organization, the International Whaling Commission, and ratified the International Convention for the Regulation of Whaling. Curiously, it sought simultaneously to achieve "the optimum level of whale stocks," the "optimum utilization of whale resources," and the "orderly development of the whaling industry." Clearly, it wasn't possible to do everything at once—priorities had to be set. For the next twenty years, the development of the whaling industry proceeded with vigor; some observed that the commission must regard the "optimum level of whales" to be a number perilously close to zero, because that seemed to be the inevitable consequence of the policies being implemented.

Only when blue whales became so rare that it was not profitable to gear up to take them—in 1965—did the International Whaling Commission finally agree to a policy of protection. Even so, Chile and Peru, not members of the commission, killed 450 that same season. A few hundred might not seem like many when compared to the thousands taken in previous years, but in 1965, this represented a significant portion of the total number of that species then alive worldwide.

Three years later, in 1968, I stood on a cliff overlooking a coastal whaling station in Chile, a picturesque cove reeking of recent death. Rusting cauldrons, lately brimming with blood and bone, were the stuff of nightmares made real. So clever, I thought, we human beings, able to reduce important creatures a thousand times our size to small chunks of meat and buckets of oil; but how much more clever it would be *not* to destroy something so magnificent—large or small—that we neither understand nor know how to replace.

Half a world away, nearly ten years later, similar thoughts seethed through the detachment I tried to maintain as six young sperm whales, adolescent males, were sliced into long slivers and strips of skin, fat, and muscle, then stuffed through a hole gaping in the blood-slippery wooden platform upon which I stood to giant heated cauldrons below. The last season of the last whaling station in Albany, at the southwestern tip of Australia, was in full motion. Clouds of steam from great steel vessels merged with gray morning mist; flensers bright in yellow boots and slickers swiftly, methodically, struck with sharp knives, like vandals stripping parts from sleek racing machines. The silver-grayness turning red, white, liquid with urgent sawing, chopping, slicing, peeling; without pause, the tools ripped through nerves connecting structures for propulsion, sight, hearing, taste, touch, reproduction, to the largest brain ever.

I wondered, *Who is doing this?* Who decides that it is all right to consign such creatures to cookpots—without even listening to what they might have to say? Am I absolved because my hand is not on a knife? Who has the right on behalf of my species to establish a relationship with these extraordinary creatures based on slaughter? Isn't anybody curious to know what they do? Doesn't anybody share my yearning to know how they manage to dive for an hour on a single breath of air, to depths beyond a mile? Or navigate over thousands of miles to destinations that have no apparent signposts? What had these young whales seen, that no human has ever glimpsed? What did they know? Might they have a sense of humor? Appreciate the music of thousands of interlacing sounds coursing through night seas? Enjoy the sensation of warm water flowing over their silky skin? What might we have learned from them, alive, that could rather exceed their worth as pounds of meat, barrels of oil? At twenty-five years of age, a male sperm whale may enter his society as a reproducing adult. Had these youngsters yet attained such stature? Surely their absence was noticed by those who shared years with them, catching squid, gliding through sparkling dark seas, touching one another with booming low sounds and signature calls eerie to our senses, but the reassuring, invisible resonating fabric of theirs.

Is there some sort of 10,000-year-old, ingrained eat-or-be-eaten habit that causes us to kill first, and wonder later if perhaps we might have tried another approach?

In Australia, at the time, the livelihood of about 150 people was at stake, two thirds of them part-time, who derived income from the sale of whale meat, teeth, oil, and an occasional lump of ambergris retrieved from a whale's stomach. Whaling interests were balanced against the mind-set of millions of Australians who believed that whales had enduring values—alive. A petition circulated throughout the country that year, 1978, calling for an end to whaling, a widely supported cry from within that helped turn Australia's focus from being a "whaling nation" to one that soon became a leader in protecting the whales that remain.

A similar revolutionary change was growing worldwide, even within the ranks of the International Whaling Commission. One by one, member countries—Australia, the U.S., New Zealand, South Africa, Britain—stopped commercial whaling altogether, while retaining their seats at the negotiating table. Nonwhaling nations joined the commission; some were brought on board by those favoring continued lethal exploitation, while others had no intention of whaling, but wanted to have a voice in determining the fate of whales. So critically depleted were whale populations by the end of the 1970s that a complete moratorium on killing whales was recommended in 1979, and again in 1980 and 1981. Finally, in 1982, a moratorium on commercial whaling was established and limits were set on those allowed to be taken by aboriginals. Not everyone was pleased. Japan, Norway, and Iceland are among the nations who not only expressed extreme displeasure, but have pressed steadily to maintain some level of killing to implement "scientific research"—despite the International Whaling Commission's rejection of the validity of the research. Their stated goal is swift resumption of commercial whaling.

Curiously, my interest in whales and whaling nearly prevented my appointment as the Chief Scientist of NOAA in 1990, which taught me some of the perils that await those who aspire to public office, and put me in the thick of issues that I had tracked from afar for years. When the Administrator of NOAA, Dr. John Knauss, asked whether I would consider having my name put forward as Chief Scientist, he added an additional incentive by saying that I might also have a chance to be involved, through the International Whaling Commission, with international as well as domestic issues concerning whales and other marine mammals. It was the nudge I needed to agree to be nominated, and it seemed that the ensuing Senate confirmation would very likely be a noncontroversial, routine matter. It was—except for objections raised by one Southern senator.

In September 1990 I was called to the senator's office and informed politely that some people were concerned about my views with respect to whales. Would I care to explain? Most particularly, would I tell him what I meant by comments made during an interview in Iceland in 1980? A fat book of clippings and other information labeled "Sylvia Earle" was produced, and there was a newspaper article—in Icelandic. A smudged, typed translation accompanied the story, roughly quoting me as believing that whales were more valuable alive than dead.

Perplexed, I said, "I might well have said something like that, but I can't vouch for a translator's account of something that a reporter said I said in a language I neither read nor speak. Can you tell me what you're concerned about?"

"Well," he said, leaning forward intently, "what were you doing in Iceland anyway, having your picture taken with these people?" He pointed to a news photo of me with two individuals notorious for helping to achieve the global moratorium on whaling.* We were featured speakers at a conference where I had been invited as a scientist to describe my experiences underwater with whales, and to show the film *Gentle Giants of the Pacific*, made during my year of research on humpback whales.

"Iceland is one of about twenty countries where I have shown the film," I told him. "One of the first was the Soviet Union. People who had worked with dead whales for years had a chance in the film to see whales as whales see whales, underwater, not just as carcasses. Other countries I visited include South Africa, Seychelles, Sweden, Switzerland, Kenya, Canada, Australia, New Zealand, Norway, Mexico . . . even Mississippi," I said, with a warm smile. "But perhaps the most fascinating audience was in China in 1980. I traveled with General Stilwell's youngest daughter, Alison, who arranged a meeting, dinner, and film-showing for Madam Soong—Madam Sun Yat Sen—and her household of about forty people."

The senator leaned closer, apparently intrigued with the concept of whales as "ambassadors" in an international arena.

"I learned something about the difficulty of communicating through translators," I continued, "but most important were the images. The photographs speak for themselves."

"Right," he said, suddenly very serious. With a hard look at the Icelandic news clipping he added, "You should watch out who you allow yourself to be

---

* Dr. Sidney Holt, world-famous fisheries and cetacean expert, and Alan Thornton, cofounder of Greenpeace-UK.

photographed next to. Photographs speak for themselves." (Happily, no photograph exists to commemorate my meeting with the senator.)

Right. By the time I left, I thought my chances for confirmation as Chief Scientist—let alone involvement with the International Whaling Commission—had slipped to zero. In fact, I had no further contact with the senator and in due course was confirmed at NOAA; later I was appointed as the U.S. deputy commissioner to the International Whaling Commission, a largely honorific position that lacks the negotiating authority vested in commissioners but gave me access to planning and led to my serving as liaison with nongovernmental groups concerned with whales and whaling. I was acutely aware that no more than 5 percent of the people doing the negotiating had ever seen a whale alive, and a fraction of that number had met a whale underwater. Many had never seen a whale, period.

When I returned to Iceland as deputy commissioner in 1991 to attend the forty-third annual meeting of the IWC, it did not take long to discover what seasoned veterans already seemed to understand full well: The focus was on whales as bargaining chips, and the IWC was regarded as a useful mechanism for nations to jostle for position, develop alliances, and gain or lose stature. For example, I was told in no uncertain terms that the United States did not intend to censure Iceland for their "scientific whaling" program, a program that was a blatant abuse of the system and led to the killing of hundreds of whales. Science was not the issue. Rather, vital U.S. military interests in Iceland might be jeopardized if we took a "get tough" position with them about whales. At the same meetings, representatives from whaling nations—Japan, Norway, Iceland—tried to find new respectability for the business of killing whales for commercial gain by suggesting not only that they could be taken on a sustainable basis, but that it was somehow irresponsible not to do so.

The publication in 1987 of *Our Common Future*, the report of the Brundtland Commission, an international commission headed by Mrs. Gro Harlem Brundtland, the Prime Minister of Norway, turned the thoughts of many to that elusive goal, so beloved by fisheries managers, of wresting from natural systems a sustained yield, of getting something for no investment, other than the costs associated with catching. The Brundtland Commission focused on the issue of "sustainable development," a term that has about it a tantalizing but inherently unattainable "perpetual motion" aura. The longer definition is more promising: "development that meets the needs of the present without compromising the ability of future generations to meet their needs." Some say, more realistically, "sustainable *use*."

The flaws inherent in the sustained commercial exploitation of wild species have been demonstrated convincingly with whales and many kinds of fish. Recently, sharks have joined the ranks of those to whom the theory is being applied. To many, the only good shark is a dead shark, a target of sport, and increasingly a popular item on menus in posh restaurants. Few seem to appreciate the role of sharks during more than 300 million years as critical links in complex ocean ecosystems. To some, it is a "macho" meal, a bit like eating lion steak—a comparison that has validity beyond image alone. Like lions, sharks are predators and require a wide territory for survival. More than 100 square miles of ocean is required to support the eating habits of a 200-pound individual, and reproductive rates are low for most of the 370 or so species. Sharks tend to mature slowly and individuals may live for decades. As with large land predators, it is easy to eliminate sharks from coastal areas through unrestricted killing, for sport or commercial use.

In the 1970s, concerns were raised about the large numbers being taken by fishermen, but little sympathy could be aroused for these extraordinary, durable creatures who, like nautiluses, coelacanths, horseshoe crabs, crinoids, and other "living fossils," survived changing times through hundreds of millions of years. However, nothing had prepared sharks for the technologically powered onslaught now brought to bear against them.

Despite knowledge that most shark species could quickly be depleted because of their top-predator position, slow growth, and low reproduction potential, in the 1980s in the United States fishermen were deliberately encouraged by the National Marine Fisheries Service to focus on sharks as the basis for a major fishery, the idea being that they were "underutilized species." The dangerous decline of "overutilized" kinds was not perceived as fair warning that the same pattern could—and would—follow with the new targets, but that is exactly what happened.

There are other, more recent, explanations for the increase in shark killing. One is the growing popularity of shark-fin soup, especially in Asia, a demand fueled by economies strong enough to pay top prices for fins imported from all over the world. The high value of fins relative to the rest of the shark led to the grotesque practice of "finning"—slicing off fins and tail, then dumping the still-living body overboard. There is also a belief that consuming the cartilage that forms the basis of a shark's skeletal structure will deter or cure certain kinds of cancer. Since 1992, hundreds of thousands of sharks have been taken from ecosystems in Central and South America, even from the presumably protected waters of the Galápagos Islands, specifically to supply the demand for dried, powdered cartilage.

In 1989, less than ten years after actively encouraging shark fishing, NOAA introduced the draft of a "recovery plan" for sharks, one, that would limit takes and attempt to restore their now seriously depleted populations. The shark industry reacted with outrage, largely because many had entered the business as a direct result of NOAA's encouragement—only to find that the raw materials for their industry were suddenly in short supply—and because restrictions on U.S. fishermen would not prevent foreign fishermen from taking sharks that freely move in and out of what humans declare to be "territorial waters." The refrain "Why should we leave them for someone else to take?" is a modest variant of the familiar "If we don't take them, someone else will." The end point of this attitude is, of course, that soon, there will be none for anyone.

Some kinds of sharks are currently escaping human predation, especially those in the deep sea, from the curious foot-long "cookie-cutter shark," a toothy creature fond of nipping neat, circular bites from large sharks, whales, and soft parts of submarines, to the giant "mega-mouth," a handsome species known for its bioluminescent lips, tiny teeth, and habit of munching on small crustaceans. Despite the possibility that some species of sharks may be hustled over the edge to extinction in the near future, chances are good that others will continue to prosper in the sea—just as in the past hundred years, there have been losses of bird, mammal, insect, fish, plant, and many other species, yet the basic categories remain. Still, because the number of species is low—less than 400—sharks are perhaps more vulnerable as a group than, say, the much more numerous and diverse bony fishes.

What would be the consequences of totally removing sharks from the sea? Or any other category of creatures? Clearly, species have come and gone through time, and so have orders, classes, perhaps even phyla. Most earth scientists accept as fact the occurrence of five major catastrophes resulting in massive extinctions since jellyfish, squids, sharks, horseshoe crabs, and the ancestors of krill first made their appearance several hundred million years ago. Various explanations have been offered as to what caused these waves of extinctions and great loss of biodiversity; suggestions include collision with meteorites or comets; major, sustained volcanic activity; continental drift and coincident climate change; and more. No one debates the cause of the present high level of losses, however. It is at once humbling and horrifying to discover that the wondrous capability humans have for altering materials to our advantage has also been applied with such devastating force as to cause change of geological magnitude, in the span of a human lifetime.

Some shrug and note that nature is obviously resilient. Many miss the crucial corollary to resilience, and that is that nature takes her time. Restoration to a healthy level of diversity following major catastrophes spans tens of millions of years. No one knows how long it might take to recover from the devastation to natural systems already wrought by human pressures, nor can anyone guarantee that in the long run there will be a place for humankind in the mix. A global experiment is in full motion, with humankind as initiators, observers, and obligatory participants.

If it were possible to start again, with the planet as it was 10,000 years ago—or even 500 years or 200 or 20—with knowledge now gained, there would be a better chance to achieve the much-discussed goal of "sustainable use" of planetary resources. Concerning whales, perhaps an entirely different relationship might have been constructed, one focused not on taking intelligent, 40- or 50-year-old creatures as a source of tangible commodities but on viewing them as a source of far more valuable intangibles, including insights into the way the world works. We can never know what the potential might have been for understanding ancient societies of marine mammals, and perhaps even for establishing a productive rapport with some, had the options not been closed by those who single-mindedly chose to turn those complex systems into chunks of raw meat.

While there may have been better times to act to preserve ecological and other options for the future, now is better than later, or never. In 1978, motivated by a growing sense of urgency, individuals from three organizations, the World Wildlife Fund International (WWF),* the International Union for Conservation of Nature (IUCN),† and the United Nations Environment Programme (UNEP) teamed up to draft a "World Conservation Strategy," a global action plan for resource conservation that preceded the Brundtland Commission by nearly a decade, and in some ways paved the way for its formation.

From 1978 through 1980, while serving as a member of IUCN's governing council and WWF's board of trustees for both the international and the U.S. organizations, I participated in several meetings with the industrialist-conservationist Maurice Strong, the World Wildlife Fund's founder and chairman, Sir Peter Scott, and several dozen others who gathered to brainstorm priorities. The widely endorsed strategy that evolved embraces the three most important objectives of conservation, namely, to achieve:

- Maintenance of essential ecological processes and life-support systems
- Preservation of genetic diversity
- Sustainable utilization of species and ecosystems

The same underlying principles provided the framework for the United Nations Conference on Environment and Development, UNCED, held in Rio de Janeiro in June 1992. Spearheaded by Strong as UNCED's Secretary General, it was an unprecedented meeting involving official representatives and the heads of state of many countries. For a while, world attention was focused on

---

* Established in 1961 and since 1988 called the World Wide Fund for Nature, WWF is associated with more than 30 national organizations. One of these, the World Wildlife Fund-U.S., retains its original name.

† Since 1992 the IUCN has been known as the World Conservation Union with a membership of more than 60 countries, 128 government agencies, and over 400 nongovernmental organizations in 118 countries. Established in 1948, the union has headquarters in Gland, Switzerland, and an environmental law center in Bonn, Germany.

the deteriorating condition of the global environment and what to do to prevent further loss of quality and productivity.

UNCED was a herculean accomplishment, one with positive repercussions that will resound through the workings of future policies forevermore. Yet some participants left dissatisfied as a result of opportunities lost, momentum that passed without an enduring legacy of results in vital areas. As an official guest and witness to the proceedings, I listened closely for strong words that needed to be said about marine topics, and there were some, especially with respect to the much-discussed Agenda 21, involving, among other things, policies concerning biodiversity.* But terrestrial and atmospheric issues dominated headlines and action plans. The oceans, many feel, have priorities less urgent than critical terrestrial and freshwater problems.

Whatever the issues, for me as a biologist it is satisfying to see a growing awareness of the importance of the creatures with whom humankind shares the planet, whether they come from wet places or dry.

I continually ask myself, deep down, *why is it* important to keep intact the elements of "ecosystem Earth"? What harm could there possibly be in the next half century or so (the time most relevant to most people) if 50 or 500 or 5,000 of the 500,000 or perhaps a million species of insects should cease to exist? Or why not 25 or 250 of the 25,000 or so kinds of fish? And who cares about jellyfish? Who even knows how many there are or what they do? Who would miss 3 or 30 or 100 of the 300 kinds of primates—as long as *Homo sapiens* were not on the list of those marked for extinction. . . .

If it were my job to justify to aliens intent on consuming every life form except those absolutely essential for human life to continue indefinitely, I would be inclined to stand firm and say, "Stop! Every scrap of diversity is essential. The system works as it is presently constructed. There is no evidence to indicate that any parts can be deleted without unfavorable consequences." Every variation on the theme of those creatures that make up ecosystems, and

---

* While the protection of marine biodiversity was not a major issue, the political aspects of bio-diversity created a stir when the United States would not sign the widely endorsed international treaty on biodiversity that includes a provision for royalties on products derived from natural sources (mostly rainforests) to be paid by users (mostly pharmaceutical companies) to countries (mostly developing) where the plants and animals originated.

the diversity of the systems themselves, enhances the likelihood that stability will be maintained throughout the ups and downs of broad physical, chemical, and geological change. Each species is part of a planetary insurance policy for maintaining gradual, not cataclysmic, adjustments to changing environmental circumstances. The burden of proof should be on those who deliberately propose eliminating diversity to demonstrate, conclusively, that no harm to ecosystem Earth, and thus human survival and well-being, will result. Once gone, no amount of high-tech engineering, wishful thinking, or mumbo-jumbo can restore a lost species, let alone a lost genus, family, order, class, phylum—or ecosystem.

Red Data Books, published by the IUCN, provide a backbone of information for the grim accounting of losses, including the following very small sample:

> *One fifth of the 11,000 or so bird species that lived on earth 2,000 years ago are gone, principally following human occupation of islands. Of the current 9,040 species remaining, eleven percent are endangered.*
>
> *Of the 250,000 species of flowering plants in the world, about 20 percent are expected to be gone by the year 2000. Of the 20,000 or so species that live in the United States, more than 200 have already become extinct and nearly 700 more species and subspecies are expected to be gone within the decade.*

E. O. Wilson addresses himself to the question, "What difference does it make if some species are extinguished, if even half of all the species on Earth disappear?" In his 1993 book *The Diversity of Life*, he responds:

> *Let me count the ways. New sources of scientific information will be lost. Vast potential biological wealth will be destroyed. Still undeveloped medicines, crops, pharmaceuticals, timber, fibers, pulp, soil-restoring vegetation, petroleum substitutes, and other products and amenities will never come to light. . . . It is also easy to overlook the services that ecosystems provide humanity. They enrich the soil and create the very air we breathe. Without these amenities, the remaining tenure of the human race would be nasty and brief.*

Wilson focuses on the land, especially his beloved rainforests, and with good reason. Within the past two centuries, more than half of these once-and-nevermore
ecosystems have been destroyed, mostly victims of clear-cutting for timber or
burning for short-term use of the land for agriculture. In the process, the planet
has incurred a staggering loss of biological wealth. Although occupying only
6 percent of the land, tropical rainforests are believed by Wilson to be home
for more than half of Earth's species, including a high proportion of the flowering plants, ferns, amphibians, birds, and the incredibly diverse variations on
the theme of arthropods, including spiders, ticks, mites, pillbugs, centipedes,
millipedes, and especially insects. Those who might be cheered by the fact that
fewer "bugs" exist now than was so two or three decades ago might think again
after hearing Wilson's assessment of the value of such creatures to humankind:

> So important are insects and other land-dwelling arthropods that if all were to dis-
> appear, humanity could not last more than a few months. Most of the amphibians,
> reptiles, birds, and mammals would crash to extinction about the same time. Next
> would go the bulk of the flowering plants, and with them the physical structure of
> most forests and other terrestrial habitats of the world. . . . The land would return
> to approximately its condition in early Paleozoic times . . . largely devoid of animal
> life.

In making a case for insects, Wilson directs attention to their crucial importance
as the cornerstone of life on Earth. In the sea, crustaceans are insect counterparts. Although not as diverse in terms of numbers of species, crustacea are
linchpins in complex ecosystems that are utterly dependent on their continued
prosperity. In effect, they are energy "middlemen," concentrating and converting
plant energy into something palatable and usable by hordes of other creatures.

An example of the importance of a single crustacean species of crustacea is
the Antarctic krill, *Euphausia superba*, a translucent pinkish-red creature about
as long, when mature, as a human thumb. Each is equipped with enormous
black eyes and the wonderful ability (shared with nearly all of the other eighty
or so kinds of krill in the sea) to emit a beautifully eerie, brilliant blue-green

bioluminescence. Antarctic krill mature rather slowly, passing through more than a dozen stages from egg to adult, each variation feeding on slightly different elements of the plankton soup surrounding them. It may take two years for a single individual to mature to the point where it can reproduce. If it isn't trapped in a net by fishermen or consumed by one of the several species of whales, or by seals, fish, birds, squid, and other creatures that rely on krill for sustenance, an individual may live for eight or nine years. This species has prospered so well in Antarctic waters over the ages that it has become one of the single most numerous kinds of creatures in the sea, and in so doing has become a critical source of food for many organisms unable to tap into plant energy directly. Krill are considered absolutely vital to the healthy functioning of the entire Southern Ocean ecosystem, an aquatic area that accounts for one tenth of the global ocean. So vital is this one species that it would seem that an all-out effort should be made to ensure that *nothing* is done to disturb its continued prosperity. If there is one creature in all of the Southern Ocean that should be left alone, it is *Euphausia superba*.

Instead, hundreds of thousands of tons of krill are removed from the Antarctic ecosystem annually. Consequently, the sun's energy and minerals fixed by plants and consumed by krill that normally would help drive the local ecosystem have been transported to distant parts of the planet as high-protein food for poultry, cattle, and other livestock as well as for direct human consumption. Removal of the krill also means not only that the local birds, squid, fish, seals, and whales have less to eat, but also that there are fewer krill to reproduce, and fewer to provide resiliency to natural fluctuations. Without knowing, and perhaps not caring, about the consequences, several nations took krill without restriction from the 1970s until 1991, when a limit was set for South Atlantic regions.[*]

The setting of quotas was intended to avoid a repeat of the devastating losses of whales, seals, penguins, and certain fish in Antarctic waters that were taken

---

[*] The Commission for the Conservation of Antarctic Marine Living Resources, an international organization formed in 1981, recommended that the take be limited to 1.5 million metric tons per annum in the South Atlantic regions. A year later, a limit was set for the southern Indian Ocean as well. This figure seems conservative to some in view of the estimated total "standing population" of krill—calculated to be more than 100 million metric tons—but there is no way of knowing whether or not the population estimates are near or far from the truth, no way of knowing what impact removal of large quantities from certain areas may have on the species or the system.

without restriction to the point of population collapse. The optimistic aim now is to pluck a few feathers from the goose while hoping for a steady supply of golden eggs. Unfortunately, the goose may need all her feathers to survive. Krill are simply too important to the stability of the Southern Ocean ecosystems to gamble that there is enough built-in slack to allow for repeated, large-magnitude losses to human predation.

The same principle applies to other species, and to other ecosystems, including the "big one," Earth itself. Without pausing to understand the consequences to the living systems that support us, humankind has over the ages relentlessly diminished genetic and ecological diversity to the point where full recovery is no longer possible. Each species lost diminishes the chance that we can "get it right," that is, find an enduring place for ourselves within the living matrix that sustains us. If humankind proves its ability to be wise as well as clever, we'll implement new and durable measures to maintain the living assets that we have too often taken for granted, and still do. Every day, the number and demands of humankind increase; the size of the planet does not.

Chapter 12

POISONING THE SEA

*With amazing arrogance we presume omniscience and an understanding of
the complexities of Nature, and with amazing impertinence, we firmly believe
that we can better it. . . . We have forgotten that we, ourselves, are just a part
of nature, an animal which seems to have taken the wrong turning, bent on
total destruction.*

Daphne Sheldrick, The Tsavo Story

Sudden death came to Florida Bay in 1993. The ripe stench
of stricken sea grasses, corals, sponges, starfish, crustaceans, and fish dying en
masse assaults the senses; blooms of algae dominate once-diverse marine meadows; hundreds of square miles fester, an infected wound in the ocean.

For forty years, I have admired this region just south of the Everglades,
world-renowned for its beauty, and a productive marine nursery and home for
billions of sea creatures. I therefore felt a sense of personal loss, anger, frustration, and infuriating helplessness when long-brewing problems reached a
critical level. It is difficult to pin down the precise cause or causes of Florida
Bay's ecosystem upheaval, but no one doubts that it is one of thousands of man-
made disasters now sweeping the oceans. High concentrations of fertilizers,

pesticides, herbicides, and other exotic chemicals flowing primarily from large agricultural developments in south Florida, especially sugar plantations, coupled with decades of diversion and manipulation of traditional sources of freshwater, are believed to be responsible for the wholesale collapse of the bay's ecosystem.

Call the problem "pollution," if you will. The sea, like a great vat of milk, can tolerate many drops of vinegar, but there comes a time—no one can predict just when—when the system goes sour. The wondrous nature of water, that it is at once the "universal solvent" and the global transport system for molecules and masses of debris alike, also makes it particularly vulnerable to debilitating, sometimes lethal, contamination.

No place in the sea, including the deepest regions, is now devoid of recent man-made pollutants, but some areas, such as Florida Bay, have "gone critical" and turned. The thought that we might unwittingly initiate huge-scale reactions over, say, large ocean basins or the entire planet, haunt the depths of my being. Worldwide, this generation's signature is written in hard trash, heavy metals, chemical wastes and spills, agricultural runoff, industrial and domestic wastes, oil spills, radioactive materials, even noise and electromagnetic fields. These pollutants individually and collectively cause modifications to the character of life in the sea, to seawater itself, and to the ocean's underlying systems and life processes. Florida Bay is just one conspicuous example, a danger flag that should serve to alert us to wider, deeper issues, and bring into focus critical questions never before facing mankind. One of the most important is: To what extent are changes caused by pollution in the sea undermining the fundamental framework of living ocean systems and underlying processes that have, through time, shaped the nature of Earth into a planet hospitable for life as we know it? And, just as significant: *What can be done about it?* Facing up to these questions, and responding to them effectively, will have much to do with our future success as a species, and some thoughts about both will be considered in the following chapters.

It is tempting to hope that there is no need to be concerned about "life-support systems" or grand Earth "processes," and to embrace the thinking of skeptics who point out that the planet is, after all, large, robust, resilient, and quite capable of taking care of itself. After all, life persisted through immense meteor-induced devastation, the impact of multiple mega-volcanoes, and more.

Surely, the argument goes, life will continue to prosper no matter how much damage we inflict on the natural systems.

If concerns about the consequences of pollution in the sea were focused only on the question of threats to *life* on the planet, we could probably stop worrying. While it is possible to imagine that humankind could set in motion a chain of events causing Earth to become truly lifeless, like the moon or Mars, it is far more likely that human actions could modify the planetary systems in ways that might be unfriendly for, say, flowering plants and air-breathing vertebrates, but still be acceptable for cockroaches, fungi, bacteria, pond scum, and many sea creatures. In any event, there is no guarantee that humankind will be among the survivors of even subtle changes to the global environment, despite our technological cleverness and the widespread belief that our species enjoys unique status that somehow absolves us from responsibility.

Mankind aside, the swift decline of highly successful, numerous creatures with large populations provides insight into the consequences of abrupt environmental changes. After millions of years of prosperity, an entire category of organisms, the dinosaurs, were eliminated from Earth, apparent victims of environmental changes to which they could not adjust. Passenger pigeons, once numbering billions, were brought to extinction within a few years of determined effort by a single predator—humans. My parents witnessed the swift decline of ubiquitous chestnut trees and elms, the victims of fungi introduced from Europe. In the 1940s, disease decimated and nearly destroyed Atlantic eelgrass meadows consisting of billions of individual plants, and another epidemic of unknown cause eliminated most of the Gulf of Mexico's numerous commercially valuable sponges. Atlantic bluefin tuna, certain kinds of salmon, and even the fecund icon of New England fishing, the cod, are recent examples of once very numerous species poised for a permanent exit.

Such swift, sweeping losses for extremely abundant and apparently durable species should make us consider our own vulnerability. Might the subtle changes we are causing in the chemistry of the sea now set in motion waves of extinction? Could certain microbes, now occupying highly specialized, restricted niches, find the conditions we are creating more favorable—and enjoy population explosions that trigger other events inhospitable to us? Changes in the sea in the past few decades should command our rapt attention—the sort of interest one might take in, say, the life-support system of a spacecraft housing all of the past, present, and future of humankind.

As a child, I was oblivious to environmental changes that were already well under way, and for years I remained largely unaware of the simmering environmental issues that would soon explode as mainstream concerns. The articulate turtle specialist, Archie Carr, helped prod me out of my complacency and evoke a personal sea change in attitude. It was unintentional on his part; he was just being himself, talking in his usual eloquent, laconically amusing style at a meeting of the American Institute of Biological Sciences in Gainesville, Florida, in 1953.

In an entertaining hour, Carr traced the past several million years of history of Florida, illustrated with sketches of the camels, mastodons, horses—and large turtles—now extinct, that once abounded in the very area where I then sat, with about three hundred other rapt listeners. *Elephants and camels in Florida?* I thought. *What a concept!* I knew something about horse history, but had not appreciated the magnitude of the numbers and kinds of large animals that had once enjoyed a very different Florida from the one I was getting to know. Moreover, Carr described the rise and fall of the sea level, alternately covering most of the peninsula, then exposing submerged land that essentially doubled the state's size. That was a sobering thought, especially when he noted that it was inconceivable that things would remain as they were for long, geologically speaking. Ice ages would come and go, sea level would rise and fall, and some species would prosper, while others would fade. It would have been enough had he stopped there, but he added a significant caveat, a single throw-away comment that jolted me into a new way of looking at the world. He said, matter-of-factly, that the millions of years of changes just described were "more than matched" by changes he had witnessed growing up in Florida! Now *that* was something to think about. Change of geological magnitude witnessed by individuals boggled my mind. It still does! I try to remember the way things were as far back as I can, compare those images with the present, and muse endlessly about the future.

The exercise, often repeated, personalized my reaction to Rachel Carson's pivotal work, *Silent Spring*, published in 1962. She opens with a chilling fable about a town in the heart of America where "all life seemed to live in harmony with its surroundings," which was changed through actions people brought on themselves to a "spring without voices . . . Everywhere was a shadow of death."

In places I knew as a child, the trend toward a world such as Carson anticipated is clear. Orchards of pear and apple trees that once mingled their blossoms in springtime breezes with petals of wildflowers, blown from surrounding untamed forests and meadows, have disappeared. Grassy fields and wooded areas are now scarce, and gone, too, are most of the foxes, hawks, songbirds, bats, flying squirrels, and wonderfully diverse assemblage of insects—not just the infamous Jersey mosquitoes!

More mesmerizing than the most ingenious toys of my early years were the ethereal, pale green–winged luna moths, elegant mossy-backed creatures wondrous enough to inspire belief in elves and fairies, or otherworldly visitors. These silent beauties mysteriously appeared on the front screen door on summer nights, sometimes two at once, perhaps drawn to the light emanating from our hallway. My brothers and I played with giant metallic blue-green June bugs that punctuated their humming flight with a solid *thwump*, landing on soft earth and leaves carpeting our yard. By unspoken agreement, some of our lima bean crop was shared with bean beetles, some of the apples, plums, and pears with enterprising birds and grubs, some of the corn with corn worms, and some of the roses with beautiful but voracious Japanese beetles, an inadvertent Asian import to North America. We even shared some of the tomato leaves with plump, jade-green caterpillars—a true concession, given my parents' passion for vine-ripened tomatoes, and the justified suspicion that there might be a few less because of the great green munchers.

In the 1940s, the wonder chemical DDT (dichloro-diphenyl-trichloro-ethane) was growing in popularity as a panacea for ridding plants of unwanted insect pests, but the concept for using poisons on crops destined for the table was not warmly received in our household. Instead, kid power was harnessed to help limit the numbers of insects by picking them off, one by one, and offering them as a welcome treat for our small flock of chickens. Soon, however, like it or not, all of the members of my family, like most other living creatures on Earth, took DDT and a wide range of other exotic chemicals into our systems. The small farms in southern New Jersey, like farms throughout the world, embraced DDT and its many successors as valuable allies that greatly enhanced the quality and quantity of production—but with an unexpected price tag. Half a century of dedicated use of pesticides has brought about the insidious legacy predicted by Rachel Carson in 1962: Land, water, and all living systems now bear a toxic load new to the planet, with unknown but surely not desirable consequences. Everywhere *is* a shadow of death, often manifested as lessened resiliency to normal day-to-day encounters with disease, accidents, and age. Entire ecosystems, too, are made more vulnerable to decay with the loss or decline of a full, balanced complement of players.

Recently, my mother, my younger daughter, Gale, and I visited the small farm where I first learned the vital importance of rain, the wisdom of taking care of systems that take care of us, and where I developed a permanent love affair with critters of all sorts. We found that most of the nearby farms had been replaced by buildings connected by a gridwork of cement and asphalt streets. During a conversation with my lively 80-year-old aunt Frances, who had spent most of her life in the nearby town as an artist and keen observer of life, I could not resist asking, "Do you remember when you last saw a luna moth?"

She brightened, as if recalling the face of a beloved but distant friend, one not heard from for some time. "Oh! Those big green moths with long tails on their wings?" she replied, and paused.

I hoped for a recent sighting, a sign that somewhere in an overlooked wooded patch or backyard garden, there might still be new generations of lunas in the making.

"Goodness, I haven't seen one of them since—well, not since your family left the farm and moved to Florida. Nineteen forty-seven, right?"

Right, I thought. Well, perhaps there are some, just not as often seen as they once were. Or maybe, of course, these ancient settlers are no longer part of the south Jersey ecosystem, having become inadvertent casualties of the quest in recent decades for perfect lawns and increased vegetable production.

I was delighted to discover wild violets still sprouting on a favorite hillside. My daughter gathered a small bouquet of blossoms to present ceremoniously to my mother, as I had myself many times as a child. Eagerly, I moved toward the base of the hill, where a small, lively creek wound through the neighborhood, ultimately flowing into the Delaware River and then to the sea. I particularly remembered a time when my mother allowed me to dip my hands in the creek ("Your fingers should be wet . . .") then form a cup with them to hold what she obviously regarded as a precious but lively treasure ("Be careful! Be gentle . . . don't squeeze . . . that's it"), a glistening, emerald and topaz-hued bullfrog, with sparkling dark eyes ringed with pure gold ("See, he's looking at you. Put him in the water, now, and watch how he kick-swims away").

There was water at the bottom of the hill, all right, but sluggish and dark, an unnatural iridescence curling ominously across its surface. Now I looked in vain in familiar places for tadpoles that should have been there. Had it just been a bad year for frogs? Or might their absence be linked to the national passion for eradicating insects, or to something sickening in the rainbow-streaked creek? Was this just a local loss? Or was it part of the worrisome global decline in amphibians noted in recent years, without certain cause but likely tied to widespread changes in chemistry felt by frogs and their kin, though not yet perceived clearly by humans. Is this one of a growing number of signals that all is not well with the healthy world I knew as a child?

Increasingly, Carson's concerns about the heavy-handed application of pesticides and other toxic chemicals to the planetary systems that support us have been vindicated, and many nagging questions have been answered the hard way—with grim statistics verifying that yes, exposure to high doses of insecticides such as DDT, aldrin, and dieldrin *does* increase the risk of cancer in many creatures, including humans. And yes, the chlordane sold to my father in the 1970s as a wonderfully safe alternative to dangerous chemicals was, in fact, not only harmful to fire ants and termites but also not good for human beings. Years after confirmation that chlordane is a neurotoxin, inducing fatigue, headaches, memory loss, and vertigo, it was finally banned in 1988. Malathion, more recently introduced as a substitute for more potent poisons, has now been proved to have unexpected drawbacks. Any pesticide that is not highly specific in its effects is likely to impact a much broader part of the ecosystem than is intended, including a damaging influence on humans. As Carson made clear years ago, "pesticides" and even "insecticides" should be called "biocides," to rightly reflect what they actually do.

A 1992 news report from the U.S. National Research Council notes that less than 10 percent of some 70,000 chemicals used in commercial and industrial products have been tested for toxicity. Despite expanded efforts to identify chemicals that could cause problems, especially to the nervous and immune systems of humans, it will be many years before straight answers can be given to basic questions concerning the safety of these new compounds. Meanwhile, more chemicals are being synthesized, used, and discarded into the atmosphere, groundwater, and the sea without clear understanding of their short- or long-range impacts on human health directly, and on the environment, which in turn is the basis for human survival and well-being.

Toxins build in natural systems, flowing through groundwater and runoff and via aerial transport into streams and rivers that ultimately join with the sea. DDT has made its way into the tissues of krill, fish, penguins, and other creatures in Antarctic ecosystems; polychlorinated biphenyls (PCB's) occur in polar bears and other dwellers of Arctic systems, far from any obvious source of pollution. Beluga whales in Canada hold in their tissues toxic loads, including heavy metals, accumulated directly from the inland and coastal waterways where they swim, and indirectly, through the fish they have consumed.

One of the first examples of problems that toxins in the sea can cause occurred in the early 1950s, when families of fishermen near Minamata, near the island of Kyushu, began to show signs of a mysterious illness whose symptoms were loss of coordination, convulsions, difficulty in speaking, and numbness in the mouth and extremities; the malady was variously diagnosed as cerebral palsy, brain tumors, encephalitis, and even syphilis. Investigation that spanned more than three years finally tracked down the cause of the strange malady: consumption of locally caught fish containing high levels of mercury. Chemical wastes from the town's principal industry were discharged directly into the sea and were then incorporated into the ecosystem, with concentrations mounting through the food chain from small crustaceans to large, carnivorous fish. Before the cause was determined and actions taken to stop further tragedy, more than two hundred people were afflicted, including mentally and physically disabled children born to mothers who had consumed contaminated fish. Fifty-three people died.

Until recently, the "deep-sixing" method of disposal, either through sewers draining into the sea or deliberate dumping offshore, has been thought to be an easy, safe way to get rid of hazardous wastes—or anything else not wanted on land. The out-of-sight, out-of-mind scheme seemed to work, until evidence (such as the poisoning of Minamata's residents) began to grow that such policies have undesirable, sometimes deadly consequences.

In the early 1970s, scientists on a routine mission to sample fish stocks in the clear waters near one of Hawaii's main islands proved the point the hard way. The research team watched as their vessel's stern winch groaned and lines on the sampling trawl snapped taut, stressed by an overly heavy catch—something none of them had seen before.

"Looked like three or four bronze cannons at first," one scientist later recounted, "but what rolled out on the deck were fat, heavy cylinders, obviously metal containers of some sort sealed at both ends. Something started leaking out of one of them that made the guys on deck cough—then they started screaming. Whatever it was was blistering their arms and legs and hurt like hell!"

No one died, but those who came in contact with the leaking containers were hospitalized and some were left with permanent burn scars, a lifetime reminder of the deadly catch—mustard gas or something equally noxious dumped into the sea after World War I, or perhaps more recently. No one knows for sure. The mysterious canisters were rolled back into the sea as soon as their lethal potential was realized.

Measures have been taken to prevent deliberate disposal of toxic wastes into the sea, especially in areas where fish and other marine life are taken for human consumption. However, toxins enter the sea through other avenues, the so-called "nonpoint sources" of pollution. Pathways include groundwater under lawns sprayed with weed killers flowing to streams and rivers and ultimately to the sea; lead-loaded exhaust fumes from automobiles, lofted into the atmosphere and later deposited with water droplets onto a lake or river—or directly into the sea; organically active agricultural chemicals (pesticides, herbicides, fertilizers) washed into rivers and ultimately making their way to the sea following heavy rains or floods.

Some regard the oceans as infinitely capable of absorbing and rendering harmless even the most noxious products of human civilization. "The solution to pollution is dilution" is a phrase commonly used by proponents of turning to the sea as the best place to dispose of whatever is not wanted on land. Over the years, however, policies favoring ocean waste disposal have developed a reputation for creating illusion, delusion, and confusion—but rarely solving the problems of safely coping with toxic substances.

The very nature of the sea as a global environment, encompassing the entire world, compounds the problems. Most nations agree not to deliberately dump toxic wastes into the sea, but if any do, all suffer. Everybody wins if everybody respects the guidelines, but all lose if even one does not.

Late in 1991, when I was serving as Chief Scientist of the National Oceanic and Atmospheric Administration (NOAA), a worried friend called to share some troubling news that illustrates the problem. Dr. Ted DeLaca, a distinguished marine scientist and veteran of numerous polar research expeditions, had just been approached by two scientific colleagues from the former Soviet Union, then visiting in Washington.

"They want me to let others know what we've just discussed—terrible reports of pollution throughout the country, most of it kept secret until now," DeLaca told me. "What they are telling me is truly frightening."

Like most concerned with environmental issues, I had followed reports about the nuclear fallout in the USSR that had spread over much of the northern hemisphere after a reactor meltdown at Chernobyl, where many made heroic sacrifices to keep the catastrophe from being much worse. But the two visiting scientists had more sobering news about at least another dozen "potential Chernobyls," widespread radioactive contamination of land and groundwater associated with various engineering projects and the dumping of nuclear wastes and disposal of "hot" equipment, including entire intact nuclear submarine reactors, into the ocean.

"And that's not all," DeLaca added. "Little attention has been focused on other toxic wastes that have been released into the rivers, lakes, and the sea. They say the Baltic is dying."

The Baltic Sea, lakelike in some ways because it is nearly enclosed by several bordering countries, is suffering from a gruesome legacy of the USSR: decades of dumping raw sewage, industrial waste, fertilizers, nuclear wastes, and other noxious substances. In St. Petersburg harbor, one thousand times the normal levels of lead and cadmium were recently found. Inland lakes and rivers have also suffered. According to one report, some three thousand factories dump nearly 13 billion cubic feet of contaminated waste and other effluents into the legendary Volga River *every year*. The river flows into the Caspian Sea, home of the famous and now gravely endangered sturgeon, source of Russia's famous black caviar.* To the north, the consequences of using the Barents and Kara seas as major dumping areas for radioactive wastes and toxic chemicals, coupled with seepage into the sea of polluted groundwater, added up to a contamination catastrophe with global implications.

The entire North Sea and Arctic Ocean ecosystems are bearing the brunt of the poisons, each contaminant in its own way blunting the resiliency and recuperative powers of the creatures living there. Many have died; some have diminished in numbers. Others, able to cope with the changing chemistry, may prosper for a while, setting in motion new cycles with unknown endpoints.

There is no fence around the contaminated waters bordering the former Soviet countries, no barrier to contain the toxins to keep them from permeating adjacent areas. In due course, through currents and living vectors, the deadly inheritance will be shared far and wide.

Stresses of this magnitude did not haunt the sea in the 1940s, when as a child I first fell in love with froth-topped waves crashing green, cold water on clean, pale sand along the New Jersey shore, and with the creatures there: sand crabs, starfish, glistening jellies, bristling tangles of red and brown sea-weed, and sometimes the strangest, most formidable-looking but gentle beasts

---

* Some jest that dining on Russian caviar is the latest version of Russian roulette, hut it is no joke, either for those consuming the toxins, or for the sturgeon, who at this critical time *need* their eggs and the fish that produce them to keep alive their species' slim hope for survival.

imaginable—horseshoe crabs, dozens of them, sometimes hundreds, jostling one another and pushing their way high onto the beach. Their smooth-domed shells, glossy as polished leather, contrasted with the barb-fringed aft end, especially the long stiff "tail" spine that was a convenient handle with which to pick them up for a closer look. I was fascinated by the crabs' determined efforts to get out of the water to the sand, where, I thought, they would surely die. I gathered as many members of the gleaming brown armada as I could and turned them back into the sea—not aware that I was intruding on a vital ritual of mating: the release of eggs and sperm in the wave-washed intertidal area. It did not occur to me then that these enchanting creatures would one day bring into sharp focus the devastating impact one species—mine—is having on the future of life in the sea, and on the planet as a whole.

I *was* aware of the clusters of shorebirds, sweeping in chattering abundance just in front of the advancing waves, apparently gorging themselves on something edible in the sand. In later years, I discovered that those small brown birds—ruddy turnstones, sanderlings, semipalmated sandpipers, dunlins, red knots, and more—were pausing to rest and eat on a journey begun many weeks before in South America. Drawn by a feast of millions of nourishing crab eggs, many thousands of birds arrive precisely when the crabs come ashore at certain locations along the Atlantic seaboard and the Gulf of Mexico. The birds stay long enough to recharge their reserves before continuing to nesting areas farther north, leaving behind fecal material that washes into the surf, priming plankton blooms that link many other forms of life to one of the longest, most slender, most mobile, and most ancient ecosystems in the world.

I later also found that the great, shiny brown crabs that gave me such delight as a child are not crabs at all but are rather creatures of great antiquity, more closely akin to spiders and scorpions. The Atlantic species, *Limulus polyphemus*, is one of just four living members of the class Merostomata, marine arachnids that represent an intact direct-line connection to an abundant and diverse group of animals that prospered nearly half a billion years ago. The other three species live in the Pacific, from Japan and Korea to the Philippines and East Indies.

When as a budding biologist I discovered something about the distinguished lineage of my old friends, I regarded them with enhanced respect, befitting their status as very senior citizens. Finding a live dinosaur would be headline news, but here were creatures whose ancestry precedes Stegosaurus, Tyrannosaurus, and all those other sauruses by hundreds of millions of years—and they are still around! Horseshoe crabs and their unique genetic information have survived numerous cataclysmic eras in the history of Earth, somehow persisting through the demise of dinosaurs and other waves of mass extinctions.

Yet in the span of one human lifetime—mine—the future of these durable, time-tested creatures has become uncertain. They have declined and in some cases disappeared from coastal areas where only decades ago they were common. The reasons are several. I joined David Cottingham, a senior scientist at NOAA, who, like the birds and crabs, is drawn a few days after the full moon tides of late May and early June to a place along the Delaware Bay where fecund crabs and feasting birds converge in an orgy of energy flow.

We were not the first to approach the sandy, sloping beach, lively with the fluttering of many wings, scuttering legs, and probing bills. Several people were clustered around a telescope, aimed at the confusion of birds along the shore and two men who zigzagged the beach near the water's edge turning crabs. They rejected some, gathered others.

"Probably, they're taking them for research," Cottingham said, mindful of my concern about the fate of the collected crabs. "Medical labs use a lot of them because of their special blood. Have you seen it?"

I had—it's bright blue when exposed to air. Where human blood is colored red by oxygen-carrying iron pigments, horseshoe crab blood turns an extraordinary blue because of an oxygen-carrying copper compound. Extracts of the crab's unique blood are used for diagnosing certain human bacterial diseases and in checking for the presence of bacterial endotoxins in drugs and intravenous solutions. Often, after some blood is taken, the crabs are released—or are retained in special holding tanks for later donations.

The crabs taken from this beach had another destiny, however. I glanced in the back of a pickup truck parked by the roadside and was astonished to see that it was nearly full of upside-down crabs, brown legs kicking uselessly. A young man arrived with a half dozen new captives, and I asked what he planned to do with them.

"Eel bait," he said. "We only take the females with eggs—freeze them, then saw them in half. This load will last us for a while. The eels love the eggs. Males are useless."

Like the eels, humans are not fond of the meat contained within the crab's thorny hide. Gulls and other birds sometimes work over upturned crabs, but in their legs-down position, the smooth shell is effective tanklike armor against most would-be predators. They are, however, an easy catch for upright primates with nimble hands. Many horseshoe crabs are gathered and ground for fertilizer or animal food, thus removing the source of egg production and making life harder for migrating birds that have few other food options available. Every horse-shoe crab removed also means the species' chances of weathering natural predation and other pressures are diminished.

Normally, a single female may release about eighty thousand eggs in a season, but it is expected that only one or two will survive the necessary eight to ten years required to reach reproductive maturity. A lucky individual may live to be twenty years old and make repeated, perilous ventures to the beaches a few days after the full moons of late spring and early summer. On beaches such as this one, many crabs—an estimated 10 percent—become stranded and die. Others never find an appropriate shore, thwarted by coastal developments featuring seawalls and jetties. But the greatest concern of all for the future of the Atlantic *Limulus*, as well as the three Asian species, is their absolute dependence on shallow, coastal waters—and in these waters pollution levels are high. Hundreds of millions of years of successful coping with changing times has not prepared this species for surviving the actions of mine.

As might be expected, pollution levels in the sea are highest close to urban and agricultural areas, and tend to decrease with increasing distance away. In 1984, NOAA initiated a program called the National Status and Trends Program (NSTP) to assess the conditions of environmental quality in the coastal zones of the United States, and to determine where pollution is increasing, decreasing, or holding steady. Mussels and other bivalve mollusks are used as "sentinel organisms" to detect environmental contamination. These small creatures have the advantage of filtering through their systems large volumes of water, and concentrating certain substances that would otherwise be difficult to detect or measure.

They also stay put, a characteristic that makes them useful subjects for observers who want to monitor specific locations repeatedly. For the NOAA assessment program, mussels and oysters are collected from more than 200 sites annually, nationwide, and are analyzed for polycyclic aromatic hydrocarbons (PAHs), an indicator of oil pollution, for PCB's, chlorinated pesticides, butyltins, sewage tracers, and toxic trace elements. Similar analyses are made of fish taken at 79 bottom locations in coastal estuaries.

Dr. Charles "Bud" Ehler, head of NOAA's division of Ocean Resources and Conservation Assessment (ORCA), has been a leading proponent of and active participant in marine monitoring programs for many years, watchdogging and documenting the ominous rise of toxins in heavily used estuaries—and sometimes, as in 1992, recording declines that may result from restrictions on the use of certain chemicals. He believes the "mussel watch" and "benthic surveillance program," as the NSTP assessments are sometimes called, are beginning to show some trends, but that long-term observations, spanning decades, are needed to establish what is really happening to substances introduced into the sea through human activity and to understand what impact they may be having on marine life.

In the ocean, levels of pollution can be very low and still have an adverse effect on reproductive potential or resistance to other stresses. Substances may also accumulate through food chains, starting with minuscule amounts absorbed by individual small planktonic creatures. Top-of-the-line predators such as swordfish, sharks, barracuda, grouper, snapper, tuna, dolphins, and certain whales are among those most likely to have high concentrations of pollutants, sometimes in sufficient quantity to cause health problems for those creatures—and for the humans who might eat them.

It should come as no surprise that substances toxic to one kind of life are likely to be toxic to others. One of the two most profoundly important discoveries about life is how similar all of it is, chemically, from bacteria to behemoth. The genetic material inherent in ferns and fungi, foxes and flying fish are composed of the same basic substances, and respond to the same underlying rhythms that govern production of the next generation. The basic "how-to" principles of life's chemistry were established by microbes billions of years ago and those patterns remain, permanently etched in the ongoing processes comprising the nature of all life on Earth.*

The surprising news is that substances lethal for some are *not* lethal for all, not even all individuals of targeted species that normally are susceptible. Some are more or less resistant than others, just as some individuals are more or less able to fend off viral or bacterial infections. The underlying reason for this relates to the other most profoundly important discovery about life, and that is, of course, how varied life is, not only in the broad categories designated as phyla or divisions, classes, orders, families, genera, species, and the like—but, most incredibly, at the very specific individual level. It is mind-stretching but understandable that every human being is distinctively different from every other, including so-called identical twins. Likewise, each individual plant, animal, and bacterium has his, her, or its own unique batch of genetic potential that is subjected to a one-time-only life span of environmental circumstances that shape the specific nature of that individual. The miracle is that somehow stable patterns emerge, despite—or maybe because of—the variability and flexibility in the system.

Because of the inherent differences among individuals of every species, some will be more likely than others to survive storms, predation, disease, poison, stress, or whatever other selective pressures may be brought to bear. This is the reason that insects can develop tolerance to substances such as DDT; these hardy survivors of spraying generate resistant offspring that in turn produce more, sometimes with enhanced reproductive success because many natural predators are destroyed in the process of attempts to eliminate the targeted pests.

In curious ways, the new environmental chemistry shaped by human activity during the past century is exerting selective pressures on all forms of life, not just insects and others regarded as "pests." The nature of humankind, too, is being

---

* A recent article by Peter Little in the British science journal *Nature* illustrates the phenomenon. The potential of the pufferfish, *Fugu rubripes*, the famous but dangerous Japanese epicurean delight, is described as an apt candidate for research on the human genome. According to Little, this fish has the smallest known genome of any vertebrate (mammals, birds, reptiles, amphibians, fish), about seven times smaller than that of humans, and he believes its use might help simplify certain aspects of genome research. He points out that much of our understanding of basic human biology stems from work on surrogate organisms used to overcome the practical, ethical, and social problems of experimenting on human beings. A few surrogates stand out as having proved to be especially valuable: the bacterium *Escherichia coli*, yeasts, nematode worms, fruitflies, clawed toads, zebra fish, and house mice: a highly diverse lot, but all valid, because of their underlying similarities to one another—and to us.

subtly influenced by individuals' greater susceptibility to or enhanced tolerance of the toxins newly introduced into everyday life. Awareness that this is so is disconcerting, but perhaps not nearly as alarming as the unintended consequences for the natural communities of life, for the land and aquatic systems, mostly marine, that together make up the life-support system upon which humankind utterly depends.

If we are intent on maintaining enough of an environmental status quo to be sure that the planet functions reliably, the one thing we do *not* want to do is to disrupt the ecosystems responsible for these essential processes. Concerns are raised from time to time about the consequences of destructive environmental changes and in response, many moves are made to try to protect and safeguard vital systems. But until quite recently, the reality of what is at stake has not been widely appreciated and there is still little public concern about the alarming changes now under way in the sea.

Recently, late on a summer afternoon, I revisited one of the beaches I loved as a child. I noticed the glint of sunlight on something small and round, and gently lifted onto my fingertip a tiny, iridescent sphere. There, immersed in clear liquid inside a space smaller than a teardrop, was a miniature horseshoe crab, with all the critical bits intact: rounded front end, oval eyes, leaflike gills, a small nub for a tail, and a priceless multimillion-year cargo of unique genetic information.

Thoughtfully, and with great care, I returned the crab-to-be to the sea, mindful that my personal chances of surviving the next few years were probably somewhat greater than those of this minute crablet. So far, this one had avoided being eaten by birds, offered to eels, turned into nutrients for a garden, or poisoned by some noxious new addition to the ocean from nearby cities, but it still had a long way to go. It would have to elude these and other perils for a decade before it might return to shore as an adult to spawn, and then have perhaps ten more years to prowl the ocean, crunching clams and other succulent seabed morsels.

But what of the future of the *category* of creatures we call horseshoe crabs? Should it not tell us something—maybe something profoundly important for our own future—that in half a century we are able through no special effort or malicious intent to bring about the decline and possible demise of such proven, durable forms of life? What plan can we possibly devise that will ensure the continued, healthy functioning of a planet suitable for a relative newcomer, ourselves, if within decades we can render the place inhospitable for those who have endured the sweeping changes of the preceding few hundred thousand millennia?

Loss of horseshoe crabs might not seem like a big deal. I can imagine some of my cynical pals, drinking beer, munching pretzels, teasing me with killer questions, including the clincher: "Who cares? *I* don't!"

I can also imagine a philosophical crab, perched on its several hundred million–year track record of success, disdainfully reviewing our meager history . . . and thinking the same of us.

Chapter 13

BEYOND FLOTSAM
AND JETSAM

Plastics are now the most common man-made objects sighted at sea.
The Center for Marine Conservation,
A Citizen's Guide to Plastics in the Ocean:
More Than a Litter Problem

A fierce glare, flashing beak, arching talons, flapping wings, and a lusty will to survive imperiled the Humane Society's Dennis Stevenson on an osprey rescue mission in my mother's Florida backyard. The feathered fury was high in a pine tree, upside down, hopelessly snared in a maze of fishing line, but not ready to yield defeat, be captured or freed from the tangle. Skillfully, Dennis gathered the big bird in his arms and managed to climb down, with his indignant passenger and his own hide miraculously intact.

The young osprey was a regular visitor to that tree, a fine lookout post where she scanned the skies, spotted fish in the water below, and—my mother hoped—might someday fashion a nest in the tree's craggy branches. But Dennis grimaced, gently touching swollen, oozing legs, gangrenous where tight wraps of monofilament rubbed against bare bone. This osprey would never again perch—on any tree. Plastic debris, an insidious new threat to ocean life, had claimed another victim.

Wherever I go, I keep a log of underwater observations and dive logistics, and along the way have kept track of changes in the kind and quantity of debris in the sea. This backbone of information has developed as a result of the lifetime habit of recording who and what I encounter while diving reefs, jetties, and wrecks; prowling waterfronts; haunting beaches, marshes, and mangrove islands; and plunging the depths in deep-diving submersibles.

For the Gulf of Mexico and Caribbean Sea, the records began sketchily in 1947 when I was 12, and continue to the present, with a high concentration of note keeping from the early 1950s through 1966, when I turned in my doctoral dissertation—*Marine Phaeophyta (brown algae) of the Eastern Gulf of Mexico*—at Duke University, ten years and two children after completing my master's degree there, *Marine Chlorophyta (green algae) of the Upper West Coast of Florida*. There have been other periods of concentrated looking underwater along the shores of all of the continents, many islands, and numerous places in the deep sea far from any shore, but the Gulf of Mexico and Caribbean Sea have commanded my attention the longest, and with the greatest degree of focus. So I can say with a high level of confidence that until the mid-1970s, debris in the Gulf and Caribbean was mostly the benign sort: chunks of wood, bottles, shoes, sometimes a piece of rotting net or tangle of line, and occasionally a real treasure—a glistening deep green or amber glass fishing float encased in a web of knotted line, crusty with salt. There were also natural castaway gems—fragile white shells of the small deep-sea squid, *Spirula spirula*; fleets of silver and indigo jellyfish, *Porpita*; or purple-violet sea snails, *Janthina*, sometimes with floating rafts of moist, silvery bubbles and egg masses still attached.

Huge mounds of *Sargassum*, a fragrant golden-brown seaweed that prospers in large, floating colonies in the Gulf, the Caribbean, and the Sargasso Sea, are sometimes swept ashore, where they provide a bonanza of food and shelter for numerous beach-dwelling creatures until the weed decomposes or is carried back out to sea. Storms often dislodge sea grasses and other bottom dwellers, and they, too, sometimes pile up in enormous nutritious windrows, as they have for millennia. I learned a lot about what lives offshore by pawing through such mounds, and often dragged home tubs of seaweed for leisurely microscopic inspection.

After a 1959 storm that left beaches near my Dunedin, Florida, home wreathed with yard-high heaps of weed, my four-year-old nephew, Eric, helped me sort the red, green, and brown marine plants from pliant piles of brilliant orange, yellow, and purple *Leptogorgia* coral, the long, U-shaped tubes of parchment worms, *Chaetopterus*, and numerous aromatic sponges. Later, while walking with his grandmother along a beach fragrant with the robust essence of rotting weed, Eric observed that it "smells just like Aunt Sylvia"—a rather nice notion, I thought at the time. Piles of seaweed on a beach have a concentrated sea essence, the potent perfume version of the tantalizing ocean cologne sensed from afar when approaching a healthy beach or bay or marsh. For much of Florida, however, and for coastal areas elsewhere, that essence has changed, and today I would not welcome the comparison to coastal aromas.

Heaps of decaying seaweed along beaches worldwide are mixed with a bizarre array of new ingredients. Some, such as tarballs and sewage, have distinctive signature smells. Most plastic goods that are cast ashore have no detectable aroma, but by tangling, snaring, or choking birds, fish, seals, dolphins, and even whales, their indirect impact can be powerful. They bring to the sea the smell of death.

The first time I picked up a piece of junk underwater, an old jar, I discovered that it was occupied by a disgruntled little toadfish who, without pause, slithered away to avoid further attention from the huge predator I must have appeared to be. The jar, snuggled down among sheltering blades of Gulf of Mexico sea grass, was definitely not one of those fine old glass pieces sometimes discovered along Florida's west coast, tossed or lost overboard during centuries of pirate-studded sea trade. Rather, it appeared to be exactly the kind of container then in the refrigerator at home, filled with mayonnaise. *This* jar, however, was a jar transformed.

Fronds of pale hydroids, small animals resembling a diminutive flowering of baby's breath, sprouted along the rim. The shaded sides bristled with clusters of a tiny plant that seemed to be constructed of pink crystal beads strung end-to-end. Firmly cemented to the smooth topside were white hollow curls of tube-dwelling polychaete worms no larger than the capital C's in this book, sharing space with a lacework of encrusting animals—bryozoans and orange sponge. A flat brown crab, several pale, curved crustaceans, and an urchin the size of my smallest fingernail clung tenaciously to a patch of green turflike algae. So robust was the community camouflaging the fish's glass cave that I might have regarded it as a curiously shaped rock, but for the anomalously perfect round entrance that caught my attention. The jar, though aesthetically out of phase in the otherwise pristine submarine forest, was arguably doing no harm. Like most human debris in the sea up until that time, from bottles and cans to sunken cities and shipwrecks, the bit of glass had experienced a sea change into a miniature metropolis . . . a natural work of art . . . a fish's lair.

I picked up that jar in 1947, three years after Allied forces during World War II blitzed Japanese military ships and planes concentrated in Truk, a cluster of Micronesian Islands in the South Pacific. The attack proved to be one of the pivotal events of the war, diminishing in one horrific day, February 17, 1944, much of Japan's sea power. Some sixty warships were sunk in Truk Lagoon, the clear blue 40-mile-wide haven of calm water in the open sea that served as Japan's naval stronghold, including nearly a dozen inhabited islands and numerous mangrove islets, encircled by a protective necklace of reefs. In a stroke, ships designed for destructive missions found a new destiny as artificial reefs, the largest concentration of man-made reefs in the world.

Thirty-one years later, I explored those ships as a scientist curious about the effects the ships might have had on the lagoon—and vice versa. I expected to find every exposed surface adopted as living space, like the jar along the Florida shore. Because the exact moment of sinking had been documented, the communities of corals and other creatures were of precisely known age, so minimum growth rates could be determined for some. All of the ships were laden with live munitions and fuel oil; some carried tanks, trucks, and supplies of highly refined aviation fuel. The remains of dozens of planes, the ships themselves, and thousands of men had lain undisturbed on the sea floor for more than a quarter of a century before divers began to visit and discover submerged guns wreathed with garlands of feathery coral, the masts of ships shaggy with sponges and lacy hydroids, the doorways and passages havens for thousands of small fish, who were a lure for larger predators.

A woman whose son died when his ship sank in the South Pacific during World War II asked me, "What is it like down there? I have nightmares, thinking of how dreadful it must be underwater."

I told her that I can imagine no monument on Earth as beautiful as the sunken warships I have explored, ships of death transformed into celebrations of life. Some resemble cathedrals, tall masts and arches framed with shafts of sunlight and hung with living masterpieces; others are reminiscent of colorful gardens, alive with blue, green, silver, and gold flashes from fish that flit like flocks of birds among trees of coral, shrubs of sponges. Without hesitation, I assured her, "If there is Heaven on Earth, it's under the sea."

I was inspired to go to Truk by a master teller of sea stories, Al Giddings, who with sweeping gestures, much suspense, and vivid phrases punctuated with "How about that, kiddo!" described his discovery of the *San Francisco Maru*, laden with mines "lined up like eggs in a box," the mysterious *Ikoku Maru*, not seen since it slipped underwater, and the giant *Fujikawa Maru*, its keel resting in 150 feet, mast projecting into sunlight, and visiting schools of silver-blue barracuda forming shimmering, living whirlpools.

This "kiddo" was hooked on the concept, and with Al, proposed a research project to *National Geographic* magazine. Their support and Al's ingenious way of turning bright ideas into mega-projects brought together a team of four, plus two part-time assistants to help deal with fifty-two boxes, bags, and cases of gear—all but four duffels (of clothing) jammed with camera and diving equipment. Thanks to a suggestion from a friend at the National Geographic that "Al and Sylvia might be able to use some help" the expedition was joined for two weeks by two novice teenage divers, John F. Kennedy, Jr., and his cousin Timothy Shriver, who soon earned their keep as expert bag carriers and keen underwater observers.

Once ensconced on the island of Moen, to get around, we engaged a thirty-foot diving support boat from Kimiuo Aisek, a robust Trukese diver and grandfather who became a regular part of the team for the duration of our stay. As a boy of 17, Kimiuo witnessed the 1944 surprise air attack on the Japanese fleet and survived by hiding in a cave on the side of Dublon Island. From his vantage point above the lagoon, he watched blue sky darken with smoke and flames, and the sea burst with the impact of thousands of exploding shells. "For more than two years afterward, oil from ships and planes covered the beaches and reefs," Kimiuo told me, then added, "but the sea is healed now."

I was intrigued by the thought. Had the lagoon fully incorporated the ships and their deadly cargo as benign elements of the underwater community? Trukese officials wanted to know, and so did I. Oil still seeped from some of the ships, creating smooth, rainbow-hued swirls on the surface above. The paint, metal, and munitions were continuing sources of toxic contamination. There was talk of trying to pump out the fuel that could be recovered and perhaps salvage and safely dispose of the explosives. Special concern focused on the potential damage that could be wrought by *San Francisco Maru*'s mines, and the huge projectiles aboard the *Yamaguri Maru*, originally intended for use on battleships with a firing range of more than 18 miles. Some fishermen had already been wounded while exploding munitions from the sunken fleet as a crudely destructive method to stun and capture reef fish.

The lagoon is shallow along the shore, sloping to 150 feet in some places, and more than twice as deep in others. Most of the bottom is smooth, carpeted with soft, calcareous sand, mud, and extensive foot-high forests featuring at least six kinds of tropical green algae of the genus *Halimeda*, each resembling some variation on the theme of diminutive cactus plants; another half dozen kinds of bright green species of the seaweed *Caulerpa* added dense clusters of lush feathery and grapelike blades. The ships, resting upright on the sea floor, introduced high-rise habitats unlike those naturally present. The passageways and staterooms were caves populated by glassy red shrimp, red squirrel fish, cardinal fish, ghostly white sponges, and fragile clusters of encrusting animals; the steep sides resembled deep, shaded cliffs, fine places for black coral, *Antipathes*, and long spirals of whip coral; the decks and other hard surfaces provided attractive places for planktonic larvae of many species of coral and sponges to settle and grow that would not normally survive on the lagoon's naturally soft, silty bottom. Plants, more than a hundred species, prospered wherever there was sufficient light and space. All in all, the underwater archipelago of ships provided favorable new sites for some species that were previously uncommon and, by adding surface area where attached plants and algae-bearing corals could grow, may have bumped the overall productivity of the lagoon up a notch.

One sign of a healthy community is the presence of large predators, the rationale being that an ecosystem generating sufficient energy to sustain itself and yield enough food to support top-of-the-line carnivores (grouper, snapper, jacks, barracuda, sharks) is probably doing all right. Other predators, local fishermen, reduced the number of large fish that might otherwise be present in Truk Lagoon, but schools of jacks are common, barracuda are frequent, and gray reef and white-tipped sharks are occasional visitors. Sharks have a Hollywood-inspired reputation for being "eating machines," constantly snatching and engulfing everything they can get their jaws around. In fact, most photographs taken of sharks engaged in eating something are staged: Bait is used to entice sharks to come close enough to where a human may be lurking, camera preset and ready. (It often works, despite claims by underwater photographers that the best shark repellant yet devised is a loaded camera.)

In many years of dedicated shark-watching, including hundreds of totally benign close encounters with the impressive jaws of many species, I have never been able to catch sharks in the act of munching, but as we explored the *Shinkoku Maru*, a six-foot-long gray reef shark nipped a live jack from a large active school, shook it, then swallowed it whole. Busily inspecting small critters, I missed the critical action, but glimpsed the dazed looks on John and Tim's faces—and the shark as it swept by at high speed. The sixty or so ships-cum-reefs do provide enhanced shelter for congregations of small and medium-size fish and probably make the job of finding a meal easier for certain large, roving predators.

But the wrecked ships are not just benign coral-sprouting lumps of metal. A dive on one of the freighters, *Amagisan Maru*, provided an example of an ongoing problem: leaking fuel. I had already noticed opalescent circles on the sea surface, marking the presence of the wreck 200 feet below. Some of the escaping oil, probably from the ship's bunkers, lingered on the surface, gradually breaking into miniature slicks dissipated by wave action. Other shimmering patches of color quickly spread, then swiftly contracted and disappeared, the signature of highly volatile aviation fuel.

During a forty-minute inspection, Al and I located a vent in the ship's hull from which greenish-gold globules of oil were escaping. No one knows how much fuel remains within *Amagisan Maru*'s dark shell, or those of the other ships in the sunken fleet, nor how much was originally released into a natural system that at the time was largely undisturbed. As it was, the massive amounts spilled in Truk Lagoon in February 1944 and in other parts of the Pacific during World War II were like small bruises on a large, robust, resilient body of water. No one knows how *many* bruises, cuts, scrapes, poisons, and other afflictions the ocean can take on without breakdown of critical functions, but it is clear that recovery of damaged areas is largely dependent on the existence of a generally healthy surrounding sea. Had Truk Lagoon been sealed off from the source of renewal—no fresh, plankton-rich seawater allowed to enter, no oil-laden water permitted to escape, no rain allowed to fall but for that generated within the perimeter of the lagoon, no fresh air allowed to dissipate the smoke and toxic fumes generated—the impact of the battle, and the dependence of local ecosystems on the surrounding broad Pacific might come into sharper focus.

As it is, natural forces *have* gracefully and effectively incorporated the introduction of grand-scale debris and oil into the lagoon's ecosystem. To remove the hulks and thus interfere with the natural processes already so successfully under way would compound problems that in time will be resolved, if left alone. I had drawn a similar leave-well-enough-alone conclusion in 1947 when I discovered the mayonnaise jar, bristling with life, along a Florida shore. I decided to put it back where I found it, and hoped the toadfish would return. I was blissfully unaware of momentum then building that would substantially alter the volume, and kind, of debris and new elements that were about to be forced upon ocean ecosystems.

The year I found the jar was the same year that scientists at Bell laboratories invented the transistor, when the first U.S. airplane was flown at supersonic speeds, and when the explorer-anthropologist Thor Heyerdahl sailed in 101 days from Peru to Polynesia on a balsa-wood raft, *Kon-Tiki*, to prove the feasibility of prehistoric navigation. During many weeks at sea, *Kon-Tiki*'s crew saw drifting logs and leaves, but not a trace of garbage or other floating civilized debris, nor was there much trash along Florida shores near where I lived other than now and then a fragment of metal or glass and pungent lumps of pulp and orange peels from a local citrus plant that discharged its daily residue directly into the bay.

Within the next two decades, however, events occurred that would forever change the nature of life for humankind—and for life in the sea. Greatly expanded use of petroleum products led to the accidental or deliberate discharge of large amounts of crude and refined oil into the ocean, and plastic, in modest use for many decades, now exploded into every aspect of everyday life as a substitute for wood, paper, metal, and glass, performing uncannily useful functions no previous material had been able to do.

Recently I returned to the place where I picked up the first piece of flotsam that I ever retrieved from the sea, that jar—half hoping that I might actually find it again. I had, after all, put it back when I discovered that it was doing all right in its new role as a fish house and miniature reef. I have found durable, centuries-old pieces of glass while diving at popular anchoring sites in the Caribbean and the Florida Keys, and have held in my hand exquisite multicolored bottles retrieved from the Mediterranean Sea that were created by Roman craftsmen more than two thousand years ago. But storms and forty-plus years of shoreline manipulation for marinas and human housing projects made the likelihood of finding my jar extremely remote. More problematic was the challenge of identifying exactly *which* might be the jar of the numerous discordant lumps now protruding from the bay's soft bottom.

Effective searching was further complicated by the nature of the water, once clear as quartz, now a succulent broth of saltwater, sediment, and organic debris. The dense meadows of sea grasses and the diverse communities they once sheltered were absent, but there was plenty of evidence that one species—mine—had been extremely active. To get to the water, I had to hop over a straight-sided seawall and scramble through heaps of shore-choking riprap, an untidy conglomeration of building scraps and chunks of concrete dumped as a futile replacement for the age-old method of effective beach holding that had preceded it: a gradually sloping shore fringed by two kinds of mangrove and patches of marsh grass, now vanished.

Thoughtfully, I disengaged a starkly barren polystyrene cup (projected lifetime, five hundred years) from where it languished in folds of soft gray-brown mud. Had a cup such as this been tossed overboard five centuries before, when Christopher Columbus was making his way across the Atlantic, it might still be around. It occurred to me that beachcombers five centuries hence might find this very one, but I decided to interfere with history and packed it home.

In 1969, the year that Buzz Aldrin and Neil Armstrong became the first men to touch the moon, Heyerdahl embarked from West Africa, this time aboard a pre-Columbian-style reed craft for a trans-Atlantic drift voyage to the New World. He had inspired his multinational crew with stories of the "beauty and purity of the marine world" observed on the first voyage, but now was stunned to find, far off the coast of Africa, water heavily contaminated with oily debris. Midway across the Atlantic, "The sea was like a soup of small black asphalt lumps." The following year, aboard a new reed raft, *Ra II*, Heyerdahl encountered tarballs during forty-three of the fifty-seven days it took to cross from Morocco to the Barbados, and every day the crew saw litter: plastic containers, bottles, cans, nylon bags, and other floating refuse.

Thereafter, Heyerdahl's view of the ocean changed. In *The Ra Expeditions*, he wrote that encountering the debris "was a lesson better than the raft voyage itself, showing me that the sea is not endless. . . . The ocean is only a big lake with no outlet, and with the human population explosion and technological boom it has become a vulnerable part of a quickly shriveling planet."

The shores of Florida, relatively free of debris when I first explored them as a child in 1947, by 1970, like beaches all over the world, were the recipients of increasing volumes of the kind of trash that so affected Heyerdahl at sea. Plastic goods began to dominate, not only because of their increasing popularity, but also because plastic items tend to float and sometimes travel many miles before being deposited high on a distant beach.

More than 20 billion pounds of plastic goods were produced in the United States in 1970; twenty years later, the volume had increased more than threefold, reflecting an increasing dependence on these attractive materials for packing, packaging, and convenient one-time-use items, from throw-away razors and cameras to hotel shower caps. Until very recently, little thought was given to the fate of these goods, once sent away in trash cans, or tossed overboard at sea.

It is difficult not to be impressed with the characteristics of plastics. They are light, strong, durable, corrosion-resistant, versatile—and inexpensive, compared to alternatives. It is also difficult to get rid of them, once used. Appealing qualities translate to appalling quantities of ocean debris: cups, lids, straws, spoons, boxes, toothbrushes, toys, rope, bags, packing material, fishing gear, and more. Much, much more.

Of the many billions of pounds of plastics produced in recent decades, most of it is still around. Plastic sandwich bags degrade in about fifty years; most garbage bags take longer. Plastic plates, utensils, bottle caps, and parts of electrical appliances may last several centuries. Until recently, six-pack rings were made of materials durable enough to last for about 450 years; many still are. Polystyrene pellets (the kind crates are packed with) and containers remain intact for about 500 years, and if burned, create still other problems through the release of ozone-damaging chlorofluorocarbons (CFC's). Among the most durable, and therefore most dangerous, of all plastic products to life in the sea is fishing gear.

Widespread use of plastic materials for fishing nets and other gear quickly replaced cotton, hemp, and other natural fibers for many good reasons. Not only were the new nets and lines resistant to rot, they were also incredibly strong, light, relatively easy to manufacture and store, and, most important, they were less costly, thus enabling fishermen to acquire and use more gear than ever before. The hidden costs of the wonderful new gear were not accounted for, because no price was charged for the unwitting casualties who died, snared in lost line and nets: seals, sea lions, dolphins, whales, birds, turtles, and, of course, many, many fish. A single derelict "ghost net" recently recovered in Alaskan waters was more than twenty miles wide and contained the remains of 350 dead seabirds and hundreds of rotting salmon.

A 1975 National Academy of Sciences study found that 14 billion pounds of garbage, much of it consisting of long-lasting plastic goods, were being deliberately dumped into the sea—*at* sea—every year, merchant ships contributing the most (close to 12½ billion pounds), then commercial fishing (almost a billion pounds) followed by recreational boating, military interests, passenger vessels, oil-drilling rigs and platforms, and various other sources.

A report from the Center for Marine Conservation, a Washington-based organization that has taken the lead in beach cleanups and the documentation of marine debris, notes that prior to 1988, the world's fleet of merchant vessels *every day* dumped at least 450,000 plastic, 4,800,000 metal, and 300,000 glass containers into the sea—enough to keep archaeologists of the future entertained for many millennia. Until recently, the United States Navy has typically thrown all wastes overboard—not a trivial contribution. A single large ship may generate more than a thousand pounds of plastic trash *a day!*

While those traveling aboard large commercial and military vessels can do little to influence the fate of the trash and garbage generated, boaters and fishermen using the more than nine million recreational vessels in the United States can control the destination of the more than 100 million pounds of their garbage tossed into coastal waters every year. Accidental or deliberate disposal of even a little bit of fishing line on each of the estimated 72 million yearly fishing trips soon translates to reefs crisscrossed with webs of tough, durable snares, many with lead weights and hooks attached. During a three-hour cleanup effort in Florida in 1988, 304 miles of such line were recovered by several thousand volunteers. But thousands of miles of monofilament remain a permanent part of the worldwide seascape, newly added during the past few decades.

No place in the world is free of such debris. During a recent return to the legendary Ras Muhammed reefs in the Red Sea, renowned as one of the "Seven Wonders of the Underwater World," I was greeted by a nightmare of plastic ribbons streaming from coral branches, bobbing fleets of plastic bags nodding with the current in a deep crevasse, snarls of plastic smothering a garden of soft corals, folds of plastic fishnet sawing through a cluster of sponges. A diver in the Mediterranean Sea showed me an impressive pile of plastic trash that he had personally recovered from the sea floor near Portofino. "The beaches are cleaned up, sometimes," he fumed, "but underwater, it's hopeless! Some of the garbage rots, but the bags—they just keep piling up." Australia's Great Barrier Reef, the epitome of ocean wilderness in the minds of many, each year becomes the final resting place for tons of lost fishing gear and dumped trash. Plastic pellets, cups, bags, and bits sail the world on ocean currents, land on distant beaches, and wherever they travel, provide a deceptive, lethal meal for birds, fish, turtles, and other sea creatures.

I sometimes wonder what our descendants might make of our present civilization if they were to try to reconstruct our day-to-day activities from beach debris. It would be a fascinating game, pawing through heaps such as those assembled by volunteer crews in Delaware in 1992. In addition to more common items (bags, butts, bottles), they found half a rake, a toothbrush, a skateboard, a headlight, boxer shorts, plastic duck, flip-flop, plastic dinosaur, doll brush, a boot, an audiotape, a sock, V-8 engine parts, some beads, and a toilet seat—among other flotsam.

In Florida, the same year's hunt yielded thousands of items, including a coffeepot, a curler, a hubcap, sunglasses, ski mask, high heels, television set, eggbeater, broom, curtains, boomerang, baby bottle, athletic cup, scissors, suitcase, barbecue grill, Barbie doll, mannequin arm, Ping-Pong ball, Christmas wreath, and, yes, a kitchen sink.

Leafing through the reports of the cleanups, I wondered if it might be possible to *guess* from beach debris where in the world I might be. Limiting the area to the United States, but otherwise without looking at the source, I pondered a list that included the following: half of a surfboard, a nine-dollar traffic ticket, a bullet cartridge, a cut-up credit card, mustache comb, black pantyhose, E.T. doll, vial of "Love Potion," three mattresses, a shopping cart, burned military clothing, a marijuana pipe, fake fingernail, tailpipe, *Playgirl* magazine cover, corncob pipe, kitchen stove, and a bathroom sink. I was distracted, for a while, by the corncob pipe, and the surfboard made me consider Hawaii, but other clues caused me to lean, correctly, in favor of my adopted home state, California. Curious, I looked at Hawaii's list, and found it to be relatively tame. Of the 65 tons of debris picked up, the most disturbing was a sea gull choked by a plastic six-pack ring.

In 1992, in the U.S. as a whole, 4,616,469 items of trash, well over half of it plastic, were recovered during a cleanup of 4,453 miles of coastline. Topping the list of the most commonly encountered items were hundreds of thousands of cigarette filtertips, followed by the usual assortment of plastic goods, paper pieces, glass bottles, and metal cans.

The Center for Marine Conservation report describing the cleanup notes, "Eliminating these items from U.S. beaches would make the beaches two-thirds cleaner than they currently are." The best way to achieve that worthy goal is to start at the source, of course. For most people though—other than the volunteers engaged in picking up accumulated beach junk—it is really difficult to connect throwing a cigarette butt, straw, or plastic bag overboard with statistics such as those noted, or with the consequences of careless tossing: litter-strewn beaches, fouled boat propellers, and animals dead from snacking on or being snagged on bits of plastic and Styrofoam pellets.

Junk on the beach translates to people avoiding such areas, however, and this comes with a price tag, measured not only in worldwide environmental degradation but also in hundreds of millions of hard-cash dollars, spent annually for beach cleanup, especially in top-dollar recreational areas. Such financial incentives have helped provoke the passage of remedial laws. But it's the outrage of individuals that has been fundamental in bringing about changes in policy to prevent ocean pollution.

Thor Heyerdahl's plea for action before the United Nations following his daily sightings of plastic debris and oil during his 1970 *Ra II* expedition triggered an immediate response. "Some people thought I exaggerated," he recalled when we recently met at a conference in Osaka. "Instead, when oceanographic expeditions from the U.S., Soviet Union, and Norway went to see for themselves, they found that the problem was *worse!*"

In 1972 and 1973, international conventions on the prevention of marine pollution from ships and aircraft were held in London and in Heyerdahl's home city, Oslo, and in 1978, international protocols to prevent ocean pollution were adopted during further meetings in London. Despite growing awareness of the problems, by the early 1980s, "death by debris" had become so acute that fifty thousand deaths a year of North Pacific fur seals were caused by entanglement with plastic, mostly discarded fishing gear. Declining numbers of seals and losses of many other species, from turtles and birds to countless numbers of fish snared in lost fishing nets inspired volunteers to take matters into their own hands. By the end of the 1980s, citizen cleanup crews began annual beach patrols in the United States and several other countries.

Finally, the United States began to enact significant legislation to deal with the plastic debris problem, starting with the Plastic Waste Reduction Act of 1986 and, in 1987, the Marine Plastics Pollution Research and Control Act, which prohibits the disposal of plastics at sea by U.S. vessels and the disposal of plastics in U.S. waters, out to 200 nautical miles.

Other bills were drafted to study the issues, establish a bounty system for recovery of lost drift nets, and mandate use of quickly degradable plastics for applications such as the ubiquitous six-pack rings. As innocuous as the ingenious and handy yokes for drinks might seem, widespread reports of an insidious "afterlife" as entanglers of wildlife such as fish, birds, even California sea lions and as beach clutter led to special consideration for these devices. The magnitude of the problem was demonstrated by the recovery of more than 15,600 six-pack rings from 300 miles of Texas coastline during a three-hour cleanup effort. In 1987, Anheuser-Busch, the world's largest brewer, voluntarily changed to photodegradable loop carriers, and other major users have followed.

The question remains, however, of how they get into the sea. In 1987, the Marine Pollution Research and Control Act (MARPOL) was passed, and the Senate also ratified Annex V of the MARPOL Protocol, thus bringing the U.S. into accord with thirty-eight other nations in prohibiting the disposal of plastics into the ocean and mandating that all vessels carry their plastic trash into port for placement in proper disposal facilities. The Senate also passed a measure that implements Annex V provisions in U.S. waters, out to 200 nautical miles. Compliance with the new laws is far from universal, and probably will remain so until it is understood that it is more costly to toss than not to. That message was received by an offender for the first time in 1993 when, alerted by a passenger, U.S. authorities fined Princess Cruise Line $500,000 for illegally dumping trash in U.S. waters.

A growing furor over the use of gargantuan plastic drift nets on the open seas came to a head in 1992, when the use of such nets—all plastic—was banned by the United Nations. Nevertheless, smaller versions are still in widespread use in coastal waters of many nations, including the U.S., and when lost, they add to the masses of trash that continue to be washed ashore, and will for centuries to come—even if nothing more is added.

Chapter 14

TROUBLED WATERS

*The truth is that we can never prevent oil spills. . . . They are the price of
an oil-driven civilization just as deaths on the highway are the price of a
mobile society.*

Frank J. McGarr, *presiding federal judge,*
Amoco Cadiz *oil-spill trial*

One whiff of asphalt warmed by hot sun, and I see, smell,
and *feel* slippery, oil-soaked rocks and viscous, bluish-brown sheens and rivers of
oil bruising the wilderness waters of Prince William Sound, or am strapped into
a Blackhawk helicopter, speeding over oceans of oil spilled in the Kuwait desert
or swimming carefully under a slick of oil along the shore of Abu Ali, Saudi Arabia, trying to absorb the significance of the largest oil spill in history—without
personally absorbing too much of the oil.

I did not intend to become an oil spill aficionada, but it is difficult to be
a modern-day marine scientist and avoid the issue. Nets dragged through the
sea anywhere are likely to come up clogged with tarballs; beaches in the most
remote areas are fringed with an oily scum or lined with sticky gobs of oil; over
the surface of much of the ocean a rainbow-hued film glistens with deadly beauty

at one time or another. No one knows the consequences to the ocean and, in due course, to ourselves, of these insidious intrusions on planetary processes, but in recent years I have been determined to learn what I can and have become increasingly eager to support measures to keep them from happening.

Oil spills have become events so familiar in modern life that little notice is taken of most of the sixteen thousand or so that occur every year in U.S. coastal waters, let alone the thousands more that flow into the sea elsewhere in the world. Like the small earth tremors that rock the residents of California so gently that they pass unheralded, most of the spills slip by unnoticed. But once in a while, spectacular "big ones" shake us enough to make us pay attention, and sometimes they stimulate precautionary actions.

Several mega-spills in the 1970s—*Torrey Canyon*, *Amoco Cadiz*, the gusher at the Ixtoc I oil well—began to raise public consciousness about the economic and environmental losses incurred when massive quantities of oil are released into the sea. Unprecedented legislation and protective policies followed, but within a few years complacency set in, perhaps born of the optimistic view that it was possible to prepare for and deal with the inevitable consequences of occasional spills. Many were convinced that oil spills—while better avoided when possible—perhaps aren't all that bad.

Swift attitude adjustment was provoked in April 1989 by the release of 11 million gallons of Alaskan crude oil from the tanker *Exxon Valdez*[*] into the Prince William Sound wilderness. When I first heard that an unprecedented amount of oil was being poured into pristine Alaskan waters, I was more than fully occupied in California at Deep Ocean Engineering, juggling roles as president and chief executive officer, wife of the cofounder, Graham Hawkes, mother and household runner for varying numbers of our combined children from earlier marriages, and simultaneously analyzing samples of marine plants recently obtained during diving operations in the Bahamas—with time out once a month to spend a long weekend in Florida with my closest friend, my then 87-year-old mother. But I was motivated to put other matters on hold long enough to go to

[*] This, the largest spill in U.S. history, dwarfed subsequent spills that otherwise might have made a more lasting impression: the *World Prodigy* spill in Rhode Island, *Presidente Rivera* in Delaware, the *Rachel B* and *Megaborg* in Texas, *American Trader* in California, and a large but unnamed spill in Florida's Tampa Bay in 1993.

Prince William Sound and try to evaluate for myself the impact of what seemed to be not only a disastrous loss of oil, but also a huge experiment with the resiliency of nature.

The immensity of the spill, and its location—a well-documented, much-loved natural area, valued for its beauty and scientific significance as well as for fishing and salmon-farming enterprises—magnified the spill's importance. During years of research on humpback whales, I had been to the area often and was aware of the growing threats to the natural health—and wealth—of the entire North Pacific region from careless shipping operations and other sources of pollution, as well as overlogging, overfishing, and aggressive development to accommodate increasing numbers of residents and visitors. Now, to these stresses was added the largest oil spill in U.S. history. By going to see for myself, I hoped to find personal answers to questions many were asking: "How bad is it?" "What's happening to the birds and otters?" "Are fish being affected?" "What might be the impact on the health of the sound, and adjacent areas?" "Is Lawrence Rawl [Exxon's CEO] right when he says that in ten years you'll see 'nothing'?" *What is the truth?* I wondered. Given a chance to go, what should I look for?

Three weeks later, thanks to arrangements made by a longtime friend and colleague, Dr. William Evans, Administrator of the National Oceanic and Atmospheric Administration (NOAA), as a scientific observer I was able to join him and several other NOAA officials and U.S. Coast Guard Commandant Paul Yost and Vice Admiral Clyde E. Robbins for an on-site review. At the time, I did not have the slightest inkling that within two years, I would be deeply embroiled in oil-spill issues as NOAA's Chief Scientist, and an early witness to the most monstrous air, land, and sea oil spill in history.

Perched near a large window of a Coast Guard C-130 aircraft, during an initial overflight of the region, I had an eagle's eye view of blinding white expanses of snow and ice, steep mountains embracing startlingly blue glaciers, surrounded by a sea that flashed molten silver or gunmetal gray, depending on the sun's angle. As the plane crossed from Cordova, at the southeast corner of Prince William Sound, we began to sight large patches of pink-and-blue sheen, windrows of the scummy brown emulsion of saltwater and heavy oil that bears a faint resemblance to the delectable chocolate dessert after which it is named—mousse—and here and there (but not everywhere), oil-slimed rocky beaches. We made passes over areas known to be critical, then paused near Naked Island, where the *Exxon Valdez* had been moved for repairs. Despite good intentions, oil was seeping past the "boom," an encircling floating fence intended to corral oil leaking from the wounded ship.

At a press conference held immediately upon landing in Valdez, Evans noted, with diplomatic understatement, "The grounding of the *Exxon Valdez* has had a significant negative impact on the environment," and, after fielding a barrage of questions with Admiral Yost, shared with him a strong commitment for swift action.

Despite the brave front, Evans was not optimistic. "There isn't much that is good that comes out of something like this," he told me. "About the only positive aspect is that we can learn how *not* to have it happen elsewhere."

NOAA scientists began mobilizing assessment research three days after the spill, some taking dismal delight in having been prepared with years of anticipatory baseline data for what *Science* called, "the predicted oil spill" (April 7, 1989). In the same issue of *Science*, Eliot Marshall noted, "When the oil companies won permission to lay a pipeline quickly from Alaska's North Slope to the port of Valdez, it was understood that ocean transport would be the riskiest part of the operation." The U.S. Department of the Interior's Environmental Impact Statement of March 20, 1972, made a spectacularly accurate prophecy, identifying as the main risk the possibility that a tanker on its way out of Valdez might break up in this remote area where little could be done to intervene, permanently changing "the solitude and wilderness aspects of this scenic area."

There is bleak satisfaction in being able to say, "I told you so," but no glee was in evidence among those whose predictions had been vindicated—only concentrated attention on how to cope, now that the deed was done.

Advice varied about how to minimize damage. To hundreds of otters, tens of thousands of sea birds, and billions of microcritters, "negative impact" translates quite simply to being dead. To those looking at damage to the Prince William Sound ecosystem as a whole, the degree of lasting impact was an open question, one that is still debated. I, too, thought about the question of how to effectively mitigate the damage wrought and restore the habitat.

Despite its toxic components, oil is an organic substance. Elements that are toxic to some creatures are tolerated by others, and even relished by a few as a source of nutrients. Spilled oil is messy and unsightly, but, in time, as much as half evaporates into the atmosphere, and the remainder, if exposed to the air, eventually breaks down. Those who regard a beach cleaned of conspicuous globs of oil as a "recovered" beach might agree with Exxon's Rawl when he suggests that after a decade, "I don't think they'll find much in terms of environmental damage." But there is much more to recovering from an oil spill than eliminating the obvious and the unsightly.

A scientist is supposed to have an open mind, and I tried to keep mine that way as I slogged along the heavily oiled beaches and slid over slippery black rocks. It was easy to maintain some detachment while contemplating the theoretical consequences of toxic substances in complex food chains, or while leafing through pages of statistics about the number of dead birds, fish, otters, and other creatures found. The detachment slipped a bit while inspecting windrows of spill victims, dead krill—glassily transparent crustaceans with enormous black eyes that have a well-earned reputation as a significant source of food for fish and whales. It seemed reasonable to conclude that a nudge to the system here, a deletion there, could well change the proportions, if not the elements, of ecosystem components, and set long-term changes into motion.

Detachment was difficult to maintain when watching thousands of individual small creatures, from crabs and barnacles to belly-up starfish and snails, struggle for survival against bleak odds in a poisoned sea. And it was impossible not to be moved to feelings of outrage when meeting the gaze of individual birds and otters at the rehabilitation center, where misery was tangible. The shrill, pained cries of sick and disoriented victims made it hard to remember that time heals many horrors, and at least the animals being treated had some chance of surviving, although their societies had been shattered.

Lingering with one listless patient, I reflected on various outcomes of a meeting between oil and otters. Some were blinded and died outright; many, inhaling toxic fumes, contracted emphysema, reducing their ability to dive and thus make a living; those with oiled fur experienced a new and chilling sensation as their legendary warm fur coat—sprouting an amazing four hundred hairs per square inch—collapsed into dark, stiff matts, useless for insulation in near-freezing water. Like cats, these bewhiskered clowns-of-the-sea groom relentlessly. Those with oiled fur suffered other avenues of contamination as they licked, tugged, and pulled their fur, poisoning their insides with toxins from the oil, which impairs normal liver and kidney functions.

Watching the distressed otter whose bright eyes met mine, then turned as the animal frantically sought a way out of confinement, I listened to one very weary volunteer who had taken leave without pay to try to help save oiled wildlife. "One remedy for oil spills," he said, "would be for those making policies about the transport of oil—tanker captains, oil company executives, and certain government officials—to be required as a prerequisite for their jobs to swim around in an oil-covered sea for a while, then, without contact with family or other familiar aspects of life, spend an indefinite period in a rehabilitation center, ideally one tended by well-meaning but ignorant otters, eagles, and murres."

While various Alaska state agencies and Evans and his NOAA staff concentrated on scientific research policies, Commandant Yost and Vice Admiral Robbins focused on how to physically deal with millions of gallons of oil on the loose. Amazingly little, approximately 4 percent, had been recovered in three weeks, and no one was sure how much of the rest had evaporated, sunk, landed on shore, or was still free and floating. Cleanup efforts had been slow to get started. As Eliot Marshall put it in his *Science* article, "The potential disaster of a major oil spill in Prince William Sound was forecast—the bumbling response was not."

Early on, when the oil was most compact and most likely to be recoverable, lack of suitable equipment and jurisdictional questions caused critical delays. Days slipped by with little action except by fishermen from Cordova to Anchorage and beyond, who were not inclined to wait for official approval. With their livelihood threatened, they took the direct approach, mobilizing people, boats, and booms, to keep the oil from critical areas.

A large handwritten sign by a restaurant owner in Cordova expressed a widely shared emotion:

FLAGS ARE AT HALF MAST DUE TO THE DEATH OF OUR ENVIRONMENT

Use of chemical dispersants was proposed by Exxon officials, the idea being that such chemicals, while as toxic to many organisms as the oil itself, would cause oil to break into small particles and dissipate or sink. There are times when such dispersants can be used to good advantage, but circumstances must be just right and the trade-offs must be weighed. As the question of whether or not to use them was debated, calm weather gave way to winds strong enough to tear the roof off the hangar at the airport, where planes carrying dispersants were standing by.

The storm, powered by 60-mile-per-hour winds, caused the oil to spread swiftly over hundreds, then thousands of square miles, influencing a vast, intricate shoreline estimated to be about as long as the entire coast of California. By the time I arrived in the sound, most of the heavy, concentrated slicks had been swept into long, slender ribbons or had piled up in thick, gooey windrows on certain islands and parts of the mainland shore.

Exxon provided a helicopter and pilot to enable me to survey, explore, and document the status of the spill two days after my arrival. While making several slow passes around the *Exxon Valdez* in Naked Harbor, I noticed that the sea along the edge of the nearby Naked Island was magically changing color from dark gray green to aquamarine blue! Like a cumulus cloud billowing from beneath the water's surface, the blue-white area expanded from a few meters across to ten, then twenty. Farther along the rocky shore, another part of the sea began to turn, then another. I had heard of the phenomenon, but this was my first chance to see an annual event of tremendous significance to the life and times of many who live in Prince William Sound: herring spawning. Tens of thousands of slim, silver-blue male fish jetted milky life-giving clouds of sperm over acres of eggs plastered by their female counterparts on rocks, shells, and seaweed—especially the smooth, golden-brown fronds of *Fucus* and crisp, red upright branches of *Odontothalia*.

Because of the spill, the area's herring season had been canceled, thus cutting off the principal source of income for many local fishermen—normally worth about $12 million—but perhaps giving the herring a better chance to recover from damage caused by the oil. To get some idea of the impact oil might be having on the newly fertilized eggs, I arranged with the Coast Guard for transportation back to the island the next day, this time to land and walk the shore.

Two friends joined the search for fish eggs, Wallace White, intrepid writer for *The New Yorker*, and the Alaskan whale biologist Charles Jurasz, a colleague of many years. I tried to pinpoint for the Coast Guard the areas so clearly evident the day before. We circled over Naked Island twice before I felt sure that we were close and a suitable patch of not-too-rocky rubble could be found for landing. The tide was low, exposing a swath of weed-covered boulders, rocks, and tidepools, glistening with oily rainbows. We were within sight of the tanker, but this beach had been spared the heavy dousing observed on other islands. Rocks along the shore of Knights, Eleanor, and Green islands had slipcovers of oil, which so softened the normally angular rock surfaces that if I held my breath and squinted my eyes, I could imagine an enormous field of slick chocolate bonbons. Rocks along Naked Island were slate gray, a tumble of boulders sloping into water clear enough to see rocky rubble extending several feet below the low-tide level.

Jurasz, the most experienced rock hopper of our trio, found the first eggs, a few glistening spheres on sprigs of seaweed. Rounding a bend in the beach, we found every blade of every plant on every rock embroidered with what appeared to be millions of perfect seed pearls—the herring eggs we sought, each holding ingredients capable of turning out, in a few weeks, a fully-finned miniature version of its parent fish.

We gathered samples to turn over to the biological assessment team for baseline documentation, and kept a few for me to hatch back in California. No one knows the real impact of the 1989 oil spill on the production of herring for the year, but I was optimistic, mostly because many of the eggs were placed well below the low-tide level, and would not likely come into direct contact with the smothering or toxic effects of an oily embrace. Reducing pressure from fishermen would also be a break for the herring that year, and for the many marine creatures dependent on them for food.

While I concentrated on a close-up view of Prince William Sound's marine life, the Coast Guard and Exxon were organizing massive cleanup operations. Admiral Yost, operating under instructions "just received from President George Bush" minced no words: "I intend to clean beaches," he said at an evening briefing attended by scientists, government officials, and others participating in the spill assessment.

Momentum began building at once to wash oil-soaked beaches with hot water. Aircraft, boats, booms, equipment, and teams of people were mobilized in an all-out effort to restore health to the sea. Within hours, the small community of Valdez was frenetic with newcomers gearing for the cleanup. Some were engaged to hand-hold high-pressure hoses to blast streams of heated water over blackened rocks. Others would wield large pom-pom mops of polypropylene to wipe softened oil, while a system of booms and skimmers was placed to contain and recover oil washed into the sea. Oil and debris would later be burned or taken to toxic-waste dumps.

"Well-organized mayhem," someone observed, watching boats loaded with gear departing for action that had more than a little resemblance to troops heading for war. Generally, there was a mood of relief that "something was being done," although there was considerable controversy about the usefulness of the costly hot-water treatments.

Dr. Jacqueline Michel, HAZMAT (short for "hazardous materials") specialist and pragmatic veteran of many of the world's worst oil spills, initially favored washing the rocks, believing that restoration might be delayed but would be more enduring on cleaned surfaces as compared to those left coated with oil. Later, she and teams of scientists engaged by NOAA to follow subsequent developments in areas cleaned as well as some sites left to nature, came to the conclusion, surprising to some, that restoration in areas treated with hot water lagged behind that of the few sites that were left alone.

Although imprecise, as most field studies necessarily are because of unavoidable and unaccountable variables, the findings appear to be straightforward and easy to explain. Oil devastated most of the life occupying the crevices and crannies on the rocky shores on both treated and untreated beaches, but some of the algae, barnacles, crabs, mollusks, and other creatures stayed alive in the left-alone areas, providing a built-in head start toward recovery. On treated beaches, however, hot water and trampling by hundreds of well-intentioned but heavy boots cooked or crushed most of whatever survived the oil. Four years after the spill, soft, even liquid, oil remains deep among the spaces between rocks and pebbles underlying both treated and untreated beaches, but wave action and natural weathering have removed most of the toxic elements and tarry residue from surface rocks—even those left alone.

The surprising conclusion is that unless action can be taken at once to contain and recover oil spills, it may be that cleanup efforts can cause more damage than good. As NOAA's Chief Scientist, it was my job two years after Exxon had spent more than $3 billion to help put things back together to announce during a jam-packed press conference that scientists had concluded that some of the efforts, particularly the hot-water treatments, were proving to be counterproductive. No one thought the expenditure was wasted, however. For one thing, a perverse yardstick of value was established for what it could cost to *try* to repair damage to a pristine wilderness. It also brought into sharp focus a clear message: No matter how much money is spent, natural forces—elements that money can *protect* but cannot *buy*—are fundamental for restoration. Not a penny of the $3 billion brought twice daily tides of clean, fresh water from the surrounding Pacific Ocean, or paid for new larval fish, plankton, otters, birds, or whales. Perhaps the most important and favorable outcome of the effort and money expended was the impact it had on the oil industry, on the government, and on the general public.

Before the spill in Prince William Sound, a patchwork of U.S. federal and state laws addressed various concerns about responsibilities, liabilities, compensation, and cleanup; after this spill, members of Congress went into high gear to respond to outcries for action from the public. On August 18, 1990, President George Bush signed into law the Oil Pollution Act of 1990, Public Law 101-380, a unanimously supported House-Senate compromise that reflected a high peak of sensitivity and concern about the consequences of oil spills. In the same month, twenty oil companies created two new organizations: the Marine Spill Response Corporation (MSRC) and the Marine Preservation Association (MPA).

MSRC, an independent organization funded with $800 million provided over five years by the MPA, is headquartered in Washington, D.C., with five regional centers in the United States. Each is equipped with gear and personnel to respond rapidly should a large spill occur. This effort was widely applauded as evidence that private industry was attentive and would be prepared to respond to distress caused by what many regarded as the inevitable spills of the future. But no one could possibly be prepared for events five months later and half a world away, when a monster slow-motion wave of oil swept over the shallow shoreline of the Persian Gulf, forever revising standards of what constitutes a "large spill."

Chapter 15

A TERRIBLE
EXPERIMENT

Wilderness is threatened everywhere. The extinction of animals is not the only danger; man faces the loss of a breathing space for all that is wild and free in his spirit. And not only his spirit, his physical welfare also, even his survival, is imperiled by the extermination of other life on this planet.

Anne Morrow Lindbergh, 1966, Earth Shine

Eyes burning, hair whipped by wind thick with oily smoke, I leaned out of the open door of a Saudi Arabian helicopter in March 1991 to survey the largest oil spill in history. Below, a black tide pressed against stark white beaches, violating the shores with a broad swath of oil as much as half a mile wide and several inches deep; narrow inlets fed oil into a lacy network of tidal streams among mangroves and marsh grass, creating a poisonous paisley of black and blue-green swirls. Aromatic compounds evaporating skyward assaulted my senses from below, while oil droplets and soot rained down from dark clouds generated by Kuwait's rich oil fields in flames. The Persian Gulf, heart and soul of the Middle East's "cradle of civilization," was wrapped in a shroud of death.

I had come to the Gulf to have a firsthand view of the impact of 11 million to 12 million barrels, some 500 million gallons of oil, or more than forty-five times the amount of oil released by the *Exxon Valdez*, crammed into less than one tenth the area. This awesome catastrophe was not an accident, but rather, a boasted act of ecoterrorism, a deliberately generated "mother of all spills" unleashed by retreating Iraqis commanded by Saddam Hussein as a closing act of defiance in Iraq's losing war against Kuwait.

This was not the first major oil spill in the Persian Gulf, but few noticed when one million barrels of oil were released into the sea during the Iran-Iraq War in 1983, nor has much been said about the approximately 250,000 barrels (one Prince William Sound–sized spill) that have flowed a bit at a time into the Gulf's warm, briny embrace each year for the past decade. No one outside the Middle East knows of such events or, if they are known, their relevance to life else-where seemed remote and thus not important. For that matter, few remember or seem concerned about the impact of the raging torrent of oil turned loose during the Ixtoc I oil-well disaster on Campeche Banks in the Gulf of Mexico, when as much as ten million barrels of oil were spread over thousands of miles for months prior to capping the out-of-control offshore gusher. And rarely is there mention of many "silent" but deadly spills, such as the one created by the Argentine tanker *ARA Bahia Priaiso* when it ran aground in 1989. Without public notice or outcry, oil leaking from that broken ship continues gradually to poison pristine Antarctic waters at DeLaca Island.

In 1991, however, the world was watching and acknowledged "Hussein's spill" as a major disaster. Grim images of dead and dying cormorants and other oil-soaked creatures provided vivid symbols of wanton slaughter, fueling outrage over the callous indifference shown for the spill's impacts on man and nature. At the time, I was trying to get used to my new role as NOAA's Chief Scientist. It was an awesome assignment. Not a natural-born bureaucrat, I was uneasily settling into the job when Dr. John Knauss, NOAA's new Administrator and a longtime colleague and friend, suggested that I might want to join a fact-finding mission to Kuwait, headed by Secretary of Commerce Robert Mosbacher. There might be a chance, Knauss said, to get a firsthand look at the oil spill, and meet with NOAA's HAZMAT team operating from a base in Dharhan, Saudi Arabia.

The main purpose of the proposed visit, sponsored by Kuwait's ambassador to the United States, Saud Al-Sabah, was to acquaint U.S. businessmen, congressmen, and a few dignitaries and officials with the challenges of restoring the economy of Kuwait in the wake of a devastating war. I would be the only scientist among a roster of politically powerful personalities including our host, Ambassador Sabah, and members of his family.

A 747 flew us nonstop from Andrews Air Force Base near Washington, D.C., to the island country of Bahrain, the smallest of the Gulf countries and notable as the place where the region's enormous oil resources were first discovered in 1932. At an elaborate formal dinner in Bahrain, I was summoned during coffee to speak with our host, the emir.

"My mother, for her health, likes to eat seafood every day, but the fish must be fresh—not frozen, and *not* canned. Tell me about the effects of the oil spill," he said. "Are the fish safe to eat?" It was the question I was to hear more often than any other about the impact of the oil spill!

I told him that no studies had yet been made on the effects of the Persian Gulf spill on edible species, but on the basis of studies elsewhere, including recent reports from Prince William Sound, toxins were not likely to appear in parts of fish usually eaten, the muscle. Whole fish might be regarded with caution, however, because of possible accumulation in certain organs, especially the liver. But to be *really* safe, I advised that for a while he might try to obtain fresh fish from distant sources. "The ecosystem as a whole is sure to be influenced," I told him. "But no one knows in what ways—or how much. It could translate to changes in the number and kind of fish living in the Gulf, reflecting changes in the ecological nature of the place."

Straight answers to straight questions are not easy to produce when so little is known about what's actually going on. I did not expect to learn much during the brief survey that would follow, but I was certain that I would have a better feel for the problems than I could gain just from looking at photographs and reading secondhand reports.

After an overnight stay in Bahrain, I reboarded the 747 with the rest of the group for a twenty-minute flight to Kuwait. Perched next to a window in the tail of the aircraft, I first noticed smoke billowing from the burning oil wells as a stark gray background under a fleet of uniformly small white clouds, a scene bearing a disconcerting resemblance to puffs of meringue floating on dark chocolate. We landed upwind of the fires in bright sunlight under a clear blue sky, and from there were caravaned aboard buses into what appeared to be an enormous thunderstorm. A billowing wall of black smoke marked the edge of the flaming oil fields; hundreds of wells spouted fountains of fire skyward.

I hugged my jacket close, chilled by the cloud-induced coolness, and shivering with apprehension about entering what General Schwarzkopf called the "gates of Hell." Some who approached too close to burning wells were killed when oil on the ground burst into flame and consumed them. Yet, in the strange way that fire fascinates humankind, I was transfixed by the roaring, flaming tornadoes and smoke-darkened sky, glowing along the horizon like burnished copper. William Reilly, Administrator of the Environmental Protection Agency, moved by the mesmerizing, contradictory beauty of the blazing columns, named my feelings: "If Hell had a national park, it would look like this."

Numerous film images have captured some of the powerful essence of those fires, and the superhuman efforts brought to bear to control them, but no words or photograph can assault one's senses with blinding, acrid smoke, shrieks of pressure-driven flames, the feel of grime darkening every surface, the grief and tangible pain emanating from people, birds, even the soot-blackened trees and grass.

Horrified, I watched one rosy-hued bird swoop into a pool of oil, lured by a struggling dragonfly. The bird barely moved, succumbing at once to shock, the slimy embrace gluing feathers, clogging nostrils, searing bright eyes, snuffing life. Nearby, deserted burrows of agile kangaroo rats and mice gaped like tombs, their former occupants mummified with legs outstretched, caught midstride as they ran for their lives.

We circled to the north to witness the infamous "highway of death," where frantically retreating Iraqis had been stopped by an allied blitz. Shattered households melded with heaps of burned-out vehicles and piles of live shells, grenades, and strips of ammunition. My chill grew stronger and intensified during a brief, numbing look at Hussein's headquarters and infamous center for torturing Kuwaiti victims.

My shock at the barbaric behavior of my species was softened slightly during a reception held by Kuwait's emir, and a flag-raising ceremony at the U.S. embassy. Along the way, smiling people waved, cheered, touched us with the joy of freedom. Feeling dazed and strangely detached, I approached a survivor who had weathered the war by hiding out at the U.S. embassy, a wobbly, bone-thin black and white cat, gray with soot and almost too forlorn to acknowledge my gesture of sympathy, half of my warm lasagna and a dish of reconstituted milk.

The following day, while businessmen and government officials evaluated the state of Kuwait's economic health, I concentrated on the underlying environmental issues. Jay Rodstein, an atmospheric scientist and member of NOAA's Seattle-based HAZMAT team, picked me up for a dawn flight via Saudi military helicopter to survey the status of the spill. Airborne, helicopter doors open for better viewing, my nose, not my eyes, alerted me to the presence of more oil than I had imagined possible—one, long, shimmering river, hugging the shoreline for as far as I could see. Rodstein said that the sheet of oil had been pushed against the shore by winds and currents, creating a smothering tide more than a foot deep in places. Tidal flats were inundated with thick, iridescent swirls of oil—beautiful, but deadly for billions of small, resident creatures who had no means of escape. Driven by winds, tides, and currents, the main mass of the slick moved south from Kuwait along the coast of Saudi Arabia and concentrated in the northwestern part of the Gulf. We hovered over Dawhat al Musallamiyah, or Musallamiyah Bay, and the adjacent lacy expanse of marshes and mangroves and the extensive web of tidal streams, looking in vain for anything green, a sign of life and hope in a morass of syrupy gray and black. We surveyed more than four hundred miles of Saudi Arabia's northern coastline, all of it awash in oil. At first, it seemed that the shores of Kuwait would escape damage, but later, as oil seeped from several sunken tankers in the north, Kuwait's southern coast and the beaches of Bubiyan Island were affected as well.

Saudi Arabia's curiously hooked island, Abu Ali, and a controversial causeway that joins it to the mainland, prevented much of the oil from flowing immediately southward into the immensely productive sea-grass meadows, coral reef and pearl oyster habitats near Bahrain and beyond. North of the causeway, oil was trapped in an immense embayment, seeping into the uppermost reaches of intricate marshes, tidal flats, and beaches; across the road, just a few feet south, beaches and surrounding waters remained nearly unscathed. The juxtaposition of these habitats was later to provide scientists with a natural laboratory for evaluating the effects of oil and various cleanup techniques.

During five subsequent visits to the area in 1991 and 1992, I was able to see the Abu Ali causeway at close range, dive among the oiled sea grasses and coral reefs, walk among oiled mangroves and marshes, view hundreds of blackened crabs, insects, birds, small mammals, and other creatures instantly fossilized in the deadly tides.

To gauge the impact of a spill, knowledge of prior circumstances is vital, wherever in the world it happens. During previous decades, scientists from Gulf countries and elsewhere had gathered enormous amounts of information about the nature of the area that now took on special value in establishing baselines against which the effects of the huge oil spill could be measured. In Saudi Arabia, scientists responded almost at once to obtain after-spill evaluations.

Efforts were made by the Saudi Arabian Meteorological and Environmental Protection Agency (MEPA), working with U.S. agencies, to track and predict the spill's behavior. Protective measures were initiated to keep oil from contaminating or, worse yet, overwhelming critical desalination plants along the Saudi Arabian coast, and extraordinary oil recovery operations were soon initiated. In time, Saudi Aramco, the Saudi oil company, and private contractors relocated more than 1.5 million barrels of liquid oil from the shoreline to holding pits carved out of nearby beaches; later, much of it was moved again to storage pits scooped out of the desert farther inland.

An attempt to set this spill in perspective begs for a comparison to the *Exxon Valdez* spill. About six times the oil spilled in Prince William Sound was *recovered* from the Persian Gulf, or about 8 percent of the total believed lost. The sheer magnitude of oil recaptured is unprecedented. An additional 5 to 6 million barrels (more than twenty times the whole Prince William Sound spill) apparently evaporated. The remainder was dispersed in shallow water, deposited in sediments, hardened into "tarmac" along beaches and marshes, or is gradually being broken down, or "bioremediated," by naturally occurring bacteria and other microorganisms. This scorecard does little to answer fundamental questions concerning the actual or potential damage wrought, but it at least helps size the problem. The Gulf spill was a spill of massive proportions. Even after the evaporated and recovered components are subtracted, 3 million to 5 million barrels remain in the ecosystem.

Using gallons rather than barrels as a measure, a 1992 report prepared by the U.S. Gulf Task Force, *Environmental Crisis in the Gulf*, offers the following statistics concerning some major spills:

		MILLION GALLONS
1991	Gulf War, Kuwait/Persian Gulf (on land)	420–42,000
1991	Gulf War, Persian Gulf (on water)	252–336[1]
1979–80	Ixtoc I, Mexico	139–428
1983	Nowruz Oil Field, Persian Gulf	80–185
1983	*Castillo de Beviller*, South Africa	50–80
1978	*Amoco Cadiz*, France	67–76
1979	*Aegean Captain*, Tobago	49
1989	*Exxon Valdez*, Alaska	10.9

1 Later figures are higher, on the order of 500 million gallons. Including oil fallout from the smoke plume that blanketed the area for many months, the amount entering the sea could easily be twice that of oil directly spilled.

Despite great interest by some in researching the effects of the spill, most efforts were overshadowed initially by the need for urgent action resulting from the burning wells and the aftermath of the war in general. The environmental focus was on the fires, smoke, and smothering effects of enormous lakes of oil, more than a hundred million barrels, flowing over Kuwait deserts. I summarized some of the findings for an assessment published by the *National Geographic* in February 1992 as a "one-year-later" report.* Concerns proved unfounded that

* Oil consumed—4.6 million barrels/day (about equal to the U.S. daily import of oil)
  Heat output—86 billion watts (about the same as from a 500-acre forest fire)

the mega-plume of smoke, ultimately extending more than a thousand miles from the burning oil fields, might impact global temperature or otherwise have a marked influence on weather or climate. Locally, however, sunlight levels were reduced, temperatures were cooler overall, and the short- and long-term influence of airborne pollutants on humans and the natural environment was real—but tough to evaluate or quantify. As with the effects of oil on a healthy ocean, smoke from an oil fire is not regarded as a prescription for well-being, but it is hard to pin down direct cause-and-effect relationships between inhalation of smoke and the incidence or onset of emphysema or asthma, or the influence of the thick layer of sooty fallout on the physical and chemical underpinnings of life in the desert and surrounding sea.

Early on, experts estimated that it would take five years to extinguish and cap the damaged and burning wells, but urgency, ingenuity, and opportunity for business combined to bring about swift control in less than ten months. Firefighters were well paid for their services, and as I watched teams in action at close range during a visit in June with EPA head William Reilly, I became convinced that every penny was earned—the hard way. Whatever the final, actual monetary price of controlling the inferno, the cost was significantly less than the daily loss of 3 million to 6 million barrels of oil in gushers or flames, the risks to human health, and the irrecoverable environmental destruction.

With Reilly, I stood behind a stiff metal shield and watched the action of high-powered hoses directing jets of water to cool the towering inferno. It was uncomfortably hot, but the metal barrier provided vital shelter from the searing, suffocating wall of heat just beyond. Briefly, I held one of the hoses and aimed it into the nearest towering blaze, a token gesture of defiance that generated fierce clouds of steam without noticeably diminishing the size or strength of my flaming adversary. To make an impact on each of the hundreds of fires required

---

Particles (30 percent soot)—12,000 metric tons/day (about 10 percent of the particles emitted daily from biomass burning worldwide)

Carbon dioxide (CO_2)—1.9 million metric tons/day (about 2 percent of the daily worldwide emissions of CO_2 from fossil fuel and biomass burning)

Sulphur dioxide (SO_2)—20,000 metric tons/day (about 57 percent of daily emissions from electric utilities throughout the U.S.)

days, not minutes, of concentrated effort, coupled with the know-how and competence to finish the job.

Snuffing out the fire was but one step in a dangerous series of maneuvers toward capping the gushers, complicated by the natural high pressure that characterizes the Kuwait fields. I glimpsed at close range the awesome power of that pressure propelling a huge, brown plume of raw oil skyward from a jagged, broken pipe . . . recoiled from the roar pummeling air spaces in my ears and chest . . . watched a feathery mist of oil merging with billowing clouds from nearby burning wells . . . tasted oil in my mouth and felt it in my nostrils, and on the surface of my eyes . . .

One of NOAA's field scientists, Dr. Will Pendergrass, gutsy meteorologist and hero to many Kuwaitis, guided me to a cluster of equipment and vehicles parked upwind of the huge, raw gusher. He was recognized by the foreman on the job, who beckoned us to come closer.

Closer! I thought, not certain that I welcomed the invitation. But when else might I have a chance to look down the throat of a gusher, a window into the very heart of Kuwait's great underground reserves. Edging carefully to the rim of a deep pit encircling the upward-thrusting jet I noticed something move. A *man* bronzed with oil was intently aiming a cutting stream of high-pressure water at the broken pipe. "Every cent," I murmured to myself. "They earn every damned cent. . . ."

With Pendergrass, John Robinson, the head of NOAA's Gulf team and for many years the head of its HAZMAT program, and an Air Force captain, I traveled the burning oil fields at night in May of 1991, something generally not allowed because of sniper activity and other uncertainties. Poisonous gases erupted from some of the wells and raw oil flowed freely, sometimes catching fire and swiftly engulfing everything combustible in its path. I had thought nighttime would seem vastly different, but mid-night seemed much the same as midday: black sky, no stars, no sun, bright flames, and all around the choking essence of hydrocarbons on the loose.

By the light of one large, vertical column of fire, I made a promising discovery, a ring of pale sand busily attended to by several industrious black ants. Grain by grain, they were effecting their version of desert restoration. Nearby was other evidence of the resiliency of life—clean round burrows of black beetles, blades of new green grass sprouting from blackened clusters, and the recent footprints of a small bird.

From the air, though the burning fields themselves looked less ominous, the catastrophic scope of an estimated hundred million barrels of oil engulfing the desert was much more evident. Joining other observers in a U.S. Black-hawk helicopter for an overflight of the burning fields and enormous lakes of oil, some more than six feet deep, I struggled to comprehend their impact on the land below, rich in human history and home to a diverse array of plants and animals—now sealed under hundreds of square miles of oil.

In the helicopter, stretching as far out as a maze of safety straps would allow, I extended my senses beyond normal constraints and imagined myself flying:

a shorebird, one of millions, drawn by ancient memories from Africa to northern nesting sites, pausing here for rest and food in the fragrant marshes and intertidal flats that lie just beyond the desert's edge. The glint of sunlight on a sparkling, calm surface lures me; I cup my wings and, too late, discover not the familiar touch of water, but of something dark and clinging . . .

Tens of thousands of shorebirds were fatally attracted to ponds, rivers, lakes, and oceans of oil, joining legions of other entrapped creatures, together forming a 1991 time capsule of animals permanently entombed in asphalt. During a return visit in January 1993, I found some of the deepest oil lakes drained and crusting over; others appeared freshly formed and were ringed by stiff, oily mummies newly snared or displaced from their burrows—sandpipers, beetles, dragonflies, and several kinds of small mammals. It is hard not to reflect that at some point, the numbers of these species will reach a critical point, and they will suddenly be gone, snuffed out, missing, and with them, links to hundreds of other creatures whose lives will be expanded, contracted, or lost altogether.

As an oceanographer, I was from the beginning most interested in the overall environmental impact of the war on the Persian Gulf and its inhabitants, and took every opportunity provided to explore the war's consequences—especially the mega-spill's—through direct observation. Several times I had the benefit of traveling to the region with Captain Francesca Cava, Chief of the U.S. National Marine Sanctuary Program, and John Robinson, on special assignment to lead NOAA's environmental assessment of the Persian Gulf War. In April 1991 we had our first glimpse of the hard-hit coastline at Abu Ali—underwater. Guiding us were scientists from King Fahd University and Dr. Ibraham Alam, a Saudi Arabian chemist and a highly respected polar explorer, but not a diver. Alam was content to watch from the shore, as was John, while two scientists from the university, Francesca, and I, outfitted with full scuba gear, clambered through an unavoidable band of thick, viscous slime to get to the bullion-colored water beyond. It was not my favorite dive, but it certainly remains one of the most memorable—and instructive.

Generally, oil floats, but in the shallow tidal flats and embayments of the Gulf, oil is brought high on the shore with incoming tides, and remains glued to the sand and mud when the tide falls. Enormous expanses of intertidal grass beds and flats are thus coated. Sand mixed with oil soon becomes heavy enough to sink and stay submerged. Sand blowing from nearby dunes also mixes with surface oil and causes some to fall to the bottom, and as light, aromatic components of oil evaporate, what remains becomes thicker, heavier, and sometimes sinks, especially when flipped and knocked about by waves. However it got there, when I plunged my hands into the muddy sand fifteen feet underwater, liquid oil oozed from among the sea-grass roots, staining my fingers with unmistakable evidence that oil was trapped below the rise and fall of the tides. It was still evident in August of 1991, when I dived there again with scientists from the university and MEPA.

Underwater, I found clusters of small clams gaping open, touched sleek, peach-colored olive shells, newly killed, and felt the crisp edge of papery brown pen shells, dead but still upright among clusters of sea-grass blades. It was impossible to gauge specific cause-and-effect relationships between the presence of oil and the dead sealife without prior knowledge of the area and controlled experiments—but there was plenty of circumstantial evidence that all was not well along this shore, and that oil was implicated. It was much easier to see the impact of oil on the marshes nearby, where a mosaic of small creatures, entombed by oil, bordered the base of blackened, dead plants.

Dr. John McCain, a U.S. scientist who has for many years researched marine life in the Persian Gulf while associated with King Fahd University, could be more precise. In the marshes near Musallamiyah and other coastal areas north of Abu Ali, studies had shown that the mudflats and shallow grass beds, long known as a rich food source for billions of birds that migrate seasonally between Africa and Eurasia, held as many as half a million small organisms . . . per square meter. Over lunch, an obviously distressed McCain described his post-spill findings: *Nothing* was alive in most oiled areas.

My quest for more widespread studies of the area materialized a month later, at a May 1991 meeting of the United Nations Intergovernmental Oceanographic Commission (IOC) in Paris. NOAA's Administrator, John Knauss, suggested that an NOAA ship might be made available as a focal point for research on the Persian Gulf oil spill by scientists from Gulf nations and elsewhere. Was anybody interested?

Having dangled the bait, Knauss clinched the concept by calling the spill a "terrible experiment," one that no one would deliberately set in motion for scientific satisfaction, but given that it was there, "the tragedy will be compounded if this opportunity to learn is missed."

Enthusiasm about the concept was contagious; even so, without a lot of good luck and an enormous amount of intelligent planning, pushing, and pulling, a project of the magnitude suggested could not have happened. I joined Captain Cava, John Robinson, and a distinguished NOAA "fish man" from Seattle, Dr. Robert Clark, who became expedition leader, to help convert Knauss's vision into operational reality as a special project of the Chief Scientist's Office. One of the greatest problems, money, was largely solved by diverting the 230-foot NOAA research vessel, *Mt. Mitchell*, for one year from a projected 150-year Gulf of Mexico survey program, so little new funding had to be found.*

Eight months later, the *Mt. Mitchell* embarked from Norfolk, Virginia, for the Gulf and in due course hosted 123 scientists from 15 countries, including representatives from the U.S., Kuwait, Saudi Arabia, Bahrain, Oman, Qatar, the

---

* All of the Persian Gulf countries—except Iraq—supported the project through contributions made to ROPME, the Kuwait-based Regional Organization for the Protection of the Marine Environment, the official sponsor of the expedition, which was coordinated through the U.N.'s IOC with financial and other assistance from NOAA.

United Arab Emirates, Iran, and several European countries. The 100-day expedition would provide the first oceanographic overview for the region since the late 1970s.

As a laboratory, the Persian Gulf is a natural. Formed about 15,000 years ago at the end of the last Ice Age, the Gulf filled as sea level rose, flooding the fertile Tigris and Euphrates river valleys with seawater flowing through the Gulf of Oman and the Strait of Hormuz. Today, it is estimated that loss of water through evaporation is ten times the volume of water added from the rivers and rain, and the Persian Gulf is therefore significantly more saline than the Gulf of Oman and Indian Ocean into which it flows. The average salinity is about 40 parts per thousand (35 parts per thousand is characteristic of the open sea); and some shallow protected lagoons may be significantly saltier, as much as 200 parts per thousand. Exchange of water through the narrow Strait of Hormuz—-i.e., a completely new batch of water—is thought to take more than five years. (In comparison, Prince William Sound has a twice-daily change of tides; full exchange of seawater takes place in about twenty-four hours.)

The Persian Gulf is shallow, with an average depth of 100 feet and a maximum depth of about 300 feet. Light penetrates to the very deepest parts of the Gulf, and productivity of phytoplankton as well as immense sea-grass meadows and other plant communities is high. Coral reefs thrive along the edges of some islands and mainland shores, hosting a rich and varied assortment of marine life including many sought after by local fishermen: pearl oysters, several kinds of shrimp, two kinds of lobsters, and numerous fish, especially hamour, a beautiful speckled grouper species.

When the expedition to explore the entire length and breadth of the Persian Gulf aboard *Mt. Mitchell* began to take shape, I could not resist getting in the thick of planning, implementing, and ultimately participating as leader of one of the six cruise segments. The opportunity to explore the Persian Gulf by diving on reefs from Qatar to Kuwait had special personal appeal because of the many similarities *this* Gulf has to another, the Gulf of Mexico, a body of water that I have come to know and love over many years. Both have coral reefs, clear warm water, sea-grass meadows, dolphins, sea turtles, and lots of tropical and subtropical algae, and both have well-established fisheries, including for several kinds of shrimp. There is even a Persian Gulf equivalent of Florida's gentle, beloved, and greatly endangered manatees, the dugongs, or sea cows. And, of course, both gulfs have been the recipients of major oil spills.

One of my objectives for the expedition was to answer the question put to me by the emir in Bahrain in March of 1991: "Are the fish safe to eat?" Or, more accurately, "To what extent is seafood quality affected by the continued presence of highly weathered oil?" As expedition leader, Robert Clark would concentrate his research efforts on this vital issue throughout the entire mission. His conclusions were roughly the same as my advice to the emir—but now with some solid reinforcing data. Other objectives included evaluating pathways and processes of oil movement from the intertidal to the subtidal (below low tide) zone, immediately after the spill and one year later; exploring the nature of current patterns and relating them to spill distribution and impact; and determining how coral reefs might be affected.

The longest and most complex studies were undertaken at Abu Ali during Leg II of the expedition, a six-week mission led by Dr. Jacqueline Michel, an oil-spill specialist and friend whom I had first encountered in a helicopter flight over Prince William Sound in 1989. Michel had her hands full, coordinating thirty-nine scientists participating in fourteen integrated projects in the shallow inshore bays north of Abu Ali island—Dawhat ad Dafi and Dawhat al Musallamiyah—and the bays and nearshore area at Tanaqip. Among other discoveries was disconcerting evidence that oil had penetrated deeply into the soft sand and mud, often channeled through burrows and passageways of multitudes of small creatures, worms, and crabs, to more than two feet below the surface. There was also evidence of previous spills, marked by layers of old oil under new sand.

In 1992, a historic exchange took place when the astronaut and oceanographer, Dr. Kathryn Sullivan, beamed down greetings to Michel, who stood on the *Mt. Mitchell*'s bridge as the shuttle swept by. This was more than a casual salute. Sullivan aimed cameras to document the nature of the area from high above to coordinate with on-site "ground truthing." Comparison of views and measurements taken from high above with those made on the spot, close up, helps calibrate Earth- and space-based remote sensing instruments and provides valuable perspectives for both.

Throughout the entire cruise, a framework of studies continued, with different areas of emphasis for each segment. During Leg V, I led a team of scientists with expertise ranging from fish, coral, and algal ecology to plankton behavior that explored the coral reefs along the northwestern Persian Gulf. Our main objectives were to try to evaluate possible impacts resulting from the spill, and to document the overall nature of the reefs and their inhabitants.

Of special interest was a chance to witness a once-a-year event, coral spawning in the moonlight—a phenomenon documented only once before in the Persian Gulf, and only recently observed in other parts of the world, from Australia's Great Barrier Reef to the Gulf of Mexico and the Florida Keys. It seems reasonable to suppose that corals have been successfully spawning for millions of years, but only in recent years have people been on the scene at the magic moment when, as one observer put it, "all Heaven breaks loose."

According to Dr. Yusef Fadlallah, an athletic zoologist who has a way with coral, a few days after the full moon in May, just after midnight, *Acropora*, one of the most common corals in the Gulf, whose branching colonies are known as "staghorn," simultaneously releases masses of eggs and sperm into the sea. He witnessed this undersea orgy for the first time the previous year near Dhahran. "Huge golden windrows of eggs formed slicks on the surface," he recalled. "They float for a while, then as larval corals start to swim as part of the plankton, eventually settling to the bottom." We speculated about the consequences of a collision between an oil slick and a raft of coral spawn, and were reasonably sure the outlook for a crop of young corals would not be good.

As one of the scientists on Leg V, Fadlallah monitored the condition of *Acropora* at every opportunity. At some places, every branch of this species was dead. Around the offshore islands of Karan, Al Arabiyah, and Jana, the coral's golden-brown branches appeared to be in good condition, and seemed about to burst with brilliant orange egg masses. Our timing was close, but sadly, the cruise ended just four days short of the blessed event in 1992 and a chance to evaluate the year's spawning was missed.

If given a list of the *natural* environmental circumstances in the Persian Gulf, I would be inclined to say that coral reefs can't grow there. As described in an earlier chapter, reef-building corals have a reputation for not living in water where the temperature falls much below about 65° F (18° C) or goes much above 86° F (30° C). Finicky creatures that they are, reef corals also flourish best, so the books say, in ultraclear water with stable salinity ranging close to that of the open ocean: 35 parts per thousand, not much more and not much less. Salinities greater than 45 parts per thousand are lethal to most corals, which literally "dehydrate" through loss of water from their tissues at higher levels.

In normal years, Persian Gulf temperatures drop to 48° F (9° C) in the winter and exceed 104° F (40° C) during summer months, although usually not remaining at these extremes for long. The Gulf's average salinity is significantly higher than that of normal seawater, averaging 40 parts per thousand, and in some places where reef corals occur, it ranges even higher than that. Exposure is another natural hazard; occasional ultralow tides cause areas that are normally submerged to be out in the open air where dehydration and exposure to high or low temperatures may be lethal. However, about sixty species of coral have not paid any attention to the books and have for thousands of years thrived around the Gulf's offshore islands and in places along the shore.

Hardy though they are, the corals in the Persian Gulf and many of the numerous invertebrates, fish, and algae associated with them live on the edge even during the best of times, and events of 1991 may have tipped the scales unfavorably, for some even permanently. Signs of stress were clear, especially in the nearshore reefs and offshore islands of Kuwait, where we found ample evidence of coral bleaching. As in the Bahamas, the Gulf of Mexico, the Caribbean Sea, and Australia, now here in the Persian Gulf I witnessed the transformation of normally golden-brown, blue, pink, or green corals to ghostly white. Most "bleached" corals were still alive, but in two areas offshore from Kuwait, many had succumbed and were overgrown with filamentous algae.

In the Bahamas and many other places, coral bleaching is linked to an increase in water temperature, typically an elevation of one or two degrees above normal, sustained for some weeks. In the Persian Gulf, however, prolonged exposure to *cold* water was implicated, perhaps brought about by lower-than-average summer temperatures caused by the "cloud-cover" plume from oil-field fires coupled with low spring temperatures and extreme low tides.

Among the recent hazards to the health of coral reefs the world over, including the Persian Gulf, are such things as smothering siltation from increased shoreline manipulation; pesticides and fertilizers applied to the land but flowing, ultimately, into the sea; sewage and pollution from industrial effluents; airborne pollutants including soot, acid rain, and heavy metals; wastes from ships; damage from anchoring and grounding; destructive fishing techniques, including lost nets; damage caused by plastic debris, especially large trash bags; tourism pressures; and of course, oil spills. The wonder is not that coral reefs worldwide are in trouble—and they are—but rather, that they have tolerated so much for so long.

In the Persian Gulf, events of 1991 exacerbated these stresses with added pollution from military ships, wartime wastes from the land, including heavy sewage discharges, and the mega-spill of oil, including an enormous fallout of soot and oil from the sky.

Whatever the cause, bleaching is bad news for these unique and resilient ecosystems. The possible good news is that world attention has been drawn to the Persian Gulf, resulting in studies that may help provide real insight into causes of deterioration worldwide—and thus lead to remedies. The "terrible experiment" may, in fact, yield some valuable answers to urgent questions.

Since the end of the war, interest has been growing in favor of providing protective measures for certain marshes, reefs, and other marine habitats in the Persian Gulf, undoubtedly enhanced by alarms over the extensive losses suffered and fear that recovery will happen slowly or not at all. In an essay for *Oceanus* magazine (Fall 1993) that I coauthored with John Robinson and Francesca Cava, we make a case for something unprecedented: to establish the Persian Gulf, *in its entirety*, as a protected area—a marine sanctuary.

Robinson, a brilliant strategist, asks tough questions: "How could this body of water, ravaged by the Gulf War and commonly thought to be among the most polluted in the world, benefit from the protection afforded by marine sanctuary status?" He continues, "Given their dependence on the Gulf for offshore oil development and transportation, how would regional governments greet an environmental initiative of this significance? How could the bordering nations, historically at war or at tenuous peace, reach agreement on the treaties necessary to establish such a joint designation?"

The questions Robinson raises, with slight variations, could be asked with regard to many places in the world where protective measures are urgently needed, if there is to be a reversal of environmental degradation, and thus survival of assets vital to all. The Gulf is, in fact, a microcosm where global issues can be seen with special clarity. As we summed up in *Oceanus*:

> *The terrifying swiftness of the 1991 damage focused world attention on the vulnerability of natural systems to human misbehavior. The before-and-after consequences of the war are immediately obvious, and the memory of the benefits of pre-war natural assets are still fresh in the minds of those now making decisions concerning the future of the region. People of several nations, diverse cultures and varying interests in Gulf use must cooperate and grasp the principle underlying the marine sanctuary concept—that everyone will win if all take care, and all will lose if even one misbehaves. On a grand scale, this is a key—perhaps the key—to achieving global environmental health.*

But there is a caveat. Despite clear evidence that damage to the environmental underpinnings necessary for life is a no-win game, population growth and immediate business interests often overwhelm good intentions. Given present trends, humankind will require twice the raw materials in 2010 that it does in 1994. To use oil in the next decade at the same rate as it is now being consumed translates to the need to discover as much in ten years as has been found in all of history.

We must come to grips with the reality that the source of our prosperity, like that of coral reefs, depends on maintaining a balance among many factors that are essential for us to survive. For corals the problem is not that they are consuming the surrounding ecosystem that supports them—but that humans are doing so. For us, it is clear that there simply isn't enough planet to go around to satisfy even the basic needs of a greatly expanding population. Our growing desires will inevitably put lethal pressure on the natural systems that sustain us.

It is not a single event—the 1991 Persian Gulf War, or the 1989 *Exxon Valdez* spill, or the advent of drift nets, or the destruction of the codfishery, or the loss of whales, reefs, or rainforests—that imperils the future of mankind. Rather, the combined effect of *all* of these, the "death of a thousand cuts," must rivet our attention, sharpen our awareness that we are a part *of*, not apart *from*, the rest. We must act while there is still time to protect the basis of our survival.

Part III

SEA CHANGE

Chapter 16

OCEAN CARE

Humpty Dumpty sat on a wall
Humpty Dumpty had a great fall.
All the king's horses
And all the king's men
Couldn't put Humpty Dumpty
Together again.

Nature Always Balances Her Books.
　　　　　　　　Arthur C. Clarke, 2061: Odyssey Three

Stunned, I stopped my car along the edge of the narrow highway to the Florida Keys and stared into a murky, gray-green sea. Heavy machinery rumbled near the shore where, five years earlier, banks of red mangroves prospered in water so clear that it seemed to merge with the sky. In their place, heat waves shimmered over stark white heaps of broken limestone rock. It was 1958 and the transformation of the Florida Keys from a lightly populated chain of islands to a crowded tourist mecca stretching from Miami to Key West was well under way.

Underwater, my worst fears were confirmed. Marly sediments and fine limestone debris had smothered the thriving reef where for the first time I had encountered the lyrical beauty of lavender sea fans, arching branches and mounds of a dozen kinds of coral, flocks of jewel-colored fish, and some of the ocean's largest, most inquisitive barracuda. It was here that I first found clusters of "baby's teeth," small snail-like creatures whose undersides made me laugh with their perfect pearly rendition of a miniature mouth. Here, too, I first met spiny lobsters, alive and well, not artfully arranged on a plate, and saw—and was seen by—the islands' signature species: pink conchs, lumbering like slow-motion bulldozers among nearby sea-grass meadows.

Today on the smothered reef there was no sign of live conchs or lobsters, only dead branches of coral and blackened, decayed skeletons of fans and sponges poking through the thick shroud of silt. Some small worms of a sort I had never seen before were taking advantage of the new seascape—just right for constructing their slippery tunnels, but lethal to the former reef residents. Siphons of small clams protruded in a few places, ambitiously filtering the clouded water for nutritious bits. *You've got your work cut out for you*, I thought, looking at the tiny tubes sucking in murky water and squirting it out clear. A few pale goatfish rummaged in the soft bottom, but the active, vibrant pulse of a healthy reef, bursting with life, had slowed to a sodden pace.

As painful as it was to witness the building project that had destroyed my favorite little patch of coral, I had to face facts. I told myself that its loss would not cause the end of the Florida Keys reef system, and there was some chance that, in time, the reef might prosper again. After all, I reminded myself, reefs recover even from hurricanes. Storms are expected events in the life and times of coral reefs everywhere in the world, forcing a continuous process of repeated disruption and repeated recovery . . . of change. After a storm, reconstruction of damaged areas, like that of war-torn cities, never follows the same pathways, but in time, storm-toppled corals sprout new branches, bits of sponge regenerate new colonies, and the scattered elements of the diverse communities reconfigure into new arrangements. Currents sweep sediment-laden waters clear more efficiently than clams and bring in young fish and planktonic larvae from distant, healthy reefs. While most hard corals grow slowly, usually less than an inch a year, shelter and sustenance for some reef dwellers is provided by fast-growing plants, red, green, and brown algae, which are gradually displaced by the slow-but-steady corals, sponges, and other more permanent reef dwellers.

But in 1958 I saw other ominous signs that did not bode well for the future of the Florida Keys. On the way to where I stood, I had passed various active projects: a marina under construction, a road being widened, a concrete sea-wall built to replace a natural marshy shoreline, a stand of mangroves cleared by motel owners to give visitors a better view of the sea, a channel dredged for straight-line access to open water. Throughout the shallow grass beds were furrows ploughed by motor boats taking shortcuts between channels; broken coral signaled impacts from carelessly tossed anchors. One of these actions by itself would not be cause for alarm, but when multiplied by thousands of similar, small, normal kinds of human behavior, ongoing through many years, the resulting force is greater than hundreds of hurricanes.

The Florida Keys are a microcosm, exemplifying the swift changes brought about by the actions of humankind globally, with causes as varied as the individuals who leave their mark. I was worried, mindful that even resilient, durable systems such as Florida's reefs might succumb to this "death of a thousand cuts." Meanwhile, the ecosystems surrounding the Keys were hit with more than the nibbling effects of physical damage; they were struck by mankind's potent double whammy: pollution and predation.

Prior to Rachel Carson's raising the alarm in *Silent Spring* about the grim consequences of lacing the natural world with deadly pollutants—and even after the alarms had been sounded—huge amounts of pesticides were used in the Florida Keys to control mosquitoes, and more flowed from residential and agricultural areas throughout southern Florida into the Everglades, from there spilling through a network of marshes and mangroves into Florida Bay and then south via the sea to the Keys. By 1994, heated controversies were raging over "Who killed Florida Bay?" and debates were held to determine what forces caused millions of fish, sponges, starfish, coral, and sea-grass meadows to die en masse, to be replaced by enormous blooms of blue-green algae and other microorganisms tolerant of the ocean's radically changed chemistry. Highly manipulated fresh-water flow from south Florida's system of dykes and canals along with high levels of fertilizers and pesticides are clearly implicated, but the specific case-by-case causes are as elusive and complex as the causes of death during a plague.

Predation came to the Florida Keys reefs in the form of fishing pressure from sportsmen and commercial interests. Both sharply increased with growing numbers of visitors who wanted to dine on the local specialties: turtle steak, conch chowder, spiny lobsters, grouper, snapper, and dozens of other tasty but vulnerable ocean residents.

Among the species most sought after commercially is my personal favorite, the Nassau grouper. These dog-sized, golden-brown creatures have great sad eyes, catlike curiosity, a hearty appetite for any grouper-mouth-sized thing that moves, and a Labrador-retriever-like personality. These characteristics make them a favorite among fish-watching divers such as I, but also cause them to be vulnerable to spearfishermen, to the dangling bait and lures of hook-and-liners, and, under special circumstances, to capture with nets.

Nassau grouper and many other commercially important fish are particularly susceptible as youngsters to unintentional wholesale taking by shrimpers, who drag their trawls over places favored by young fish as well as grown-up shrimp. Even more insidious and dangerous to the future of Nassau grouper and several other related species is the impact of fishermen who have discovered spawning aggregations—groupers grouping—and have netted tons of breeding fish in a single haul. In tropical seas worldwide, normally solitary grouper species assemble near the time of the full moon in midwinter. In a few days of focused action, they repeatedly rise in mating pairs, spilling into the sea the vital ingredients for the next generation of their kind. Large concentrations of fish have been destroyed in a season or two by fishermen who slaughter not only the adults who have traveled many miles for their time in the moonlight, but also kill the future for the species—and for the grouper fishery itself.

Within two decades of my discovery of the damaged reef in the Keys, the number of Nassau grouper had dropped sharply in Florida and the Caribbean. They are now so scarce that they are fully protected in Florida waters, and there is concern that the species may never recover to its former robust numbers. At the Caribbean Marine Research Center in the Bahamas, scientists are raising grouper in tanks with the idea that someday these fish might be cultivated for food like trout, thus taking the pressure off wild populations. There is also a vision of reintroducing young fish to places in the sea where they once were abundant, thus restoring depleted reef systems and perhaps providing the means to make possible a modest level of fishing.

The director of the research center, Bob Wicklund, has an uncanny ability to "think like a fish," a valuable asset in understanding and overcoming the problems of raising notably intelligent, independent reef carnivores in high-density tank culture. I visited the facility in May 1993 and found Wicklund eager to show off the results of years of patient research in the life and times of Nassau grouper, culminating in the successful rearing and maintaining of several thousand young. In the wild, most by now would have been munched by other predators; here, a large percentage were not only surviving and getting plump, but were also showing signs of typical independent grouperlike behavior.

Nearby were shallow tanks filled with hundreds of small members of another species targeted for benign scientific scrutiny, the famous queen or pink conch, *Strombus gigas*. Citizens of the Florida Keys are known as "conchs" because at one time many made their livelihood from gathering, selling, and consuming these big and beautiful snail relatives. In 1958 I saw enormous piles of football-sized conch shells in the Florida Keys, eloquent evidence of years of successful hunting by people who did not foresee a time a few decades later when the species would be so scarce that it would be accorded full protection throughout the Florida Keys and in many nations of the surrounding Caribbean Sea. The loss of a single species from an ecosystem featuring thousands of kinds of creatures might not seem a cause for concern, but to some, no pink conchs means no conch fritters, ceviche, chowder, and no conch "icon" to symbolize the way the Florida Keys should be . . . like an Oakland without oaks, a Bear Mountain without bears, Cape Cod with no cod.

To certain hermit crabs, octopuses, young grouper, and adult damselfish, no conch means no source of empty-shell shelter. To the ecosystem, it means the loss of a critical component, an abundant grazer on sea grass and other plants, a translator of plant energy into a source of food for hundreds of predators that dine on the vulnerable young conchs, and an occasional source of sustenance for a few creatures able to deal with the chunky limestone shell of the adults. To some humans, no conch means no income.

Economic losses more than ecological concerns have caused many to ask whether or not it is possible to restore conchs to depleted areas by culturing some young ones, then putting them where they can grow up for later capture. It sounds reasonable: Go gather up a few adults from areas where some still remain, put them in a tank, wait for them to reproduce, collect the young, feed them until they are old enough to fend for themselves, then pour them back into the ocean to restock as needed. I was excited at the prospect that restoration of conchless grass flats might be imminent.

But scientists who are involved in conch cultivation have encountered some unforeseen problems. "Raising young conchs is fairly easy," Wicklund told me. "Getting them to survive in the ocean is not. They don't know how to behave. They seem to be missing the know-how that makes it possible to escape predation. So far, we've lost one hundred percent of those turned loose. None have survived."

I have had some personal experience with attempts to reintroduce wild creatures raised in captivity to their natural habitats because throughout my childhood, I helped feed and care for various wild birds, squirrels, rabbits, opossums, and other injured or abandoned creatures brought by neighbors to my mother, "the bird lady," for recovery. Her greatest challenge, perhaps, was the successful raising of five scrawny, featherless snowy egrets, long-legged fish-eating waterbirds brought to her Dunedin, Florida, doorstep by my younger brother, Evan, who explained, "Some boys had them in a wagon along the side of the road. They said the mother had been killed."

My mother took over the job, stuffing ground fish down their gullets, egret style, and later releasing minnows and shiners in a dishpan where the young birds could succeed in the fine art of fishing. Known as "golden slippers" because their slender black legs end in conspicuously brilliant yellow feet, the birds seemed to dance in place, quivering their bright toes in the bottom of the pan, alarming the fish, who then were artfully speared. Later, the young birds fed along a natural shore near a dock in the backyard, but they continued to get handouts contributed by the local fish market. It was a startling sight to see Mother's gangling young protégés vying for priority treatment, crowding around a large table in the backyard where fish were chopped, sliced, and offered.

Snowy egrets are not currently regarded as endangered, but the species has never fully recovered from the time, years ago, when hunters killed many thousands, to pluck and sell the silky breeding feathers for adornment on women's hats. More recently, their numbers have fallen because of the widespread loss of wild wetland habitats needed for feeding, nesting, and roosting. Thus my mother's determined efforts to try to save and restore the orphans in her care. As she said, "Every egret counts."

We worried that despite her best efforts the semicivilized youngsters might choose to become permanent panhandlers, rather than rejoin the society of wild egrets and do their part for the future of their kind. Soon, however, wild egrets flew down to join them at the dock, and even became cautious participants in the once-daily backyard free-fish feeding sessions. For a while, Mother hosted not five but twelve snowy egrets, plus an assortment of other not-so-dumb species: American egrets, little blue herons, and one bossy great blue heron.

No one witnessed the hoped-for moment of departure of the young egrets, but as the weather cooled in the fall, the number of visiting birds diminished, and one day, no one showed up at feeding time. Three months later, with a rustle and flash of white wings, they were back, not to the natural shore by the dock, but to the feeding table in the backyard. Never caged, the birds nonetheless were not completely wild, but that may always be a price paid for foster care, whether of snowy egrets or of small, pink conchs.

It is generally easier to figure out what to do, or not do, to maximize the chances for successful rearing and release of bird species than of most marine creatures, especially invertebrates. I have great sympathy with the problems experienced by Bob Wicklund and others who have tried to restore conchs and other ocean dwellers to places where they were once abundant. Some successful transplants have been made of live coral to damaged reefs, and sea-grass meadows have sometimes, but not consistently, been reestablished by careful replanting of scoured areas with clumps taken from nearby healthy grassbeds. Mangrove and salt-marsh species have been planted in places where coastal development has destroyed the natural systems, a concept comparable to replanting trees in a logged forest. Such attempts to work with nature to speed an otherwise extremely slow process of natural recovery are encouraging, but the focus on individual species highlights the magnitude of what is involved in efforts to restore entire ecosystems. So much effort and skill are required to master the art of cultivating even one of the species involved!

For example, in 1958, the year my favorite little reef in the Florida Keys was destroyed, biologists Walter and Jo Starck carefully explored a nearby offshore area, Alligator Reef, and discovered 517 species of fish, including 45 previously unknown in the United States and 18 that had not been named before. Thousands of other species make up that reef system, each one with its own individual quirks, strategies, and requirements for survival. Even if all the species that such a system comprises were known and could be obtained in exactly the right numbers and stages of development, knowledge of how to put them together so they would prosper remains a mystery as elusive as the origin of life itself. The magnitude of the challenge, even for much simpler habitats, was recognized in a 1991 National Research Council Report, *Restoration of Aquatic Ecosystems*:

> It is axiomatic that no restoration can ever be perfect; it is impossible to replicate the biogeochemical and climatological sequence of events over geological time that led to the creation and placement of even one particle of soil, much less to reproduce an entire ecosystem. Therefore, all restorations are exercises in approximations and in the reconstruction of naturalistic rather than natural assemblages of plants and animals with their physical environments.

That does not mean that it is no use trying, but the magnitude of ignorance concerning how to improve, let alone re-create, a healthy ecosystem from the broken or poisoned remains cannot be overestimated.

Some people seem to miss the point entirely, such as the city officials in Los Angeles who in the 1970s authorized $75,000 to purchase and "plant" plastic replicas of trees to replace their genuine, living predecessors along a highway, a restoration solution comparable to offering a rag doll to replace a real child.

Others attempt "quick-fix solutions" to restoration of cleared areas by planting fast-growing ornamentals, sometimes with disastrous consequences. Examples include the use of Australian "pines," *Casuarina*, and the bright-berried shrub known as Brazilian pepper as replacements for native trees. The "cure" has caused the displacement of significant areas of natural vegetation in Florida, and necessitated very costly belated control measures.

Still others wisely recognize that effective restoration efforts are comparable to the actions of doctors treating a sick patient: Recovery can be accelerated with some skillful help, but true healing requires time and active, natural processes that are beyond human understanding. It is sometimes possible to manipulate those processes, but they are frighteningly easy to disrupt and destroy. And impossible to create.

Lake Erie, once one of the most polluted and ecologically damaged lakes in the world, has improved through significant "doctoring"—a combined approach of reducing pollution and restoring and protecting natural habitats. It has not been restored to the robust, healthy system it was two centuries ago, but it is better than it was two decades ago. Similarly, Chesapeake Bay, ounce for ounce one of the most productive bodies of water in the world, has been subjected to enormous stresses during the past century, greatly reducing the natural wealth of shellfish, crabs, and other marine life that greeted early European settlers in the 1600s. Efforts to address the many causes of destruction and their cures are beginning to pay off, although "total recovery" is not achievable, nor is it expected.

Environmental health is at the heart of awesome problems now facing Russia and other Eastern European countries in the wake of efforts to swiftly industrialize, without due regard for the long-term costs. The price for not respecting the environment is now being paid with blighted lives of citizens who had little to say about the policies that have altered their future. Most directly affected are those living nearest to the sources of pollution, but indirectly and inescapably, all of us are touched.

Who was consulted before the recently revealed sinking of the nuclear submarine *Komsomolets*, or the dumping of radioactive wastes by the Soviets in the

Arctic and North Atlantic? And who was asked before the blatant dumping of nuclear wastes by Russia in waters near Japan in 1993, in violation of the present international moratorium on dumping radioactive wastes at sea? Who really knows how much of what radioactive substances were thrown where in the ocean during years of Soviet secrecy? Or what is presently taking place in the former Soviet Union and in other countries eager to relocate noxious materials from their immediate backyards? No one really knows what the impact of using the ocean for disposal of nuclear wastes has been. Nevertheless, some favor the sea over the land as a dumping place, and pressures are mounting to lift the moratorium on ocean dumping, coincident with the growing volume of radioactive wastes—and the shrinking number of "safe" places to put them. Answers to urgent questions about the magnitude and location of ocean dumping, nuclear and otherwise, past and future, are dangerously slow in coming, but they must be accounted for as a vital part of ocean care.

Once the health of the land, air, and water, and thus the people, is impaired, the expense of restoring it is staggering—but ignoring the problems is sure to be even more costly. This is the conclusion of the U.S. Environmental Protection Agency with respect to the formidable work involved in cleaning up and restoring to a reasonable state of good health areas used for disposal of toxic chemicals and nuclear wastes during the past fifty years. Despite the EPA's readiness to spend billions of dollars to detoxify disposal areas, especially those designated as "Superfund" sites, it is increasingly clear that while money can be used to make improvements, money alone cannot restore health. Those elusive but vital "natural processes" are absolutely crucial to successful restoration. Medical doctors are quick to point out that it is much more desirable and cost-effective to try to prevent, rather than cure, sickness. In this spirit, *protecting* and *maintaining* existing healthy ecosystems should command our highest priority.

In the 1950s, in Florida, not many people thought that measures to protect the ocean habitat were necessary, but nevertheless, growing concern for the destructive consequences of human activity in the Florida Keys led to an inspired action: the establishment in 1960 of the John Pennecamp Coral Reef State Park, a twenty-one-mile stretch of productive reef systems offshore from Key Largo, Florida, where spearfishing and certain other damaging practices were prohibited. Some were skeptical about the prospects for success, since the park had fluid boundaries that were difficult to patrol, and there was no way to prevent pollutants from entering, or fish from leaving. Nonetheless, the concept caught the imagination of many who could readily see the parallels between protected areas in the sea and those on land.

The need to take special action to set aside natural areas on land for wildlife and for people was recognized long ago. The U.S. network of national parks, which began in 1872 with the designation of Yellowstone Park, is sometimes referred to as the "nation's crown jewels," and the "best idea America ever had." Their underwater counterparts are arguably just as significant but about a century slower in getting under way.

Coincidentally, the science of oceanography was also born in 1872, with the departure of the British research vessel HMS *Challenger* on the first global expedition to explore the oceans; one hundred years later, legislation passed in the U.S. Congress authorizing development of a National Marine Sanctuary Program. At the same time, plans were brewing in Australia that would result in the establishment of the largest marine park in the world, the Great Barrier Reef Park Authority.

By 1975, the Barrier Reef had been designated a national park, and two sites had been given protection in the United States as national marine sanctuaries. The first is a 1.3-square-mile area around the wreckage of a U.S. Civil War ironclad vessel, the USS *Monitor*, which is located in about 220 feet of water, 16 nautical miles (18.4 miles) southeast of Cape Hatteras, North Carolina. Sunk in a storm in 1862, *Monitor* was searched for and located with modern side-scan sonar during a scientific expedition in 1973, and soon thereafter was proposed for designation as a sanctuary. The designation of a second site, the Key Largo National Marine Sanctuary, adjacent to the John Pennecamp Coral Reef State Park, brought an additional 100 square miles under special protection.

Within fifteen years, the number of national marine sanctuaries had grown to nine, from tiny Fagatele Bay, a one-third-square-mile piece of coral reef in American Samoa, to the Channel Island Sanctuary, encompassing 1,655 square miles of near-shore waters, open ocean, and deep ocean surrounding several of southern California's most significant offshore islands—San Miguel, Santa Rosa,

Santa Cruz, Ana Capa, and Santa Barbara. There were discussions about designating the 2,600 square miles of ocean surrounding the Florida Keys, incorporating the Key Largo sanctuary as well as Looe Key, a small but pristine site designated for protection in 1981. The proposal of the Florida Keys sanctuary in 1990 by the U.S. Congress had a lot to do with my decision to take a temporary leave from Deep Ocean Engineering, the company that had dominated my time and life for ten years, and go to work for NOAA.

As soon as I announced my plans, several friends questioned my sanity. "Are you in your right mind? You're leaving Deep Ocean Engineering to go to Washington to work for the *government?* You'll never get out from behind a desk! Dry rot will set in!"

My stock response to such jibes was "The spouting whale gets the harpoon!" For many years I had been a "spouter," a critic of ocean policies—or lack of them—in the U.S. and elsewhere. I was happily "speared" in 1990 with the NOAA appointment. Instead of poking at policies from the outside, I thought it might be the best chance I would ever have to help implement efforts from within the system, especially those concerned with ocean exploration and research, restoration of damaged ocean ecosystems, and, most important, protection of the healthy areas that remain. I had come to believe that healthy natural ecosystems are the most important legacy we have from the past, above the ocean or below, and their protection is the single most important legacy we can bequeath to those who follow. Of particular interest was the National Marine Sanctuary Program, especially in light of the proposed designation of the entire Florida Keys reef system, and an even larger area, 5,327 square miles offshore from Monterey, California, that would eventually be incorporated late in 1992.

The program presently encompasses more than 15,000 square miles of ocean in fourteen sanctuaries, and is complemented by a related National Estuarine Research Reserve System, a nationwide network of more than twenty protected areas dedicated to research and education. From 1992 to 1994, the combined program prospered under the leadership of Francesca Cava (with whom I had worked on the *Mt. Mitchell* expedition), formerly manager of the Channel

Islands sanctuary and a captain in the NOAA Corps.* The Center for Marine Conservation, the Nature Conservancy, and local universities and research and educational institutions such as the Monterey Bay Aquarium in the west and the New England Aquarium in the east have greatly magnified the effectiveness of the sanctuary program. But demands are increasing for use of the sanctuaries, for development of new sites, for enforcement, for amenities, for research and educational programs, and for effective response to conflicts arising from the many ways that people want to make use of the same small patch of water, often simultaneously.

In some ways, much more is expected of the ocean than the land, and that applies to ocean sanctuaries as compared to their terrestrial counterparts. How large would the national parks have to be, for example, to enable an unrestricted number of people to take commercially and for sport buffalo, trees, birds, shrubs, snails, rabbits, and trophy species such as bears, wolves, and bighorn sheep? How large must the marine sanctuaries be to enable an unrestricted number of people to take commercially and for sport fish, clams, lobsters, seaweed, seashells, and trophy species such as sharks, billfish, and barracuda?

In the 1970s, when few protected areas existed in any ocean, human population was significantly smaller and the demands on natural resources far less than at present—or in the future. Moreover, materials did not then exist to make durable drift nets and other modern catch gear, resulting in massively destructive fishing techniques. Acoustic fish finders, airborne observers, sophisticated communications systems, and powerful new catch techniques have taken much of the guesswork out of commercial fishing. As a consequence, it has been possible to remove staggering quantities of life from natural ocean ecosystems that were formerly protected by their inaccessibility. One important function of marine sanctuaries, therefore, is to provide havens for species through habitat protection and perhaps, in due course, restrictions on taking fish in specific protected areas.

---

* The NOAA Corps is a uniformed branch of service involving about four hundred people who operate NOAA's ships and aircraft and otherwise participate in the agency's scientific research and monitoring programs.

The U.S. National Marine Fisheries Service (NMFS) oversees a coastal habitat protection program that is aimed specifically at protecting and restoring commercially valuable species. Marine sanctuaries and NOAA's system of Estuarine Research Reserves complement the NMFS efforts, but even these combined efforts are far from enough to solve the problem of declining fisheries. After all, unlike in our national parks, where all of the resident and migratory creatures are protected, visitors to national marine sanctuaries face few restraints on killing the most abundant and influential vertebrates present—the fish.

Many, of course, think of such creatures only as a commodity. It is less obvious, perhaps, but no less important, to consider fish as vital components of an ecosystem. Yet, as on land, removal of large predators and other key players from ocean ecosystems is certain to trigger unfavorable chain reactions in the communities from which they have been taken. Hooking squirrels, spearing hawks, shooting owls, netting songbirds and butterflies, slaughtering the deer and bears—all would be permissible if "marine sanctuary" rules applied to national parks.

Given this fact, some people are confused by the term "sanctuary" with its implication of something sacrosanct—a special haven off limits to all but a privileged few. In fact, marine sanctuaries in the United States and elsewhere in the world take many forms, but rarely are there constraints as great as those that now apply to national parks. Officially, U.S. marine sanctuaries are vaguely defined as "areas of the marine environment of special national significance due to their conservation, recreational, ecological, historical, research, educational or aesthetic qualities." This generalization of sanctuaries allows for many interpretations—and misinterpretations. Perversely, some recommend changing the term "sanctuary" to something like "management zone," rather than tackle the politically hot issues about fishing. However, as the value of designating fully protected areas for fish is demonstrated, including their importance in maintaining healthy stocks in adjacent areas available to fishermen, attitudes—and policies—may change.

"Despite the problems, what to do is much clearer than how, given the program's extremely limited resources," Cava observed recently. Currently, $11 million must be stretched to cover the cost of about seventy people managing and monitoring aquatic space larger than the land mass of the Bahama

Islands, but spread from the South Pacific to New England. The budget is about the same as that provided just for Yellowstone Park's yearly operation, one of the many parks in a system that comprises nearly 95,000 square terrestrial miles and has a staff of more than 16,000 people and a total budget of $1.4 billion, or 127 times the budget for marine sanctuaries.

Comparisons between the sanctuary program and the nation's fine system of national parks are tempting, but not entirely appropriate. One major difference is that the ocean is three-dimensional, and the space designated for management is actually significantly larger than the square-mile figures suggest. Terrestrial parks often are blessed with varied terrain, but except for birds and flying insects, the perspective enjoyed by most visitors—and residents—is essentially two-dimensional and earth-based. Not so in places such as the Monterey Bay Marine Sanctuary, where life abounds throughout the water column, from the surface to depths greater than 10,000 feet, and as high above as seabirds fly.

Another conspicuous difference concerns how property is treated on land versus the sea. On behalf of the people of the United States, the National Park Service "owns" the land it manages, but everyone—and no one—owns the ocean. Complex and overlapping local, state, national, and international laws provide a framework for what may and may not be done, as well as many years of tradition, vested interests, and a widespread passion for the vision of "freedom of the seas"—even within territorial waters. The National Marine Sanctuary Program has jurisdiction over the designated areas it manages, but cannot exclude many activities that are, in fact, damaging to the continued good health of the areas involved.

Just as polluted air moving with a breeze cannot be fenced away from national parks, toxic wastes carried by currents cannot be kept out of protected areas in the sea. It is possible to restrict deliberate dumping in marine sanctuaries, but difficult to keep out whatever comes from adjacent land and sea. The decline of reefs in the Florida Keys is linked to an overload of noxious substances, suspended debris, and chemicals, all flowing from the land into the sea, as is the case for reefs in Hawaii, Australia, and elsewhere. In fact, as ecologists have long observed, all of Earth's systems are interrelated. Whatever happens to damage the health of one area in due course has an impact on the entire earth. Likewise, keeping any area, land or water, in good condition, benefits all.

Given the diversity of size and composition of the areas designated as marine sanctuaries, it is not easy to find descriptive terms that fit all of them. Perspectives vary, too. From the standpoint of the whales who know a little patch of the western North Atlantic called Stellwagen Bank, the place might be regarded as pleasing because of its natural beauty: a craggy sea floor topped with a magnificent column of sparkling clear water loaded with small creatures. There are always interesting things to do, good things to eat, and in the later part of the twentieth century, amusing encounters with those terrestrial mammals, human beings, who stare into the sea from their various perches on the surface. Small as it is, this little patch of the western North Atlantic is valuable for far more than whales and the recreation-tourism industry, however. Protecting that pristine piece of the sea is like banking an ecological asset that yields high dividends in terms of fish, genetic diversity, and a source of materials for restoring overexploited areas. This aspect of marine sanctuaries might generally be defined as part of the planet's "environmental capital," the ultimate source of our wealth and health.

Ask any ten people "What is a marine sanctuary?" and you might get ten different answers, all of them true, probably none of them complete. To a diver, gliding among the branches of one of the last healthy stands of elkhorn coral in the United States, in the Florida Keys, a marine sanctuary is a vital national insurance policy that helps protect the future of that coral and of magnificent diving locations for millions of people every year, forever.

To a scientist, the sanctuary is a natural laboratory, one with the added appeal of long-term protection of precious study sites, and opportunities for integrated monitoring and research.

To schoolchildren, a sanctuary is a playground and a most desirable destination for field trips, a neat place with no end of appealing critters to look for in the shallow reefs and the lush sea-grass beds and along the rocky or sandy shores. It may mean a glimpse of dolphins at home in the sea or a chance to discover and be surprised.

To a genetic engineer, the waters surrounding the Florida Keys may be defined as a priceless, protected "resource bank" crammed with precious biodiversity, thousands of recipes for life, of complex chemistry developed over millions of years of Earth history, now secure.

To an environmental engineer, charged with restoring damaged ecosystems, the sanctuary is the yardstick against which "good health" can be gauged, and a source of the ingredients needed for renewal.

To a government economist, marine sanctuaries generally might be defined as one of the nation's greatest bargains, considering the territory covered and the amount that is accomplished with the means available.

Savvy visitors from Chicago or San Juan or London or Moscow to a sanctuary such as the Florida Keys may see it as part of this nation's commitment to protect natural areas that contribute importantly to the basic ingredients of a hospitable planet. The oxygen produced and carbon dioxide absorbed in the protected sea-grass meadows, like in the rainforests of Brazil, help offset oxygen consumed and carbon dioxide produced by automobiles, by industry, by people throughout the country—and the world. Billions of natural microorganisms in the sanctuary act on excess fertilizer, pesticides, herbicides, and other contaminants from air- and land-based sources, gradually breaking them down into less harmful components. Even people who never venture within several thousand miles of the Florida Keys sanctuary might regard the sanctuary there—or the four in California, the reefs offshore from Georgia, or even the little speck of reef in American Samoa—as a "priceless national asset," each area doing its part in maintaining planetary health vital to all people everywhere.

Fishermen, however, might regard a marine sanctuary as a threat to traditional freedoms, the thought being that while fishing is not now restricted in most parts of the sanctuary, it might be in the future. Yet, upon reflection, a fisherman might also see a marine sanctuary as the best hope that fishermen have for maintaining their way of life. Like human beings and all other forms of life, fish require certain things to live and grow and produce more of their kind. Protection of habitats for fish will translate to more fish, and thus protection for fishermen as well.

An oil company executive might say that marine sanctuaries are a mechanism used by some to keep areas from being developed for exploitation of offshore oil and gas resources. It is true that concern about the possible impacts on ocean health of activities associated with the oil and gas industry has provoked some to embrace openly the marine sanctuary concept as a way to inhibit future development. In fact, the one activity that is universally not allowed in U.S. marine sanctuaries is drilling for oil and gas. Most other uses continue without restraint.

However, one of the smallest of the U.S. sanctuaries is actually perched right in the midst of a field of actively producing oil rigs. The Flower Garden Bank National Marine Sanctuary consists of two small rocky outcroppings, totaling 58 square miles, about 120 miles east of Galveston, Texas, and 120 miles south of New Orleans. Despite its diminutive size, the place has been a showcase for an industry-conservation partnership, after an initial period of resistance by oil companies in the area who were concerned about potential restrictions on their activities. The area has also drawn large numbers of divers, especially eight days after the full moon in August, when several species of coral, sponges, starfish, and other reef creatures simultaneously release into the sea critical ingredients needed to establish a new reef.

The manager of the Flower Garden Bank sanctuary, Dr. Steve Gittings, becomes enormously animated as he describes with eloquent gestures the way tiny yellowish spheres—coral eggs—are cradled within the coral's tentacles, then lofted all at once into the water column while corals nearby emit smokelike clouds of sperm. Other species that have males and females in the same colony release egg packets already fertilized. Small brittle starfish slither from crevices, stand on the very tips of their five slender arms, then arch and twist with rhythmic rock-star vigor while sending a haze of small packages of genetic data into the dark sea. This planktonic blizzard concentrates at the surface, where it may travel for many miles beyond sanctuary boundaries before settling down to fill niches opened during the past season, or establish new reef systems elsewhere. Similarly, creatures inhabiting the Flower Garden Bank sanctuary are likely to have come from far away. Terrestrial counterparts are airborne seeds, insects, spiders, and spores.

In fact, marine sanctuaries cannot—and should not—be isolated from surrounding systems, and it has become increasingly clear that the best hope for maintaining areas in good condition, especially those selected for their ecological significance, is to embrace large areas. This awareness is reflected in a trend toward designation of entire, broad ecosystems rather than small, pristine places, although the Flower Garden sanctuary is a notable exception. But large protected areas are vulnerable, too. The United States and other nations can establish sanctuaries only within their Exclusive Economic Zones (EEZ). Approximately 60 percent of the ocean is in that jurisdictional never-never land, the "global commons," where policies for ecosystem protection are largely in the discussion stage.

In 1991, in an effort to explore what it might take to create open-sea "Wild Ocean Reserves," I called together a group of scientists and ocean policy experts from NOAA as well as the Environmental Protection Agency, the Smithsonian Institution, the World Wildlife Fund, the U.S. Center for Marine Conservation, and others concerned with "ocean care." There was underlying agreement that protection for open ocean ecosystems was desirable, but there appeared to be no clear mechanism for action. We did make some progress in identifying the challenge, however, and several meetings followed, each reflecting a growing certainty that we were onto something important—but that we might have identified the twitching tail of a very large tiger.

Many are concerned that measures to protect open ocean ecosystems might impact the traditional "freedom of the seas." Translated, it might mean that some who now take advantage of the "anything goes" policies on the high seas beyond national jurisdiction might have to modify their behavior. For example, some sensitive areas might be declared through international agreements to be off-limits to fishing or dumping of trash or use of explosives or other potentially disruptive activities. After the first meeting of the "Wild Ocean Reserve" group, I realized that some very large toes were being stepped upon when NOAA received worried calls from people in the State Department and the U.S. Navy. At subsequent meetings of our group, several "watchers" were present to keep track of developments.

Despite the inherent difficulties, protection for ocean ecosystems, both within national jurisdiction and in the wild ocean beyond, may be one of those important concepts whose time has come. Concern for the health of the open sea is reflected in *Caring for the World: A Strategy for Sustainability*, prepared by the World Conservation Union,*The United Nations Environment Programme, and the World Wide Fund for Nature. It states: "The ecosystems and resources of the open ocean beyond 200 miles from the coast are open access resources; and there is no effective comprehensive legal regime to regulate their use." The report goes on to recommend integrated international management and the development of an effective legal regime.

The World Commission on Environment and Development identified three imperatives for ocean management, recognizing that the sea is, after all, one continuous and interacting system. There first must be effective global management regimes, and because resources in regional seas are shared by several nations, there also must be effective regional management regimes. Finally, since the major land-based threats to the oceans require effective national actions, there must also be international cooperation.

The World Conservation Union is helping to promote all of these objectives, including development of a vitally important international environmental database. A comprehensive inventory of the world's protected areas has been assembled, ranging from small, privately held wildlife reserves to internationally renowned "World Heritage Sites" and "Biosphere Reserves," large regions identified by panels of specialists working with UNESCO. The results, published in four hefty volumes, complement a more detailed global review of hundreds of marine protected areas being compiled by ocean specialists worldwide under the leadership of Graeme Kelleher, director of Australia's Great Barrier Reef Marine Park Authority (GBRPA).

The GBRPA, with Kelleher at the helm, has set the pace for effective, integrated management and multiple use of a very large and complex marine ecosystem. Their domain covers thousands of square miles of reef, peppered with numerous islands, vast sea-grass meadows, sandy flats, deep-water channels,

---

* Formerly the International Union for Conservation of Nature (IUCN).

and steep, mostly unexplored dropoffs into depths well below scuba-diving range. Mindful of the many, sometimes conflicting, interests already established through years of use prior to establishing the park, managers and policy-makers have made efforts to develop win-win policies that will result in sustaining the good health of the natural systems involved, and thus the sustained use for all concerned.

It may seem farfetched to some to imagine a time when sharks, barracuda, grouper, and other fish may be actively protected anywhere in the sea, but deep in the heart of the Great Barrier Reef there is one place popularly known as the Cod Hole, where this apparently radical concept is already functioning.

Largely through the efforts of Ron and Valerie Taylor, naturalists, conservationists, and divers who long ago exchanged their spearguns for cameras, a part of the reef was officially designated as off-limits to fishing, specifically to protect those giant, good-natured fish known in Australia as "potato cod," which are a kind of giant grouper whose curiosity matches that of visiting human beings. Now one of the most popular dive sites in the world, the Cod Hole is one area where pound for pound, the fish are acknowledged to be far more valuable alive, generating sustained revenues for boat operators, dive shops, and photographers than they would be if sold—once—as fillets.

In protected areas, nonconsumptive uses of the resources present are increasingly being emphasized. Fish-watching is one of the prime attractions in the Florida Keys, now rivaling fish-catching in terms of economic importance. Whale-watching is the basis of a booming business worldwide. Within the recently designated Stellwagen Bank Sanctuary off the coast of Massachusetts, revenues exceeding $100 million a year are attributed to whale-watching enterprises; an even greater income is generated in the Hawaiian Islands Humpback Whale Sanctuary. Whale-watching is an economically attractive "sustained-use" alternative to the one-time sale of these large mammals as bacon, steak, or sushi—even in Japan, Norway, and Iceland, countries that are actively engaged in killing and selling dolphins and whales for meat.

Some question the need for special "whale sanctuaries," since there is a world-wide moratorium on the commercial taking of large cetaceans, and the killing and harassment of all marine mammals in U.S. waters is already prohibited by special legislation, the Marine Mammal Protection Act (passed in 1972 and reauthorized in 1992). Some whales, such as humpbacks, are additionally insured by their status as endangered species, and thus are covered by strongly protective measures embodied in the Endangered Species Act. But the National Marine Sanctuary legislation complements the other laws by protecting entire ecosystems, and thus addresses critical aspects of the creatures' life history, notably breeding and feeding, as well as safe living space. One of the outcomes of facing the issue of what it takes to truly look after the well-being of whales, or any other species, is the discovery that thousands of other creatures must be protected as well if the selected species is to live.

Whales, potato cod, coral polyps, eagles, and grizzly bears share with all other forms of life the need for habitat protection. The U.S. National Marine Sanctuary Program and other national and international protective measures provide a mechanism whereby critical elements necessary for the healthy functioning of "habitat Earth" can be safe-guarded. If we are successful, our descendants one hundred years hence—and forever after—will salute those who now have the foresight to wrestle with the tough policy issues, take on the vested interests, seek win-win solutions, and emerge with protection for special places where coral can spawn, undisturbed, by the light of an August moon, for places where kelp forests and deep-sea canyons shelter a mix of life undiminished in diversity and abundance, for places that provide safe havens for fish and whales—and a planet that is a safe haven for all.

Chapter 17

SECOND CHANCE
FOR PARADISE

Unless someone like you cares a whole awful lot, nothing is going to get better. It's not.

Dr. Seuss, The Lorax

"Mom! The Oakland hills are on fire and we are going to have to evacuate in about an hour. What should I take?"

With one breathless telephone call from California in October 1991, my daughter Elizabeth transformed a quiet dinner at my mother's home in Florida into a scramble for action—and a swift appraisal of what really matters in life. Fire is a known hazard in the area north of San Francisco Bay—known, but rarely uppermost in the minds of the residents who also philosophically live with imminent earthquakes and the everyday high risks associated with heavy freeway traffic and other threats to a serene life.

"Make sure you get out," I said, wasting no time with questions. "Take the animals, if you can."

I paused, then flashed for a moment on my lifetime collection of thousands of professional books and papers, photographs, films, files of correspondence, travel mementos, and 20,000 or so cataloged specimens of marine plants. I skipped without hesitation over thoughts of furniture, clothing, office and laboratory equipment, but lingered on images of numerous family treasures that are in my care: my mother's all-"A's" school report cards, a small jewelry box from *her* mother, six silver spoons from her *mother's* mother, and files of schoolwork and notes from my own children . . . and one-of-a-kind messages from my one grandchild, Russell. I thought of the one time out of hundreds of encounters with sharks when it seemed that I was being sized up dispassionately, not as a professional scientist or scholar or mother or shark-sympathetic human being, but simply as a piece of meat. Fire, too, is hungrily indiscriminate, engulfing with equal equanimity fibers of wood, scraps of paper, living tissue.

In an instant, priorities focused. *Life* was most precious, and could not be replaced, no matter how great the compensation. There is but one *Elizabeth!* And no dog like Blue, or each of the several cats, or Elizabeth's three birds or bright green geckos. Next in line were photographs, readily retrievable—and totally irreplaceable. I did not want to think about the rest.

"If you can, take the family albums in my office," I said. "But just go! I'll get there as soon as I can."

"What shall I do about Charlie?" Elizabeth asked.

Charlie, a four-foot-long alligator, a foundling given to San Francisco's Steinhart Aquarium, was being kept in our backyard in a special pen until he grew large enough to be safely placed in the aquarium's "gator gallery."

"How about putting him in the swimming pool. But hurry! Call me again when you can. . . . I love you."

Rarely have I felt so helpless. Even if I were to board a flight at that moment, I could not get to my doorstep, or what would be left of it, for at least seven hours. My daughter and the family belongings, my office and laboratory, would survive the approaching fire—or would not—independent of anything I could do.

On the way to the airport, I listened on the car radio for news and heard that hundreds of houses in Berkeley and Oakland already had been consumed and that the fire was out of control. On the flight, I learned that a quirky shift in the wind had sent flames moving away from the area where I live, and the evacuation alert was put on hold—time enough for my friends at Deep Ocean Engineering to mobilize an all-out effort to pack and move several truckloads of files and photographs, and soak the roof of the house to make it less vulnerable to flying sparks and debris. Many had no such leeway and barely escaped with their lives while their lifetime possessions were swiftly transformed into pyres of roaring raw energy, smoke, ash, twisted metal, melted glass, smoldering memories, living nightmares. When I arrived, I was immensely relieved to find a household intact, although still threatened by fires raging uncontrolled. Nearby, the hills of Oakland and Berkeley were brutally disrupted, forever changed by an act of arson or human carelessness.

This personal brush with catastrophe jolted me, sharpened my senses, provoked me to think about what *really* matters in life, and yielded a refined perspective concerning the nature and consequences of change. As I walked solemnly over the blackened, blistered remains of what had been a friend's home, it was clear that those with whom I share the present time matter most: family, friends, acquaintances, members of my species throughout the world, and ranging beyond humankind, to other creatures I know personally or have met and admired on their own terms, from great humpback whales who turned their eyes to meet mine to the three Stellar's jays that recently fledged from a nest over my backdoor. I wish them well, want them to prosper, will take measures to protect them, if I can . . .

Reaching back in time, I care also about those I have known who are now gone: my beloved father; dear aunts and uncles who were my friends as a child; my mother's mother and father, who died when I was very young but who live in my memory as members of the family whose being touched my life directly. I value the tangible evidence of their lives: my grandmother's small jewelry box with leaves etched on the lid, my grandfather's volumes of Mark Twain stories. But most of all, I value another inheritance that will be mine as long as I live, a legacy of memories that no fire can burn, no thief can take away, heirlooms of knowledge that I intend to pass along intact. My grandparents and parents have provided me with access to times that preceded mine; through them, I can almost hear the rustle of passenger pigeon wings, sense the wondrous aura of New Jersey marshes before they were awash with strange chemicals, anticipate the excitement of travel beyond the normal range of a horse and wagon, and vicariously experience the miraculous dawning of technological change that shapes the present. Their eyes viewed the seas of Eden. And they witnessed the start of an era of unprecedented global consumption of living resources, the transformation of an earthly paradise 4.6 billion years in the making skimmed for the short-term service of a single insatiable species.

It is difficult for me to relate with personal feeling to times earlier than the late 1800s, although vivid accounts in books and re-creations via films, museums, and my own imagination infuse a sense of understanding—and sometimes inspire a profound longing to intrude, to modify the course of history. *If only* I could reason with those who deliberately slaughtered the societies of great whales! *If only* the last Great Auks and Stellar's sea cows and Atlantic walruses had not been killed! Perhaps other choices would have been made, *if only* the dire consequences of draining wetlands, cutting primeval forests, polluting rivers, filling productive bays, had been known to those who forever closed options for me, for all who share the present, and for all who will come in the future. I care very much about that future, my time and all that follows, now compromised by actions taken long ago.

"But wait!" I counsel myself. "Such things are still happening, and the pace has picked up considerably since the eighteen hundreds—and is accelerating!" I can imagine my grandchildren and great-grandchildren of the future looking back and asking, "G-Mom*, why didn't you do something? You were there! You actually saw blue whales and horseshoe crabs, coral reefs and thousand-year-old trees! Why didn't you save them?"

I long for answers. How *can* individuals stop the actions that are degrading the quality of life, closing doors not only for future generations but also for those now alive? Whales are still being killed by wealthy nations as luxury food, despite worldwide outrage and awareness of their alternative value—economic,

* The name bestowed by my grandson.

aesthetic, scientific, moral. Thousands of once-and-nevermore species are being rendered extinct each year, perhaps as many as four a minute, through rainforest destruction and loss of other unique ecosystems. With brazen indifference to world opinion and international agreements, some nations still dump highly toxic materials into the sea and sky, a deadly legacy for the future, with immediate consequences for those around here and now. Despite clear evidence that ocean ecosystems are collapsing and fish populations cannot sustain commercial taking, huge nets, trawlers, and factory ships are still being deployed, and more are being built.

Clearly, it is not possible to go back and redirect history. But now—not for long—there is a chance, a brief window of opportunity to restore and protect the remaining healthy ecosystems that support us. Most important, most urgent, we must protect the principal substance of the biosphere: the sea. In *Earth in the Balance*, Al Gore describes the influence volcanoes have had on planetary climate and weather, then notes, "In the course of a single generation, we are in danger of changing the makeup of the global atmosphere far more dramatically than did any volcano in history, and the effects may persist for centuries to come." Even greater, more far-reaching anthropogenic changes are sweeping Earth's *aquatic* atmosphere, chemically, physically, and biologically. In the past few decades—my lifetime—the sea has changed; with each passing year, pressures on ocean resources and ocean ecosystems increase; the size of the ocean does not.

Part of human impact on the earth relates to our swiftly growing numbers. If we do not take deliberate, conscious action to maintain a reasonable balance between the numbers of people and the environmental wealth required to sustain us, nature will make appropriate adjustments, and famine, disease, and wars—the predictable outcomes of living beyond one's environmental means, of overspending environmental capital—will ultimately force a cruel discipline.

Excessive numbers alone are not the greatest problem, however; a small fraction of the present world population could cause a lot of mischief unless guided by positive policies, by an ethic of taking care of the planet that heretofore has taken care of us without much effort on our part. A few individuals, armed with modern technology, can wield enormous destructive power. One person with a bulldozer can, in a few days, eliminate a forest that has been quietly building for millennia; one fisherman deploying miles of drift nets can wipe out entire seagoing societies; one Saddam Hussein can—and did—use his power to transform the cradle of human civilization into an ecological and cultural graveyard.

But just as individuals can and do negatively impact the course of history, so can and do individuals make a positive difference. All my life I have watched my parents apply commonsense "take-care-of-the-planet" principles long before it became stylishly urgent to do so. Motivated to live lightly on the land, to value and protect what they have and leave the world better than when they arrived, they just knew that it made sense to plant native trees and wildflowers to restore woodlands that had been cut. They knew the importance of protecting the source of wellwater that we consumed, of minimizing and, when possible, making use of trash and compost generated. They respected and gave safe haven to wild creatures and willingly shared with them the land and water on their small farm. Such attitudes, as natural and important to my parents' way of life as breathing, provide the backbone of an environmental ethic that, if widely adopted, will bode well for the human future.

Some individuals have taken deliberate, courageous actions when confronted with ethical, sometimes unpopular choices concerning environmental issues. One of my personal heroes is William Perrin, a biologist with the National Marine Fisheries Service in San Diego, California. Perrin was disturbed when he became aware in the early seventies that hundreds of thousands of wild dolphins were being killed each year when U.S. tuna fishermen surrounded schools of yellowfin tuna with purse seines, and in the process of rounding up the fish captured and inadvertently destroyed dolphins who, for reasons unknown, associate with the fish. When Perrin expressed concern, he was reprimanded and warned that he should mind his own business. Unintimidated, he persisted in showing films to his colleagues of dolphins being captured, tangled, bruised, then, of dolphins panicked, stunned, dying. It took several years and the forceful involvement of many who were inspired to help, but policy changes were finally made that sharply reduced the numbers of dolphins killed. I met Perrin in 1972 when I organized a course concerning the biology of dolphins and whales at the University of California in Los Angeles; in that year an estimated 200,000 dolphins died in the nets of U.S. tuna fishermen.

Nearly twenty years after Perrin's first observations of the tuna-dolphin dilemma, another brave individual, Sam LaBudde, deliberately signed on to a tuna boat as a cook to document the continued take of as many as 20,000 dolphins a year. His resulting video of carnage educated many millions who had previously munched tuna sandwiches oblivious to the hidden costs in terms of dead dolphins and shattered dolphin societies. A newly provoked public insisted on further policy changes, and this time, business interests were compelled to guarantee "dolphin-free tuna" or face economic penalties in terms of significant lost sales.

Some businessmen and -women have been motivated to "do the right thing" environmentally, without being required to do so—even when higher costs result. Samuel C. Johnson, pilot, diver, wildlife photographer, conservationist, and president and CEO of the S. C. Johnson Company, confounded some of his colleagues when, despite increased expense, he banned the use of aerosols in the company's products, long before laws mandated such action. This and other environmentally friendly measures embraced by Johnson caused him to become in 1994 the sixteenth individual honored with the Charles A. Lindbergh Award for efforts to find the balance between environmental preservation and technological development. Lindbergh himself at one time said, "If I had to choose between airplanes and birds, I would choose birds." But according to his youngest daughter, Reeve, the real trick is not to have to choose, but to find and hold that elusive—but attainable—balance.

Other individuals given the Lindbergh Award over the years for having discovered their own pathways to a harmony of values include Jacques Cousteau, Thor Heyerdahl, Sir Edmund Hillary, Maurice Strong, Hugh Downs, Paul MacCready, Edwin Link, Russell Train, Arthur C. Clarke, Robert M. White, Murray Gell-Mann, and more, each one very different from the other, and each providing an eloquent response to the plaintive voices asking, "What can I—just one person—do?"

When Johnson was asked to tell what caused him to make choices that some thought were crazy, he said that he was concerned about the world his grandchildren would inherit, and told of taking his eleven-year-old grandson to visit a pond that he had known as a boy. Both were thrilled when the bright eyes and craggy form of an immense snapping turtle lifted out of the water, then disappeared beneath the surface. "Maybe it was the very one I saw in that pond when I was eleven years old," he said. If Johnson (his grandchildren know him as "Grand Sam") has his way, snapping turtles will prosper in that pond for eternity, there for generations of grandchildren to come—"for this sixty-six-year-old man—and for the eleven-year-old boy who lives inside."

Most individuals who change the course of history are not recognized publicly for their good or bad deeds. Most decisions that matter, in fact, seem too trivial to be noticed, but each and every choice made by all the world's individuals, taken together at the end of a day, week, year, or millennium shape the direction of society, cause the actions that collectively build or destroy civilizations.

Individuals decide:

"Shall I or shall I not provide a market for shark steak or swordfish by ordering them in a restaurant?"

"Shall I or shall I not drop a bottle or cup or bag of trash into the ocean?"

"Shall I or shall I not vote for this person—or that?"

"Shall I spend ten dollars on a movie—or send ten dollars to an organization that is trying to clean up the coastline?"

"Shall I bend over to remove a piece of junk from the beach?"

The current state of affairs, however you characterize it, is the distillation of millions of responses to questions such as these, every minute of every day. Not only do the actions of individuals matter; *only* what individuals do matters.

Sometimes I am asked what I believe to be the greatest threat to the oceans, and thus to human survival and well-being. Is it the huge amount of trash and toxic chemicals that are dumped into the sea? Pesticide-, herbicide-, and fertilizer-laden runoff from the fields and lawns of the land? Acid rain and other toxins falling from the sky? Dredging and filling of shoreline marshes and productive, shallow sea-grass meadows? Oil spills? Overfishing? The introduction of exotic species? These are all problems, and all are contributing to the changing sea we are now witnessing, and causing. But if I had to name the single most frightening and dangerous threat to the health of the oceans, the one that stands alone yet is at the base of all the others is *ignorance*: lack of understanding, a failure to relate our destiny to that of the sea, or to make the connection between the health of coral reefs and our own health, between the fate of the great whales and the future of humankind. There is much to learn before it is possible to intelligently create a harmonious, viable place for ourselves on the planet. The best place to begin is by recognizing the magnitude of our ignorance, and not destroy species and natural systems that we cannot re-create nor effectively restore once they are gone.

The biologist Lewis Thomas observed in 1979:

> *The only solid piece of scientific truth about which I feel totally confident is that we are profoundly ignorant about nature. . . . It is the sudden confrontation with the depth and scope of ignorance that represents the most significant contribution of twentieth-century science to the human intellect. We are, at last, facing up to it . . . we are getting glimpses of how huge the questions are, and how far from being answered. . . . But we are making a beginning, and there ought to be some satisfaction, even exhilaration, in that.*

Traditionally, the sea has been regarded as the common heritage for all mankind; now its care must be acknowledged as a common responsibility. To ensure a decent quality of life for the rest of our lives, as well as for all those who follow, we must develop global policies that recognize the interdependence of life and the need for nations to agree on mutually beneficial measures to protect and maintain the basic elements of life support, on a planetary scale. But it is difficult to be concerned about what we do not know, just as most of us have trouble planning for future generations that do not have specific faces attached to them, that never-never point beyond the time of one's own children, grandchildren, and, sometimes, great-grandchildren. Such planning requires a leap of faith but perhaps we can be aided by technology that helps us understand the consequences of choices made now.

Through computer imaging and sophisticated number crunching, it is possible to gain better insight than ever before possible concerning future realities. Scientists can generate predictions, convincing visualizations, and alternative simulations. The consequences to the planet of a slight warming trend are seen to be profound: The polar ice caps melt, sea level rises, seaside cities are inundated, climate and weather change . . . people die. The consequences of too many people vying for too few fish can be anticipated, as can the consequences of dumping toxic pollutants into places near and far. Perhaps with knowing will come caring, and with caring, an impetus toward the needed *sea change of attitude*, one that combines the wisdom of science and the sensitivity of art to create an enduring ethic.

As a child, I was intrigued with Aesop's fable of the goose that laid the golden eggs. Once upon a time, the story goes, a farmer went to the nest of his goose and, to his surprise, found . . . an egg of solid gold! Every day thereafter, the goose laid an egg of pure gold. The farmer was joyous, but in time, as he grew rich, he also became greedy. One day, thinking that if he killed the goose, he could have all her treasure at once, that inside her there must surely be many golden eggs, he cut her open only to end up with . . . a dead goose.

Like that farmer, we are blessed with a wondrous source of wealth, the oceans. Presently, however, humankind is insisting on taking more than a sustainable supply of golden eggs; feathers are being plucked, and some have begun to carve into meat and bone. If we care for the source and aren't too greedy, we might yet have a continuing supply of valuable resources, perhaps modest yields of fish, shrimp, clams, crabs, and seaweed, in addition to the more fundamental benefits such as oxygen to breathe, a benevolent climate, a hospitable environment, an infinite source of knowledge to be derived from the lives of the diverse sea creatures. With care, we should be able to get the goose to do all sorts of things: provide transportation, shelter, entertainment, and perhaps even consume certain kinds of garbage.

This is the time as never before and perhaps never again to establish policies—on a small personal scale as well as on a broad public scale—to protect and maintain planetary health. To be effective, actions must be taken before ecosystems are further traumatized or destroyed, before vested interests become too firmly established, before the arrival of that worrisome "point of no return."

We have an opportunity, now, to achieve for humankind a prosperous, enduring future. If we fail, through inability to resolve thorny issues, or by default born of indifference, greed, or lack of knowledge, our kind might well be a passing short-term phenomenon, a mere three or four million—year blip in the ancient and ongoing saga of life on Earth.

APPENDIX

The growing number of marine parks and sanctuaries, from 0 to more than 1,200 worldwide in less than twenty-five years, is an indication of an increasing awareness of the economic and ecological benefits of protecting marine ecosystems. The degree of protection varies widely from some small areas where no fishing or other taking of marine life is allowed to the nominal protection afforded in the great majority of the parks and sanctuaries where, ironically, most commercial, sport fishing, and other exploitation of marine resources continues with little restraint. In effect, policies typically could be compared to allowing sport hunting and commercial taking of songbirds, owls, eagles, snails, beetles, squirrels, bears, wolves, and their many relatives from Yellowstone National Park. The recent catastrophic decline of fisheries worldwide is causing many to rethink the definition and treatment of marine protected areas, but there remains a widespread belief that management of such areas can and should include the persistent removal of large numbers of fish and certain invertebrates. The reasons for such a stance are varied—and clearly political as they relate to special-interest groups.

In 1988 at the General Assembly of the IUCN in Costa Rica, the following role of "protected areas" in marine conservation was defined and endorsed: "to provide for the protection, restoration, wise use, understanding and enjoyment of the marine heritage of the world in perpetuity through the creation of a global representative system of marine protected areas and through the management, in accordance with the principles of the World Conservation Strategy, of human activities that use or affect the marine environment."

Through its Commission on National Parks and Protected Areas, IUCN has developed a program to promote the establishment and management of marine protected areas throughout the world and appointed Graeme Kelleher, Director of the Great Barrier Reef Marine Park Authority, as coordinator. A network of people worldwide was established in 1990 to implement the program, and a global inventory initiated of marine protected areas representing eighteen geographical regions involving more than forty countries.

I am grateful to the Center for Marine Conservation and to Graeme Kelleher and Chris Bleakley of the Great Barrier Reef Marine Park Authority for providing much of the material included in the following summary.

SUMMARY OF MARINE PROTECTED AREAS

REGION 1. ANTARCTICA
NUMBER OF PROTECTED AREAS: 17

The Antarctic marine environment is characterized by high diversity and high productivity induced by upwelling of nutrient-rich water. In 1994, after years of debate, the waters surrounding the Antarctic continent were established by the International Whaling Commission as an international sanctuary for whales, a move that suggested to some an unwelcome trade-off to those who would like whaling to resume elsewhere, but to others signaled a timely interest in protection for the ecosystems required by whales for their survival. The taking of krill, squid, and fish continues, but the entire oceanic region surrounding Antarctica is covered by a comprehensive environmental protective regime involving complex international agreements known collectively as the Antarctic Treaty System. These provide a basis for future management and protection of the marine and terrestrial environment for the entire region.

REGION 2. ARCTIC
NUMBER OF PROTECTED AREAS: 16

This region includes the Arctic seas of Scandinavia, Iceland, Greenland, northern Russia, and Canada, an area that encompasses some of the most pristine waters in the world, and offshore from Russia in the Kara and Barents seas, some of the most dangerously polluted. An example of one of the areas designated for protection is in the Kara Sea, where the Russian Federation, Great Arctic Zapovednic (Nature Reserve) provides for a program of research and monitoring to detect environmental changes.

REGION 3. MEDITERRANEAN MARINE REGION
NUMBER OF PROTECTED AREAS: 53

The swift degradation of the Mediterranean Sea and adjacent semienclosed waters has inspired numerous attempts to safeguard that rich and productive environment, but complex cultural and historical issues greatly complicate efforts to effectively manage the system as a whole. Encouraging progress is being made, especially through the IUCN's Mediterranean "Blue Plan." Among the areas designated for protection in the region is the Scandola Nature Reserve, which combines 1,000 hectares* of ocean and a similar area of the rugged, mountainous northwest coast of Corsica. Significant enough to be declared a World Heritage Site, valued for research and recreation, fishing nonetheless is permitted. Another example is the small (24.6 hectares) Miramare Marine Reserve and Biosphere Reserve, located on the northern part of the Adriatic Sea in the Gulf of Trieste, Italy. Although tiny, it has high significance for research and education. The much larger Tunisia, Zembra, and Zembretta National Park (4,700 hectares) in the Gulf of Tunis embraces two mountainous islands (Zembra and Zembretta), presently uninhabited and maintained as a military zone.

---

* A hectare is a metric measurement equal to 10,000 square meters or 32,500 square feet; or 6.1 square miles.

REGION 4. NORTH WEST ATLANTIC MARINE REGION
NUMBER OF PROTECTED AREAS: 89

From the middle of the Atlantic Ocean west to North America, north to the Arctic, and south to the mid-Atlantic region of North America at Chesapeake Bay, this region embraces some of the richest fishing grounds in the world. Few areas are strictly off limits to fishing, but numerous places have been designated for limited protection. In Canada, the Saguenay Marine Park covers 113,800 hectares, including a deep-water fjord, estuary, upwelling, and open water environments extending from the high-water mark to the middle of the St. Lawrence estuary in Québec. One of the newest areas designated for protection is in the U.S.: The Stellwagen Bank National Marine Sanctuary involves 218,000 hectares vital to numerous marine mammals and various commercially sought-after fish. In a division of NOAA that includes the National Marine Sanctuary Program, a portion of Chesapeake Bay has been designated as a National Estuarine Research Reserve where studies are conducted on saltwater, wetland, and open-water environments.

REGION 5. NORTH EAST ATLANTIC MARINE REGION
NUMBER OF PROTECTED AREAS: 41

From the middle of the North Atlantic east to Gibraltar and north to the Arctic, numerous small areas have been designated for protective measures. One of the best known is the Waddenzee Natural Monument in the Netherlands, an internationally important area that includes shallow coastal embayments with rich mud and sandflats vital as a nursery for fish and invertebrates and as a feeding, resting, and migration stopover for millions of waterfowl. In the United Kingdom, plans are under way to have zones for fishing and other activities at the Lundy Marine Nature Reserve, and in France, management of land and sea are combined in the Iroise Regional Nature Park and Biosphere Reserve, where nesting seabirds and colonies of gray seals are protected together with subtidal channels and sea-grass meadows.

REGION 6. BALTIC MARINE REGION
NUMBER OF PROTECTED AREAS: 43

Recent high levels of pollution have greatly endangered marine life, as well as human life, in this region. Among the places designated for protection are the Nemunas Delta Regional Park in Lithuania at the mouth of the Nemunas River, as it flows into the Kursiu Marios Lagoon, and the Gullmar Fjord in Sweden, a park that supports a variety of marine habitats and breeding areas for salmon and serves as a monitoring site for water quality in the Baltic Sea.

REGION 7. CARIBBEAN MARINE REGION
NUMBER OF PROTECTED AREAS: 92

From the middle of the North Atlantic west to the Chesapeake Bay and south to the northern border of Brazil, this region encompasses the Gulf of Mexico and numerous islands including more than 700 in the Bahamas alone. Coral reefs, lagoons, sea-grass beds, shallow bays, mangroves, and estuaries are included in one of the largest marine sanctuaries in the world, the 2,600-square-mile Florida Keys National Marine Sanctuary. Similar environments are contained in Reserva de la Biosfera Sian Ka'an in Mexico and the much smaller Parque Nacional Natural Tayrona, part of a Biosphere Reserve offshore from Colombia. Commercial and sport fishing are allowed in both, and both are of great economic importance to tourism. However, at Saba and Bonaire, Netherlands Antilles, even the fish are protected as creatures more valuable alive to the lucrative diving and tourism industry than as a commodity for fishing.

REGION 8. WEST AFRICA MARINE REGION
NUMBER OF PROTECTED AREAS: 42

The West Africa region extends from Gibraltar in the north, the middle of the North Atlantic to the west, and south around the tip of Africa to the northern border of the Republic of South Africa on the Indian Ocean. This enormous area ranges from cold open sea and coastal environments to subtropical reefs, mangroves, and sea-grass beds. One of the largest marine parks in the region is Banc d'Arguin National Park in Mauritiana, 1,173,000 hectares, half of which is marine, including coastal islands, mangroves, mudflats, and areas of upwelling. Another, noted as an area important for fish spawning, is the Senegal and Gambia Delta du Sime/Saloum National Park on the border between Senegal and the Gambia. Many marine areas have been designated for protection in South Africa, including some where full protection is given to commercially valuable species.

REGION 9. SOUTH ATLANTIC MARINE REGION
NUMBER OF PROTECTED AREAS: 19

The area offshore from South America from the mid-Atlantic to the northern border of Brazil south to Tierra del Fuego is encompassed in this region. One of the most celebrated protected marine areas in the world is in this region: Argentina's Parque Marino Golfo San José, a provincial park located in Chubut province, which provides protection for an important breeding area for southern right whales and feeding areas for migratory shorebirds. In Brazil, the Reserva Biologica Federal Atol das Rocas, 200 miles offshore from the Rio Grande do Norte state, embraces the only coral reef atoll (a lagoon surrounded by coral) in the South Atlantic.

REGION 10. CENTRAL INDIAN OCEAN MARINE REGION
NUMBER OF PROTECTED AREAS: 15

This region embraces the waters around India and Sri Lanka, south to the middle of the Indian Ocean, and includes numerous islands and important reef habitats. The Gulf of Mannar Marine National Park, on the southern coast of India, was established in response to concerns over deterioration of the marine environment caused in part by mining of coral. Heavy fishing pressure, mostly by foreign fleets, has increased stress on marine ecosystems in the area, but most protected areas continue to allow commercial and sport fishing operations.

REGION 11. ARABIAN SEAS MARINE REGION
NUMBER OF PROTECTED AREAS: 19

The Red Sea, Persian Gulf, Arabian Gulf, and immediately adjacent waters are included in this region. Numerous small areas have been designated for protection here, but all are stressed by growing pressures relating to increasing tourism, fishing, and even war. One special site, Egypt's Ras Mohammed Marine National Park, located near the entrance to the Gulf of Aqaba in the Red Sea, is known as one of the Seven Wonders of the Undersea World. Its spectacular underwater scenery and diverse marine life may be "loved to death" by increasing numbers of admiring visitors; the Persian Gulf, proposed for protection "in its entirety" for its rich cultural and biological heritage, has suffered as the scene of recent wars.

REGION 12. EAST AFRICA MARINE REGION
NUMBER OF PROTECTED AREAS: 54

This region encompasses the area along the eastern coast of Africa from Somalia south to the northern border of the Republic of South Africa and eastward to the middle of the Indian Ocean, including Madagascar, the Comoro Islands (home of the celebrated "living fossil fish," the coelacanth), the Maldives, Aldabra, the Seychelles, and numerous small islands. Greatly increased fishing pressure, military operations, tourism, and other activities have brought swift change to this region in recent decades, stimulating action to protect some of the most critical areas. One example is the Malindi Marine National Reserve in Kenya, a portion of the mainland coast noted for its fringing offshore coral reefs, sea-grass beds, mangroves, mudflats, and significant shorebird populations. Another protected area along the mainland coast is the Bazaruto Marine National Park, an archipelago in Mozambique prized for its high diversity and relatively undisturbed condition. Pelagic areas, coral reefs, rocky intertidal areas, sandy beaches, tidal sandflats, sea-grass meadows, and mangroves provide habitats for large numbers of marine mammals and sea turtles and breeding, nursery, and recruitment areas for numerous marine organisms. Offshore, one of the best-known protected areas is the Aldabra Strict Nature Reserve of the Seychelle Islands. This is an undisturbed coral atoll with mangroves and sea-grass meadows noted for its endemic birds, giant tortoises, nesting green turtles, and seabirds.

REGION 13. EAST ASIAN SEAS
NUMBER OF PROTECTED AREAS: 92

Well known for supporting extraordinarily high diversity of marine life, this region includes the ocean surrounding thousands of small islands and reefs north of Australia to the southern part of China, including Indonesia, Malaysia, Thailand, and the Philippine Islands. Marine life in this region has been greatly depleted in recent years by the commercial taking of many creatures, from fish and turtles to shells and sea snakes. Among the areas designated for some form of protection is the Ujung Kulon National Park in Indonesia (also a World Heritage Site) located along the western tip of Java. Coral reefs, mangroves, sandy beaches, and other rich habitats support nesting sea turtles, birds, and numerous fish and invertebrates and provide the basis for a thriving tourist industry based on scuba diving. Commercial and artisanal fishing activities continue, however. In 1987, a multiple-use marine park (fishing included) was established at Pulau Redang in Malaysia, one of the Terengganu Group of islands, located off the eastern coast of the Malaysia peninsula. The state and federal governments of the area recognize the ecological and economic (tourism) value of the rich coral reefs in the area, and are giving high priority to their protection.

In the Philippines, one of the most notable protected areas is the Tubbataha Reefs Marine National Park (a World Heritage Site) located in the Sulu Sea about 150 miles southeast of Palawan. The park includes relatively intact and extensive coral reefs, algal and sea-grass meadows, and significant turtle nesting sites. Tourism and recreational uses are high during the calm summer months, but natural storms, while sometimes destructive, afford protection from human pressures during much of the winter. Two national parks in Thailand, Mu Ko Surin and Mu Ko Similan, provide havens in the Andaman Sea for well-developed fringing coral reef and mangrove habitats about thirty miles offshore from the mainland.

REGION 14. SOUTH PACIFIC
NUMBER OF PROTECTED AREAS: 66

This region encompasses the broad open sea of the South Pacific Ocean, including the extraordinarily diverse marine fauna and flora associated with many small islands, atolls, and reefs. So far, only nearshore places have been designated for protection, but consideration is being given to possible areas in the open ocean. In Tonga, Fanga'uta and Fangakakau Lagoons Marine Reserve protects a shallow, almost enclosed estuarine embayment separated from the open ocean by a complex system of reefs and channels.

The sea around Papua New Guinea has until recently been exploited only for limited local uses, but widespread logging, agriculture, rapid land development, and foreign commercial fishing are swiftly degrading formerly pristine, productive areas. In response, some places, such as the Maza Wildlife Management Area off the coast of the Western Province in the Torres Strait, have been designated for protection with an emphasis on "sustained use" by the local populations. The value of tourism—diving and snorkeling—motivated the protection of the Palolo Deep Marine Reserve in Western Samoa, a small reef system on the north coast of Apia that includes a hole about 600 feet in diameter and 30 feet deep within a fold in the fringing reef. Sustainable subsistence fishing and tourism development are management goals for another protected area of reefs in this region along the coast of Kosrae Island in the Federated States of Micronesia. If successful, such multiple-use reserves are likely to increase in number and size.

REGION 15. NORTH EAST PACIFIC
NUMBER OF PROTECTED AREAS: 168

The area included in this region extends along the entire western coast of North America from the Aleutian Islands south along the shores of Canada, the U.S., and Mexico. Five of the U.S. National Marine Sanctuaries are in this region, including the second-largest marine protected area in the world, the Monterey Bay National Marine Sanctuary (about 5,000 square miles of the sea offshore from California). Another kind of marine protected area is represented by the Alaska Maritime National Wildlife Refuge, extending from Forrester Island in southeastern Alaska to Attu Island at the tip of the Aleutian Chain and almost to Barrow on the Arctic Ocean (nearly 2 million hectares of coastal land and the surrounding marine area). Designed to conserve fish and wildlife populations including marine mammals, birds, and the marine resources on which they rely, special allowances are made for native people of the area. Among the protected areas along the coast of Canada is the Pacific Rim National Park, about 15,540 hectares offshore from the coast of West Vancouver Island. Examples for Mexico are Reserva de la Biosfera El Vizcain on the coast of the Sonora near Isla Tiburón, where an area is managed as a refuge for commercially valuable species of fish and crustacea, and several lagoons along the Pacific coast of Baja California where gray whales gather to breed and bear their young.

REGION 16. NORTH WEST PACIFIC REGION
NUMBER OF PROTECTED AREAS: 172

The area from the middle of the North Pacific westward to Asia and south to the southernmost part of China is encompassed in this region. Examples of marine protected areas in China include the Sanya Coral Reef National Marine Nature Reserve, in the southern Hainan province off the tip of Hainan Island, where coral reefs are becoming increasingly popular as a tourist attraction, and the Nanji Archipelago National Marine Nature Reserve, offshore from Zhejiang Province, a zone of transition between temperate and subtropical waters. In Taiwan, the Kenting National Park, on the southernmost point of the island, provides protection for coral reefs and surrounding habitats. One of Japan's marine parks is in Okinawa, the Taketomijima Takidonguchi Marine Park, where a visitor center promotes recreational activities associated within the park's coral reefs and nearby habitats.

REGION 17. SOUTH EAST PACIFIC
NUMBER OF PROTECTED AREAS: 18

Encompassed within this region are the waters lying west of Central and South America. Two well-known protected areas are included here, Costa Rica's Parque Nacional Isla del Coco (Cocos Island) located about 350 miles southwest of mainland Costa Rica, and Ecuador's Reserva de Recursos Marinos Galápagos, the 7 million hectares of ocean surrounding the Galápagos Archipelago. While restrictions apply in both areas, much more protection is afforded to the fauna and flora of the land than to the surrounding sea. Commercial fishing has greatly impacted many marine species in recent years, notably sea turtles and sharks.

REGION 18. AUSTRALIA/NEW ZEALAND
NUMBER OF PROTECTED AREAS: 260

The area from the middle of the Indian Ocean eastward around Australia and New Zealand and across the southern Pacific Ocean is included in this region. The largest marine protected area in the world is here, the Great Barrier Reef Marine Park (a World Heritage Site), extending along the north coast of Queensland for nearly 2,000 miles and encompassing about 35 million hectares—the largest existing system of corals and associated plants and animals. Established in 1975 (one of the first marine protected areas), the park is managed to accommodate multiple uses (sport and commercial fishing, tourism, artisanal taking, research) and achieve a balance between such uses and conservation of the natural resources. Another of Australia's notable marine protected areas is Cobourg Marine Park, situated at the confluence of the Indian Ocean and Pacific Ocean off the north coast of the Northern Territory. Here, the land and adjacent sea are managed in concert to protect both.

In New Zealand, one famous marine protected area is Poor Knights Island near the eastern coast of the North Island. The profusion of fish, plants, and invertebrates draws divers and snorkelers underwater to glimpse a part of Earth's fast-disappearing 4.6-billion-year legacy.

U.S. NATIONAL MARINE SANCTUARIES
Since 1972, when enabling legislation set in motion the formation of a system of National Marine Sanctuaries, the following areas have been established:

Monitor, designated 1975
A one-and-one-third-square-mile protected area surrounds the site of the wreckage of the USS *Monitor*, located in about 220 feet of water sixteen nautical miles southeast of Cape Hatteras, North Carolina. This U.S. Civil War ironclad vessel was the prototype for a class of warships that significantly altered

nineteenth-century marine architecture and technology. Sunk in a storm in 1862, it was searched for and located with modern side-scan sonar during a scientific expedition in 1973. Depth limits the number and kind of dives that can be made to visit the site, and permission must be obtained to go, but film documentation has been made during nearly a dozen expeditions, and some of the artifacts, including the anchor and a lantern, have been recovered for exhibit and study.

Key Largo, designated 1975

One hundred square miles of coral reef three miles offshore from Key Largo, Florida, and extending eight miles seaward was established in an effort to protect a portion of the largest living coral reef in North America. It also includes numerous shipwrecks and Caryfort Lighthouse, one of the oldest functioning lighthouses in the U.S. Mile for mile, it is perhaps the most heavily used site in all of the NMS system, hosting more than a million visitors a year, including many divers.

Florida Keys Reef Tract, designated 1990

Twenty-six hundred square miles of ocean surrounding the Florida Keys from Key Largo to beyond the Dry Tortugas are included in this second-largest of the U.S. marine sanctuaries. Economically driven by tourism, the Florida Keys provide a clear example of the "sound environment/sound economy" rationale underlying much of the current conservation policymaking, worldwide. The value of protecting marine ecosystems has won the support of many commercial interests, including some fishermen once opposed. Others have joined a group of treasure salvors in resisting constraints on what they regard as traditional freedoms. Grounding of ships on protected reefs has been a continuing problem. A program is under way to designate part of the reef for special protection and others for various kinds of use in an effort to respond to multiple—and sustained—use objectives.

Flower Garden Banks, designated 1992

The waters surrounding two small rocky outcroppings, totaling fifty-eight square miles, protect the northernmost coral reefs in the United States, located approximately 120 miles offshore from the coasts of Texas and Louisiana. Set amid a field of producing oil rigs, and the subject of great controversy concerning conflicting uses, this sanctuary now is viewed as an admirable example of successful industry-private-government cooperation.

Monterey Bay, designated 1992

Designation of 5,327 square miles offshore from Monterey, California, as a sanctuary brought into being a marine protected area second in size, worldwide, only to the territory administered by Australia's Great Barrier Reef Park Authority. This sanctuary includes extensive nearshore undersea forests of giant kelp, regarded by many as the "sequoias of the sea." Like redwood forests, these enormous plants provide the cornerstone of complex, diverse resident communities and haven for many transients.

Just offshore, the subsurface terrain drops precipitously, creating submerged canyons that rival the Grand Canyon for size and grandeur, and are filled from sea surface to ocean floor with life representing greater genetic diversity than is present in any terrestrial environment. Public interest in marine life in the area has tended toward large mammals, birds, and fish, but through activities at the Monterey Bay Aquarium and several research institutions in the area, concern has been extended to include the role—and fate—of the entire ecosystem. Controversies continue concerning commercial and sport fishing, recreational activities such as jet skis, offshore dumping, shipping traffic, and oil and gas development that conflict with scientific, environmental, and aesthetic issues. Resolution of such problems is pivotal to the success of the multiple-use concept that underlies the management plan for this sanctuary, and for the U.S. marine sanctuary program as a whole.

Channel Islands, designated 1980

One thousand six hundred fifty-five square miles of nearshore, pelagic, and deep ocean surrounding several of southern California's most significant offshore islands—San Miguel, Santa Rosa, Santa Cruz, Ana Capa, and Santa Barbara—were designated to protect marine ecosystems that support major fisheries and recreational interests as well as important ecological, cultural, aesthetic, and educational values. Located squarely within a region well known for oil production and active shipping traffic, this sanctuary provided the first test cases for many hotly debated multiple-use issues. Despite modest funding, this sanctuary was also the first to initiate a strong and now widely acclaimed marine education program, "Los Marineros," aimed at fifth- and sixth-grade students in Santa Barbara schools.

Grays Reef, designated 1981
Due east from Sapelo Island, this twenty-two-and-a-half-square-mile protected area encompasses a limestone-based reef and is notorious as the most popular sport fishing area along the Georgia coast. Host to rich and diverse communities of plants and invertebrates, this site is prized for its scientific, ecological, and educational value as well as being a significant habitat for commercially valued species.

Looe Key, designated 1981
A small but a pristine part of the Florida reef system, the seven-square-mile region around Looe Key is especially esteemed as a research and conservation area. Special restrictions have been in effect for many years to protect the resident fish and other reef creatures. It is a very popular destination for snorkelers and divers in search of a glimpse of the way many of the reefs "used to be"—and may be again, with time and care.

Gulf of the Farallones, designated 1981
One thousand two hundred thirty-five square miles of nearshore and offshore waters ranging from wetlands and intertidal areas to pelagic and deep-sea communities are included in this sanctuary, located just north of San Francisco. It is large enough to embrace a fair cross section of marine life characteristic of the cold temperate eastern Pacific region that extends from British Columbia

to Point Conception, including conspicuously large and diverse populations of seabirds, marine mammals, and fish. Less obvious are the enormously abundant and diverse microorganisms, plants, and invertebrates in the water column and benthic regions that form the underlying structure of the ecosystem. Multiple uses abound, ranging from recreational and commercial fishing to sailing and surfing as well as bird-, whale-, and people-watching. It also contains some of the busiest shipping lanes in the world.

Fagatele Bay, designated 1986
One third of a square mile of reef bordering Fagatele Crater in American Samoa has been set aside for protection and study following a major outbreak of crown-of-thorns starfish that killed much of the coral in the area. Although primarily focused on research and ecological values, it is also a popular place for divers and snorkelers.

Cordell Bank, designated 1989
Five hundred twenty-five square miles of ocean about forty miles due west of Point Reyes, California, were designated to protect the unique assemblage of creatures that have developed in association with California's northernmost sea mount—in effect, an underwater island. Known to fishermen for many years as a place where certain kinds of fish were likely to be concentrated, it became championed by a small group of individuals who recognized the bank's unusual scientific and ecological values, and developed a compelling case for sanctuary status.

Stellwagen Bank, designated 1993
Located thirty miles off the coast of Massachusetts, between Cape Cod and Cape Ann, the 521-square-mile area around Stellwagen Bank recently designated as a marine sanctuary has been recognized for ages by more than thirty kinds of seabirds, seals, and cetaceans as *the* place to go in the Gulf of Maine for good fishing. Discovered by humans in 1854 when hydrographer Henry S. Stellwagen conducted soundings in the area, the biologically diverse area on and around the

bank generates approximately $100 million annually for fishermen. In recent years, whale-watching has provided comparable revenues—and an attractive nonconsumptive use of the bank's resources.

Hawaiian Islands Humpback Whale Sanctuary, designated 1993

An area near Maui, Hawaii, has been designated to protect the ecosystem vital to the survival of humpback whales, an endangered species and one of Hawaii's most popular natural residents. Humpbacks travel thousands of miles from their summer feeding grounds in Alaska to be in Hawaii in time for the birth of calves for that year. Other national and international policies and laws protect the whales themselves from harm; the sanctuary protects a critical part of the whale's environment that is vital to their life history.

Olympic Coast National Marine Sanctuary, designated 1994

This sanctuary encompasses an area off the Olympic Coast of Washington that is twice the size of Yosemite National Park, stretching from Cape Flattery to Copalis and running seaward thirty to forty miles. The pristine coastline, rugged offshore islands, lush kelp forests, and other habitats are home to millions of birds, fish, invertebrates, and at least twenty species of marine mammals. Multiple uses create great challenges for those charged with protection of the natural assets of this sanctuary, while promoting economic development. It is a recurrent theme, one that will increasingly be a cornerstone of conservation measures everywhere. Undersecretary of the U.S. Department of the Interior Timothy Wirth maintains critical perspective, however, with his observation that "the economy is one of the subsidiaries of the environment." Only by protecting environmental assets can there be economic dividends.

SELECTED REFERENCES

Adams, Douglas, and Mark Carwardine. 1990. *Last Chance to See*. London: William Heinemann, 208 pp.

Anderson, L. G. 1986. *The Economics of Fishery Management*. Baltimore: Johns Hopkins U. Press.

Anonymous. 1994. "Fish. The Tragedy of the Oceans." *The Economist* 330 (7855): 13–14; 21–24.

Anonymous. 1994. *An Overview of the Sanctuaries and Reserves Division*. Washington, D.C.: National Oceanic and Atmospheric Administration, 104 pp.

Anthony, V. C. 1990. "The New England Ground Fishery after 10 Years of the Magnuson Fishery Conservation and Management Act." *The North American Journal of Fishery Management* 10 (1990): 175–184.

Backus, Richard H., and Donald W. Bourne, eds. 1987. *Georges Bank*. Cambridge, Massachusetts: MIT Press, 593 pp.

Bardach, John. 1968. *Harvest of the Sea*. New York: Harper & Row, 301 pp.

Bardach, John, John H. Ryther, and William O. McLarney. 1972. *Aquaculture. The Farming and Husbandry of Freshwater and Marine Organisms*. New York: Wiley-Interscience, 868 pp.

Bartholomew, C. A. 1990. *Mud, Muscle and Miracles. Marine Salvage in the U.S. Navy*. Washington, D.C.: Dept. of the Navy, 505 pp.

Bascom, Willard. 1928. *The Crest of the Wave. Adventures in Oceanography*. New York: Harper & Row, 318 pp.

Bean, M. J. 1984. "United States and International Authorities Applicable to Entanglement of Marine Mammals and Other Organisms in Lost or Discarded Fishing Gear and Other Debris." Final Report for the Marine Mammal Commission, Contract MM26299943-7. National Technical Information Services, NTIS P885-160471. Springfield, Virginia, 65 pp.

Beddington, John, and R. M. May. 1977. "Harvesting Natural Populations in a Randomly Fluctuating Environment." *Science* 197 (4302): 463–465.

Beebe, William. 1934. *Half Mile Down*. New York: Harcourt, Brace and Co., 344 pp.

Bennett, Peter, and D. H. Elliott, eds. 1982. *The Physiology and Medicine of Diving*. San Pedro, California: Best Publishing, 570 pp.

Berger, John J. 1990. *Environmental Restoration. Science and Strategies for Restoring the Earth*. Washington, D.C.: Island Press, 398 pp.

Beverton, R. J. K., and S. J. Holt. 1957. "On the Dynamics of Exploited Fish Populations." Fishery Investigations of the Ministry of Agriculture and Fisheries. Food (G. B.) Ser. II, vol. 19, 533 pp.

Bohnsack, James A. 1993. "Marine Reserves: They Enhance Fisheries, Reduce Conflicts, and Protect Resources." *Oceanus* 36 (3): 63–71.

Boraiko, Allen. 1980. "The Pesticide Dilemma." *National Geographic* 157 (2): 144–183.

Brown, Lester. 1985. "Monitoring World Fisheries." In Brown et al. *State of the World*. Washington, D.C.: Worldwatch, pp. 73–96.

Bulloch, David K. 1986. "Marine Gamefish of the Middle Atlantic." Highlands, New Jersey: The American Littoral Society Special Publication No. 13, 83 pp.

Butler, Michael J. A. 1982. "Plight of the Bluefin Tuna." *National Geographic* 162 (2): 220–239.

Campbell, Todd. 1991. "World Fisheries Management Earns Poor Marks." *National Fisherman Yearbook*, vol. 71 (13): 42–45.

Canby, Thomas Y. 1991. "After the Storm." *National Geographic* 180 (2): 2–35.

_____. 1994. *Our Changing Earth*. Washington, D.C.: National Geographic Society, 200 pp.

Carr, Archie. 1984. *The Sea Turtle: So Excellent a Fishe*. Revised edition. New York: Charles Scribner's Sons, 280 pp.

Carson, Rachel. 1951. *The Sea Around Us*. New York: Oxford U. Press, 230 pp.

_____. 1962. *Silent Spring*. Boston: Houghton Mifflin, 368 pp.

Center for Environmental Education. 1988. "1987 Texas Coastal Cleanup Report." Washington, D.C., 105 pp.

Center for Marine Conservation. 1988. "A Citizen's Guide to Plastics in the Ocean: More Than a Litter Problem." Washington, D.C., 143 pp.

_____. 1989. "California Marine Debris Action Plan." Washington, D.C., 89 pp.

_____. 1992. "1992 National Coastal Cleanup Results." Washington, D.C., 336 pp.

_____. 1992. "1992 International Coastal Cleanup Results." Washington, D.C., 217 pp.

Chin, Edward. 1970. "*Anton Bruun* Reports. Scientific Results of the Southeast Pacific Expedition." Galveston, Texas: Texas A&M University Press.

Clark, Eugenie. 1953. *Lady with a Spear*. New York: Harper & Brothers, 243 pp.

Clayton, David, and Keith Wells. 1987. *Discovering Kuwait's Wildlife*. Kuwait: Published by Fahad Al-Marzouk, 253 pp.

Cleveland, Harlan. 1990. *The Global Commons. Policy for the Planet*. Lanham, Maryland: The Aspen Institute and University Press of America, 118 pp.

Collette, Bruce B., and Cornelia E. Nauen. 1983. "Scombroids of the World." *FAO Species Catalogue*, vol. 2. FAO Fisheries Synopsis, no. 125, vol. 2, 137 pp.

Collette, Bruce, and Sylvia A. Earle, eds. 1971. "Results of the Tektite Project: Ecology of Coral Reef Fishes." *Science Bulletin*, Natural History Museum of Los Angeles County, 14:16–47.

Conniff, Richard. 1993. "From Jaws to Laws—Now the Big Bad Shark Needs Protection from Us." *Smithsonian* 24 (2): 32–43.

Conservation Foundation. 1987. "State of the Environment: A View Toward the Nineties." Washington, D.C., 614 pp.

Cottingham, David. 1988. "Persistent Marine Debris: Challenge and Response: The Federal Perspective." Washington, D.C.: National Oceanic and Atmospheric Administration Report, 41 pp.

Council on Environmental Quality. 1990. "Environmental Quality. Twentieth Annual Report." Washington, D.C.: The Executive Office of the President, 494 pp.

Cousteau, Jacques. 1953. *The Silent World*. New York: Harper & Brothers, 266 pp.

Cunningham, J. T. 1896. *The Natural History of the Marketable Marine Fishes of the British Islands*. London: Macmillan, 375 pp.

Cuyvers, Luc. 1993. *Sea Power: A Global Journey*. Annapolis: Naval Institute Press, 254 pp.

Daily, Gretchen C., and Paul R. Ehrlich. 1992. "Population, Sustainability, and Earth's Carrying Capacity." *Bioscience* 42 (10): 761–771.

Dicks, Brian, ed. 1989. *Ecological Impacts of the Oil Industry*. New York: John Wiley & Sons, 316 pp.

Dr. Seuss (Theodor S. Geisel), and Audrey S. Geisel. 1971. *The Lorax*. New York: Random House.

Earle, Sylvia A. 1969. "Phaeophyta of the Eastern Gulf of Mexico." *Phycologia* 7 (2), 182 pp.

————. 1971. "Science's Window on the Sea (part two): All-Girl Team Tests the Habitat." *National Geographic* 140 (2): 290–296.

————. 1971. "The Influence of Herbivores on the Marine Plants of Great Lameshur Bay, with an Annotated List of Species." In Collete, B., and Sylvia A. Earle, eds., "Results of the Tektite Project: Ecology of Coral Reef Fishes." *Science Bulletin*, Natural History Museum of Los Angeles County, 14: 16–47.

————. 1976. "Life Springs from Death in Truk Lagoon." *National Geographic* 149 (5): 578–613.

————. 1979. "Humpbacks: The Gentle Giants." *National Geographic* 155 (1): 2–17.

————. 1983. "Will Robots Replace Man in the Sea?" *Sea Technology*, May, 69 pp.

————. 1989. "Troubled Waters." *Ocean Realm*, Summer 1989: 30–37.

————. 1990. "Ocean Everest—An Idea Whose Time Has Come." *Marine Technology Society Journal* 24 (2): 9–12.

————. 1989. "Sharks, Squids and Horseshoe Crabs: The Significance of Marine Biodiversity." *Bioscience* 41 (7): 506–509.

————. 1992. "Persian Gulf Pollution. Assessing the Damage One Year Later." *National Geographic* 181 (2): 122–134.

Earle, Sylvia A., and Al Giddings. 1980. *Exploring the Deep Frontier.The Adventure of Man in the Sea*. National Geographic Society, 296 pp.

Eckert, Scott. 1992. "Bound for Deep Water." *Natural History* March: 28–35.

Edwards, Steven F. 1987. *An Introduction to Coastal Zone Economics*. New York: Taylor & Francis, 135 pp.

Egan, Timothy. 1994. "U.S. Fishing Fleet Trawling Coastal Water Without Fish." *The New York Times*, March 7, pp. 1, A10.

Eicher, Don L. *Geologic Time*. 2nd edition. Englewood Cliffs, New Jersey: The Prentice-Hall Foundation of Earth Sciences Series, 150 pp.

Eiseman, Nathaniel, and Sylvia A. Earle. 1983. "Johnson-Sea-Linkia Profunda, a New Genus and Species of Deep Water Chlorophyta from the Bahama Islands." *Phycologia* 22 (1): 1–6.

Ellsberg, Edward. 1929. *On the Bottom*. New York: Dodd, Mead, 324 pp.

Eno, Amos. 1990. *Needs Assessment of the National Marine Fisheries Service*. Washington, D.C.: National Fish and Wildlife Foundation, 315 pp.

Ernst, W. G., and J. G. Morin, eds. 1982. *The Environment of the Deep Sea. Rubey Volume II*. Englewood Cliffs, New Jersey: Prentice-Hall, Inc., 371 pp.

Food and Agriculture Organization of the United Nations (FAO). 1991. "Environment and Sustainability in Fisheries." Rome: FAO.

———. 1993. "Marine Fisheries and the Law of the Sea: A Decade of Change." FAO Fisheries Circular no. 853. Rome: FAO.

Fowle, Suzanne. 1993. *Fish for the Future. A Citizens' Guide to Federal Marine Fisheries Management*. Washington, D.C.: Center for Marine Conservation, 142 pp.

Gore, Al. 1990. *Earth in the Balance: Ecology and the Human Spirit*. New York: Houghton Mifflin, 408 pp.

Gore, Rick. 1990. "Between Monterey Tides." *National Geographic* 177 (2): 2–43.

Gould, Stephen Jay. 1987. *Time's Arrow, Time's Cycle*. Cambridge, Massachusetts: Harvard U. Press, 222 pp.

Grassle, J. Frederick. 1991. "Deep-Sea Benthic Biodiversity." *Bioscience* 41 (7): 464–469.

Gray, Jane, and William Shear. 1992. "Early Life on Land." *American Scientist* 80 (5): 444–456.

Griffin, Nancy. 1993. "Northeast Shrimpers Finish Up a Poor Season." *National Fisherman* 74 (3): 16.

Griggs, Tamar. 1975. *There's a Sound in the Sea*. San Francisco: Scrimshaw Press, 96 pp.

Grun, Bernard. 1963. *The Timetables of History*. New York: Simon & Schuster, 664 pp.

Hanson, Lynne Carter, and Sylvia A. Earle. "Submersibles for Science." *Oceanus* 30 (3): 31–38.

Hardin, Garrett. 1968. "The Tragedy of the Commons." Science 162 (3859) 1243–1248.

Hass, Hans. 1951. *Diving to Adventure*. Garden City, New York: Doubleday & Co., Inc., 280 pp.

_____. 1973. *Challenging the Deep. Thirty Years of Undersea Adventure.* New York: William Morrow and Co., 266 pp.

Hass, Lotte. 1972. *Girl on the Ocean Floor.* London: Harrap, 166 pp.

Hastings, J. W. 1971. "Light to Hide By: Ventral Luminescence to Camouflage the Silhouette." *Science* (173): 1016–1017.

_____. 1983. "Biological Diversity, Chemical Mechanisms and the Evolutionary Origins of Bioluminescent Systems." *Journal of Molecular Evolution* (19): 309–321.

Hawkes, G., and P. Ballou. 1990. "The Ocean Everest Concept: A Versatile Manned Submersible for Full Ocean Depth." *Marine Technology Society Journal* 24 (2): 79–86.

Herring, Peter J., A. K. Campbell, M. Whitfield, and L. Maddock. 1989. *Light and Life in the Sea.* Cambridge, UK: Cambridge U. Press.

Hessler, R., and H. Sanders. 1967. "Faunal Diversity in the Deep Sea." *Deep Sea Research* 14: 65–78.

Heyerdahl, Thor. 1977. *Kon-Tiki,* special edition for the Oceanic Society. New York: Pocket Books, 238 pp.

_____. 1971. *The Ra Expeditions.* London: Chronica Botanica India, 334 pp.

Hodgson, Bryan. 1990. "Alaska's Big Spill." *National Geographic* 177 (1): 5–43.

Holliday, Mark C., and Barbara K. O'Bannon. 1991. "Fisheries of the United States, 1990." U.S. Department of Commerce, National Oceanic and Atmospheric Administration, Current Fishery Statistics No. 9000, May.

Hollister, Buell, and H. Arnold Carr. 1992. "1991–1992 Annual Review of Developments in Marine Living Resources, Engineering and Technology." Massachusetts Division of Marine Fisheries and the MTS Living Resources Committee in cooperation with the National Marine Fisheries Service Northeast Regional Office. August, 73 pp.

Holt, Sidney J., and Lee M. Talbot. 1978. "New Principles for the Conservation of Wild Living Resources." Wildlife Monographs No. 59. The Wildlife Society, 33 pp.

Horton, Tom, and William M. Eichbaum. 1991. *Saving Chesapeake Bay.* Washington, D.C.: Island Press, 325 pp.

Intergovernmental Oceanographic Commission. 1992. "Expedition Report: Mt. Mitchell Expedition to the ROPME Sea Area, February–June, 1992." 58 pp.

International Union for Conservation of Nature. 1991–1992. *Protected Areas of the World: A Review of National Systems,* vol. 1–4. Cambridge, UK: IUCN Publication Services Unit.

_____. 1988. *Coral Reefs of the World,* vols. 1–3. Gland, Switzerland: IUCN, and Nairobi, Kenya: UNEP.

IUCN-The World Conservation Union et al. 1991. *Caring for the Earth: A Strategy for Sustainable Living.* Gland, Switzerland, 228 pp.

Jannasch, Holger. 1990. "Marine Microbiology: A Need for Deep-Sea Diving?" *Marine Technology Society Journal* 24 (2): 38–41.

Joseph, James, and Joseph W. Greenough. 1979. *International Management of Tuna, Porpoise, and Billfish—Biological, Legal, and Political Aspects*. Seattle: Washington Press, 253 pp.

Joseph, James, W. Klawe, and P. Murphy. 1988. "Tuna and Billfish—Fish Without a Country." La Jolla, California: Interamerican Tropical Tuna Commission, 65 pp.

Kahari, Victoria A. 1990. *Water Baby: The Story of Alvin*. New York: Oxford U. Press, 356 pp.

Kaufman, Les, and Kenneth Mallory. 1987. *The Last Extinction*. Cambridge, Massachusetts: MIT Press, 208 pp.

Kooyman, G. L. 1989. *Diverse Divers. Physiology and Behavior*. New York: Springer-Verlag, 200 pp.

Larkin, P. A. 1977. "An Epitaph for the Concept of Maximum Sustainable Yield." Transactions of the American Fisheries Society 106 (1): 1–11.

Lean, Geoffry. 1990. *World Wildlife Fund Atlas of the Environment*. New York: Prentice Hall.

Levenson, Thomas. 1989. *Ice Time. Climate, Science and Life on Earth*. New York: Harper & Row, 242 pp.

Link, Marion. 1973. *Windows in the Sea*. Washington, D.C.: Smithsonian Institution Press, 198 pp.

Little, Peter. 1993. "Nature. Small and Perfectly Formed." *Nature* 366 (6452):204–205.

Lovelock, James E. 1979. *Gaia. A New Look at Life on Earth*. New York: Oxford U. Press.

McGowan, John A. 1991. "The Role of Oceans in Global Change and the Ecosystem Effects of Change." In *Proceedings of the National Forum on Ocean Conservation*. Washington, D.C.: Smithsonian Institution.

MacInnis, Joseph. 1993. "The World Columbus Never Saw." *Ocean Realm* April: 48–49.

McIntosh, William C., and Arthur Thomas Masterman. 1897. *The Life Histories of the British Marine Food Fishes*. London: Cambridge University Press, 516 pp.

McKinnon, Michael, and Peter Vine. 1991. *Tides of War*. London: Boxtree Ltd., 192 pp.

McMahon, B. 1994. "Homo Aquaticus and the Eight-Liter Lung." *Outside*, March, pp. 75–80, 134–35.

Margulis, Lynn. 1978. *Evolution of Cells*. Cambridge: Harvard U. Press.

Massachusetts Offshore Groundfish Task Force. 1990. "New England Groundfish in Crisis—Again." Boston: Massachusetts Offshore GroundFish Task Force.

Maurizi, Susan, and Florence Poillon, eds. 1992. *Restoration of Aquatic Ecosystems. Science, Technology, and Public Policy*. Washington, D.C.: National Academy Press, 552 pp.

Maury, Matthew Fontaine. 1857. *The Physical Geography of the Sea*. London: Samson Low & Co., 360 pp. + 13 pls.

May, Robert M., John R. Beddington, Colin W. Clark, Sidney J. Holt, and Richard M. Laws. 1979. "Management of Multispecies Fisheries." *Science* 205 (4403): 267–277.

Moseley, H. N. 1892. *Notes by a Naturalist on HMS* Challenger. London, 540 pp.

Murphy, G. I. 1966. "Population Biology of the Pacific Sardine (Sardinops Caerulea)." *Proceedings of the California Academy of Sciences* 34 (1): 1–84.

Nakamura, Izumi. 1985. "Billfishes of the World." *FAO Species Catalogue*, vol. 5. *FAO Fisheries Synopsis*, no. 125, vol. 5, 65 pp.

Nash, Ogden. 1937. "A Beginner's Guide to the Oceans" in *Good Intentions*. Boston: Little, Brown and Company, p. 53.

National Academy. 1992. *Conserving Biodiversity: A Research Agenda for Development Agencies*. Washington, D.C.: National Academy Press, 127 pp.

National Marine Fisheries Service. 1991. "Our Living Oceans: The First Annual Report on the Status of U.S. Living Marine Resources." Washington, D.C.: U.S. Department of Commerce, National Oceanic and Atmospheric Administration, 123 pp.

National Research Council. 1992. *Marine Aquaculture. Opportunities for Growth*. Washington, D.C.: National Academy Press, 290 pp.

_____. 1992. *Oceanography in the Next Decade. Building New Perspectives*. Washington, D.C.: National Academy Press, 202 pp.

Nicol, Stephen, and William de la Mare. 1993. "Ecosystem Management and the Antarctic Krill" *American Scientist* 81 (1): 36–47.

Norse, Elliott, ed. 1993. *Global Marine Biological Diversity: A Strategy for Building Conservation into Decision Making*. Washington, D.C.: Island Press, 383 pp.

O'Conner, Thomas P., and Charles N. Ehler. 1991. "Results from the NOAA National Status and Trends Program on Distribution and Effects of Chemical Contamination in the Coastal and Estuarine United States." *Environmental Monitoring and Assessment* 17: 33–49.

O'Hara, K. J., N. Atkins, and S. Iudicello. 1986. *Marine Wildlife Entanglement in North America*. Washington, D.C.: Center for Marine Conservation, 219 pp.

_____. 1987. *Plastics in the Ocean: More Than a Litter Problem*. Washington, D.C.: Center for Marine Conservation, 128 pp.

Ono, R. Dana, James D. Williams, and Anne Wagner. 1983. *Vanishing Fishes of North America*. Washington, D.C.: Stone Wall Press, 257 pp.

Piccard, J., and R. Dietz. 1961. *Seven Miles Down*. New York: G. P. Putnam's Sons, 249 pp.

Pitt, David. E. 1993. "Despite Gaps, Data Leave Little Doubt That Fish Are in Peril." *New York Times*, August 3, p. B-7.

Pope, Gregory T. 1991. "Deep Flight." *Popular Mechanics*, April: 70–72 + cover.

Potter, Frank, ed. 1993. "National Marine Sanctuaries: Challenge and Opportunity." A report to the National Oceanic and Atmospheric Administration. Washington, D.C.: The Center for Marine Conservation, 25 pp.

Ray, G. Carleton. 1976. "Critical Marine Habitats." In *Proceedings of an International Conference on Marine Parks and Reserves, Tokyo, Japan, 12–14 May, 1974.* IUCN Publications New Series, No. 37.

_____. 1991. "Coastal-zone Diversity Patterns." *Bioscience* 41 (7):490–498.

Revelle, Roger. 1969. "The Harvest of the Sea and the World Food Problems." *Oceanus* 14 (4): 1.

Robinson, John, Francesca Cava, and Sylvia A. Earle. 1993. "Should the Arabian (Persian) Gulf Become a Marine Sanctuary?" *Oceanus* Fall: 53–62.

Robison, B. 1990. "Biological Research Needs for Submersible Access to the Greatest Ocean Depths." *Marine Technology* 24 (2): 34–37.

Rose, George A. 1993. "Cod Spawning on a Migration Highway in the North-west Atlantic." *Nature* 366 (6454): 458–461.

Ross, David A. 1982. *Introduction to Oceanography*, 3rd edition. Englewood Cliffs, New Jersey: Prentice Hall, 544 pp.

Ross, R. M., and L. B. Quetin. 1986. "How Productive Are Antarctic Krill?" *Bioscience* 36 (4): 264–269.

Rudloe, Jack, and Anne Rudloe. 1994. "Sea Turtles in a Race for Survival." *National Geographic* 185 (2): 94–121.

Ryther, J. H., and G. C. Matthiessen. 1969. "Aquaculture, Its Status and Potential." *Oceanus* 14 (4): 2–14.

Salm, Rodney V., and John R. Clarke. 1989. *Marine and Coastal Protected Areas: A Guide for Planners and Managers*, 2nd. edition. Gland, Switzerland: International Union for Conservation of Nature and Natural Resources, 302 pp.

Sanders, H. L. 1968. "Marine Benthic Diversity: A Comparative Study." *American Naturalist* 102: 243–282.

_____. 1977. *Evolutionary Ecology and the Deep-sea Benthos.* Academy of Natural Sciences special publication 12: 223–243.

Sanders, H. L., and R. R. Hessler. 1969. "Diversity and Composition of Abyssal Benthos." *Science* 166: 1074.

Sayles, Richard E. 1951. "The Trash Fishery of Southern New England in 1950." *Commercial Fisheries Review* XIII (7): 1–4.

Scammon, Charles. 1874. *The Marine Mammals of the North-western Coast of North America and the American Whale Fishery*, 1969 facsimile edition. Riverside, California: Manessier Publishing, 319 pp.

Service, Robert. 1993. "Under the Sea: A New Order." *Newsweek* September 20: 5.

Sharp, Gary D., and Andrew E. Dizon, eds. 1978. *The Physiological Ecology of Tunas.* New York: Academic Press, 485 pp.

Sheppard, Charles, Andrew Price, and Callum Roberts. 1992. *Marine Ecology of the Arabian Region. Patterns and Processes in Extreme Tropical Environments.* London: Academic Press, 359 pp.

Sherman, Kenneth. 1992. "Where Have All the Fish Gone?" *Nor'easter: Magazine of the Northwest Sea Grant Programs* 4 (2): 14–19.

Shomura, R. S., and H. O. Yoshida, eds. 1984. *Proceedings of the Workshop on the Fate and Impact of Marine Debris. 27–29 November, 1984*. Honolulu: Dept, of Commerce, NOAA, technical memo. NMFS NOAA-TM-NMFS-SWFC-54. Washington, D.C., 580 pp.

Society of the Plastics Industry. 1987. *Proceedings of a Symposium on Degradable Plastics, 10 June*. Washington, D.C., 55 pp.

Spry, W. J. J. 1877. *The Cruise of Her Majesty's Ship* Challenger. Toronto: Belford Brothers Publishers, 388 pp.

Stern, Paul C., Oran R. Young, and Daniel Druckman, eds. 1992. *Global Environmental Change*. Washington, D.C.: National Academy Press, 308 pp.

Stone, Gregory S. 1992. "Japanese Ocean Research and Development." *Marine Technology Society Journal* 26 (3): 11–19 + back cover.

Sullivan, Walter. 1993. *Continents in Motion*, 2nd edition. New York: American Institute of Physics, 430 pp.

Talkington, Howard R. 1981. *Undersea Work Systems*. New York: Marcel Dekker, 165 pp.

Taylor, Harden F., et al. 1951. *Survey of Marine Fisheries of North Carolina*. Chapel Hill: University of North Carolina, 555 pp.

Thomson, Wyville. 1874. *The Depths of the Sea*. London: Macmillan and Co., 527 pp.

Thorne-Miller, Boyce. 1993. *Ocean*. San Francisco: Collins Publishers, 240 pp.

Thorne-Miller, Boyce, and John Catena. 1991. *The Living Ocean*. Washington, D.C.: Island Press, 181 pp.

Travis, John. 1993. "Deep-sea Debate Pits Alvin Against Jason." *Science* 259: 1534–1536.

United Nations Environment Programme. 1990. Group of Experts on the Scientific Aspects of the Marine Environment (GESAME). "The State of the Marine Environment. UNEP Regional Seas Reports and Studies No. 115." Nairobi.

U.S. Gulf Task Force. 1992. "Environmental Crisis in the Gulf. The U.S. Response." Washington, D.C., 22 pp.

Walford, Lionel A. 1958. *Living Resources of the Sea: Opportunities for Research and Expansion. A Conservation Foundation Study*. New York: Ronald Press, 321 pp.

Walsh, Don. 1990. "Thirty Thousand Feet and Thirty Years Later: Some Thoughts on the Deepest Ocean Presence Concept." *Marine Technology Society Journal* 24 (2): 7–8.

Ward, Peter Douglas. 1992. *On Methuselah's Trail*. New York: W. H. Freeman & Company, 212 pp.

Weber, Peter. 1993. "Abandoned Seas. Reversing the Decline of the Oceans." Worldwatch Paper 116, 66 pp.

———. 1994. "Net Loss: Fish, Jobs and the Marine Environment." Worldwatch Paper 120, 76 pp.

White, Donald J. 1954. *The New England Fishing Industry*. Cambridge, Massachusetts: Harvard U. Press, 205 pp.

Whitehead, Hal. 1990. *Voyage to the Whales*. Post Mills, Vermont: Chelsea Green Publishing, 195 pp.

Wieland, Robert. 1992. "Why People Catch Too Many Fish: A Discussion of Fishing and Economic Incentives." Washington, D.C.: Center for Marine Conservation, 56 pp.

Wille, Chris. 1993. "The Shrimp Trade Boils Over." *International Wildlife* 23 (6): 18–23.

Wilson, Edward O. 1992. *The Diversity of Life*. Cambridge, Massachusetts: Harvard U. Press, 424 pp.

Wise, John P. 1991. "Federal Conservation and Management of Marine Fisheries in the United States." Washington, D.C.: Center for Marine Conservation, 378 pp.

Wolf, Edward C. 1985. "Conserving Biological Diversity." In *State of the World*. Washington, D.C.: Worldwatch, pp. 124–146.

Wolfe, D. A., ed. 1987. "Plastics in the Sea." *Marine Pollution Bulletin* 18 (6B).

World Resources Institute. 1992. *World Resources 1992–1993: A Guide to the Global Environment*. New York: Oxford U. Press, 385 pp.

———. 1994. *World Resources 1994–1995. People and the Environment*. New York: Oxford U. Press, 403 pp.

Yergin, Daniel. 1991. *The Prize*. New York: Simon & Schuster, 885 + 32 pp.

Young, Nina, ed. 1992. "Understanding the Revised Management Procedure." Washington, D.C.: Center for Marine Conservation, 72 pp. + appendices and index.

INDEX

ABOUT THE AUTHOR

Called "Her Deepness" by *The New Yorker* and *New York Times*, "Living Legend" by the Library of Congress, and first "Hero for the Planet" by *Time* magazine, Sylvia Earle is an oceanographer, explorer, author, and lecturer with experience as a field research scientist, government official, and director for various corporate and nonprofit organizations. She is founder of Mission Blue, honorary chair of the Harte Research Council for Gulf of Mexico Studies, founder of Deep Ocean Exploration and Research, Inc., a founding Ocean Elder, a founding patron for the International Union for the Conservation of Nature, and a lifetime trustee of the Aspen Institute and the Woods Hole Oceanographic Institution, and currently a National Geographic Explorer in Residence.

Formerly chief scientist of the National Oceanic and Atmospheric Administration and member of various commissions, she has also served as curator of phycology at the California Academy of Sciences; research associate, Los Angeles County Museum of Natural History; research associate, University of California, Berkeley; research fellow, Harvard University; Radcliff Scholar; and resident director of the Cape Haze Marine Laboratory (now the Mote Marine Laboratory).

She has a BS degree from Florida State University, has an MS and a PhD from Duke University, has 33 honorary degrees, and has authored more than 240 scientific, technical, and popular publications—including nine books about the ocean and five books for children. She has lectured in more than 90 countries and appeared in hundreds of radio and television productions, including being the subject of the 2017 NGS/True Blue Films *Sea of Hope* and the 2014 Emmy Award Winning Netflix documentary *Mission Blue*. Leader of more than 100 expeditions, she has logged more than 7,500 hours underwater—including leading the first team of women aquanauts during the Tektite Project in 1970—participated in ten saturation dives, and set a record for solo diving in 1,000 meters depth. Her research concerns marine algae, wildlife, and ecosystems with special reference to exploration, conservation, and the development and use of new technologies for access and effective operations in the deep sea as well as developing a global network of marine protected areas, "Hope Spots," to safeguard species, ecosystems, and planetary processes.

Her more than 150 honors include the Princess of Asturias Concordia Award (Peace), Trident d'Or Prize, Lewis Thomas Prize, the EARTHx award, 2016 Harris World Ecology Medal, IUCN Icon of Nature, Prince Albert II Award, UNEP Champion of the Earth, Glamour Woman of the Year, Walter Cronkite Award, the Carl Sagan Award, the National Geographic Hubbard Medal, UNESCO E-Award at Rio+20, the Royal Geographical Society Patron's Medal, the Explorers Club Medal and Lowell Thomas Medal, the Medal of Honor from the Dominican Republic, the TED Prize, the Netherlands Order of the Golden Ark, Australia's International Banksia Award, Italy's Artiglio Award, the International Seakeepers Award, the International Women's Forum, the National Women's Hall of Fame, American Academy of Achievement, Los Angeles Times Woman of the Year, UN Global 500, the US Department of the Interior Conservation Service Award, and medals from the Philadelphia Academy of Sciences, Lindbergh Foundation, National Wildlife Federation, Sigma Xi, Barnard College, and the Society of Women Geographers.

OTHER BOOKS IN THE HARTE RESEARCH INSTITUTE FOR GULF OF MEXICO STUDIES SERIES

Marine Plants of the Texas Coast
Roy L. Lehman

Beaches of the Gulf Coast
Richard A. Davis

Texas Seashells: A Field Guide
John W. Tunnell, Noe C. Barrera, and Fabio Moretzsohn

Benthic Foraminifera of the Gulf of Mexico:
Distribution, Ecology, Paleoecology
C. Wylie Poag

The American Sea: A Natural History of the Gulf of Mexico
Rezneat Milton Darnell

Birdlife of the Gulf of Mexico
Joanna Burger

It's More Than Fishing: The Art of Texas Trout and Redfish Angling
Patrick D. Murray